T0226218

Communications in Computer and Information Science   **789**

Commenced Publication in 2007
Founding and Former Series Editors:
Alfredo Cuzzocrea, Xiaoyong Du, Orhun Kara, Ting Liu, Dominik Ślęzak,
and Xiaokang Yang

More information about this series at http://www.springer.com/series/7899

Andrey Filchenkov · Lidia Pivovarova
Jan Žižka (Eds.)

Artificial Intelligence and Natural Language

6th Conference, AINL 2017
St. Petersburg, Russia, September 20–23, 2017
Revised Selected Papers

 Springer

Editors
Andrey Filchenkov
ITMO University
St. Petersburg
Russia

Lidia Pivovarova
University of Helsinki
Helsinki
Finland

Jan Žižka
Mendel University
Brno
Czech Republic

ISSN 1865-0929   ISSN 1865-0937   (electronic)
Communications in Computer and Information Science
ISBN 978-3-319-71745-6   ISBN 978-3-319-71746-3   (eBook)
https://doi.org/10.1007/978-3-319-71746-3

Library of Congress Control Number: 2017960865

Printed on acid-free paper

This Springer imprint is published by Springer Nature
The registered company is Springer International Publishing AG
The registered company address is: Gewerbestrasse 11, 6330 Cham, Switzerland

Preface

The 6th Conference on Artificial Intelligence and Natural Language Conference (AINL), held during September 20–23, 2017, in Saint Petersburg, Russia, was organized by the NLP Seminar and ITMO University. Its aim was to (a) bring together experts in the areas of natural language processing, speech technologies, dialogue systems, information retrieval, machine learning, artificial intelligence, and robotics and (b) to create a platform for sharing experience, extending contacts, and searching for possible collaboration. Overall, the conference gathered more than 100 participants.

The review process was challenging. Overall, 35 papers were sent to the conference and only 17 were selected, for an acceptance rate of 48%. In all, 56 researchers from different domains and areas were engaged in the double-blind reviewing process. Each paper received at least three reviews, in many cases there were four reviews.

Beyond regular papers, the proceedings contain six papers about the Russian Paraphrase Detection shared task, which took place at the AINL 2016 conference. These papers followed a slightly different review process and were not anonymized for reviews.

Altogether, 17 papers were presented at the conference, covering a wide range of topics, including social data analysis, dialogue systems, speech processing, information extraction, Web-scale data processing, word embedding, topic modeling, and transfer learning. Most of the presented papers were devoted to analyzing human communication and creating algorithms to perform such analysis. In addition, the conference program included several special talks and events, including tutorials on neural machine translation, deception detection in language, a hackathon for plagiarism detection in Russian texts, an invited talk on the shape of the future of computational science, industry talks and demos, and a poster session.

Many thanks to everybody who submitted papers and gave wonderful talks, and to whose who came and participated without publication.

We are indebted to our Program Committee members for their detailed and insightful reviews; we received very positive feedback from our authors even from those whose submissions were rejected.

And last but not the least, we are grateful to our organization team: Anastasia Bodrova, Irina Krylova, Aleksandr Bugrovsky, Natalia Khanzhina, Ksenia Buraya, and Dmitry Granovsky.

November 2017

Andrey Filchenkov
Lidia Pivovarova
Jan Žižka

Organization

Program Committee

Jan Žižka (Chair)	Mendel University of Brno, Czech Republic
Jalel Akaichi	King Khalid University, Tunisia
Mikhail Alexandrov	Autonomous University of Barcelona, Spain
Artem Andreev	Russian Academy of Science, Russia
Artur Azarov	Saint Petersburg Institute for Informatics and Automation, Russia
Alexandra Balahur	European Commission, Joint Research Centre, Ispra, Italy
Siddhartha Bhattacharyya	RCC Institute of Information Technology, India
Svetlana Bichineva	Saint Petersburg State University, Russia
Victor Bocharov	OpenCorpora, Russia
Elena Bolshakova	Moscow State Lomonosov University, Russia
Pavel Braslavski	Ural Federal University, Russia
Maxim Buzdalov	ITMO University, Russia
John Cardiff	Institute of Technology Tallaght, Dublin, Ireland
Dmitry Chalyy	Yaroslavl State University, Russia
Daniil Chivilikhin	ITMO University, Russia
Dan Cristea	A. I. Cuza University of Iasi, Romania
Frantisek Darena	Mendel University in Brno, Czech Republic
Gianluca Demartini	University of Sheffield, UK
Marianna Demenkova	Kefir Digital, Russia
Dmitry Granovsky	Yandex, Russia
Maria Eskevich	Radboud University, The Netherlands
Vera Evdokimova	Saint Petersburg State University, Russia
Alexandr Farseev	Singapore National University, Singapore
Andrey Filchenkov	ITMO University, Russia
Tatjana Gornostaja	Tilde, Latvia
Mark Granroth-Wilding	University of Helsinki, Finland
Jiří Hroza	Rare Technologies, Czech Republic
Tomáš Hudík	Think Big Analytics, Czech Republic
Camelia Ignat	Joint Research Centre of the European Commission, Ispra, Italy
Denis Kirjanov	Higher School of Economics, Russia
Goran Klepac	University of Zagreb, Croatia
Daniil Kocharov	Saint Petersburg State University, Russia
Artemy Kotov	Kurchatov Institute, Russia
Miroslav Kubat	University of Miami, FL, USA
Andrey Kutuzov	University of Oslo, Norway
Nikola Ljubešić	Jožef Stefan Institute, Slovenia

Contents

Social Interaction Analysis

Semantic Feature Aggregation for Gender Identification in Russian Facebook

Polina Panicheva[(✉)], Aliia Mirzagitova, and Yanina Ledovaya

St. Petersburg State University,
Universitetskaya nab. 7-9, 199034 St. Petersburg, Russia
ppolin86@gmail.com, amirzagitova@gmail.com, y.ledovaya@spbu.ru

Abstract. The goal of the current work is to evaluate semantic feature aggregation techniques in a task of gender classification of public social media texts in Russian. We collect Facebook posts of Russian-speaking users and apply them as a dataset for two topic modelling techniques and a distributional clustering approach. The output of the algorithms is applied as a feature aggregation method in a task of gender classification based on a smaller Facebook sample. The classification performance of the best model is favorably compared against the lemmas baseline and the state-of-the-art results reported for a different genre or language. The resulting successful features are exemplified, and the difference between the three techniques in terms of classification performance and feature contents are discussed, with the best technique clearly outperforming the others.

1 Introduction

Data on verbal and behavioral patterns in social networks can provide insight into numerous sociological and psychological characteristics [14]. Open-vocabulary approach to social media data is widely used to predict demographic and psychological characteristics of users [37]. However, in recent years the language-based features are aggregated in various ways, with meaningful groups of highly correlated features identified in English data [2,3,16]. This allows to increase the features' impact by combining similar units together, dramatically decrease computational costs, and gain greater interpretability comparing to individual term or linguistic category usage.

Current study is a part of a larger research project aimed to explore the relations among behavioral data, personality traits and the language a person uses in online communication. We perform 3 feature aggregation techniques using public Facebook post data by Russian-speaking users, and evaluate the aggregated features in an author profiling task of gender identification.

The paper is organized as follows. Section 2 presents a short overview of topic modelling and distributional clustering algorithms, and feature aggregation techniques applied to author profiling tasks in social media. In Sect. 3 we describe the procedure of obtaining the dataset of Russian Facebook posts. Section 4 is a recount of the techniques used for feature aggregation and labeling. In Sect. 5 we

© Springer International Publishing AG 2018
A. Filchenkov et al. (Eds.): AINL 2017, CCIS 789, pp. 3–15, 2018.
https://doi.org/10.1007/978-3-319-71746-3_1

present the experiment, with both performance results and exploratory analysis. The conclusions are outlined in Sect. 6.

2   Related Work

2.1   Feature Aggregation for Author Profiling in Social Media

In traditional closed-vocabulary approaches [32] features are aggregated manually into supposedly meaningful categories, thus forming a look-up vocabulary for word-count statistics. Feature aggregation for author profiling relies on automatic identification of meaningful categories: topic modelling and distributional semantic techniques. Thus, Latent Semantic Analysis modelling has been successfully compared to the traditional LIWC dictionary approach in predicting author's age and gender in multi-genre English texts, including social media [2]. User Embedding algorithms allow learning user-specific aggregated features, rather than just co-occurrence based, reportedly accounting for personal verbal and behavioral patterns: verbal information is aggregated to predict mental health outcomes (depression, trauma) in Twitter [3]; Facebook likes are used to model a behavioral measure of impulsivity [9].

Authors of [16] apply Factor Analysis to identify factors of lexical usage by English-speaking Facebook users. They evaluate the obtained language-based factors in terms of Generalizability and Stability, by correlating them with the Big5 Personality Traits and comparing their performance with Big5 in terms of predicting some behavioral (income, IQ, Facebook likes) and psychological (satisfaction with life, depression) variables. Thus the language-based factors are established as proper latent personality traits based on large-scale behavioral data rather than questionnaire self-reports.

2.2   Topic Modelling

Topic modelling is a statistical technique widely used in the field of natural language processing for analysing large text collections. One of the first and most commonly used methods for fitting topic models is Latent Dirichlet Allocation (**LDA**), a probabilistic graphical model regularised with Dirichlet priors [7]. LDA presupposes that each document is a finite mixture of a small number of topics and each word in the document can be attributed to a topic with a certain probability.

The author-topic model (**ATM**) is an extension of LDA which accounts for authorship information and simultaneously models the document content and authors' interests [36]. While LDA models topics as a distribution over words and documents as a distribution over topics, ATM models topics as a distribution over words and authors as a distribution over topics. Thus, LDA is seen as a special case of ATM where authors and documents have a trivial one-to-one mapping and author's topic distribution is the same as document's topic distribution. The case of one-to-many relationships, with authors owning multiple texts, is referred as the single author-topic model [33]. To the best of our knowledge, there are no reported results of applying ATM to Russian corpora.

Resulting topics are conventionally represented as a simple enumeration of topics together with the top terms from the multinomial distribution of words [7]. For better and easier interpretation, experts can manually assign these word lists a textual label. Since manual annotation is a costly and time-consuming task, there have been proposed numerous methods for automatic topic labelling. These can either rely solely on the content of the text corpus [15, 19, 24] or use external knowledge resources like Wikipedia [18], various ontologies [11, 22] or search engines [1, 27].

2.3   Distributional Clustering

Distributional semantic models allow for representing word meanings in a multi-dimensional vector space [10, 26]. The representation effectively captures semantic relations [28] and can be used to obtain clusters of related meanings in an unsupervised way [5]. We apply a Russian National Corpus-based semantic model [17], and automatically obtain Distributional Semantic Clusters (**DSC**) of words using K-Means clustering [6]. K-Means clustering over word-embeddings has been successfully applied to topic and polarity classification in English [38, 39]. DSC has also been recently utulized as a feature aggregation technique on a smaller Russian Facebook dataset in a study on content correlates of personality traits of users [30].

3   Dataset

8367 Russian Facebook users participated in the study by completing a questionnaire with an instant feedback about their personality traits and providing consent to share their publicly available posts. The application with the questionnaire had been advertised on Facebook. The public posts by the users have been gathered, with text cited or written by the users themselves, repost information being out of scope of the current work.

The basic data collection procedure and the questionnaire details have been described in [8, 30]. However, the described data were obtained in 2015, while the current dataset is generated by a different set of users and collected in October 2016. There were also a number of important changes introduced in the questionnaire, including the "outlier" criteria, and in the text collection procedure, allowing to download a larger sample by every user.

Out of the 8367 initial participants, 3973 users (47%) have written more than 10 posts in Russian (as identified by the *langid* library [21]). These data are used as raw texts for topic and distributional modelling.

The data was filtered according to the following criteria, so that only the 3341 (40%) users who performed the questionnaire properly were included in the final sample:

- users who finalized the questionnaire;
- correctly answered a trivial "trap" question;
- did not score too high on the social desirability scale;
- did not answer too many questions too shortly (less than 5 s).

1684 users (20%) have both written more than 10 posts in Russian and have performed the questionnaire properly. There are 807 male (48%) and 872 female (52%) authors; 5 authors have not indicated their gender and are excluded from the current experiments. The final dataset consists of **130** posts on average for each participant, standard deviation = 126. This is on average **401** sentences (std = 748) or **5395** tokens (std = 11185) per author.

4 Feature Aggregation Models

In order to obtain semantically interpretable aggregated features, we apply 3 semantic models: LDA, ATM and DSC. The dataset used for topic modelling and clustering experiments consisted of 343492 posts written by 3973 users, with the overall word count being 6248565. Prior to fitting the topic models, the data had been preprocessed: after removing stop words and hapax legomena, the vocabulary contained 100 K unique tokens. For direct comparability of features we set the number of topics/clusters $K = 500$ in all cases. $K = 500$ was chosen as it results in on average 200 words per cluster, which is the maximal cluster size allowing for cluster coherence and interpretability, according to a preliminary manual analysis of the resulting clusters.

4.1 LDA

We have performed LDA on the dataset using the Python *gensim* library [35]. We deployed the multi-core implementation of LDA which allows to develop topic models much faster and efficiently than the simple one-core version. We selected the default symmetric Dirichlet priors $1/K$, the number of iterations was 10 with 20 passes.

We did not pool the documents for LDA, so the model treated each post as a separate document. The average length of the preprocessed posts was 22.4 words, which was quite short and thus posed a challenge for LDA, as there could have been insufficient term co-occurrence statistics in each document.

4.2 Author-Topic Model

The second model, namely the single ATM, was intended to reflect the authorship information contained in the data. The single ATM is effectively equivalent to the author-wise pooling strategy, i.e. aggregating the documents written by the same author into a new longer document [23]. This way, the model could utilize the most of the given data and presumably better identify the features immanent in different authors' combined texts. For this purpose, we took advantage of the *gensim's* ATM module [36]. The chosen hyperparameters were the same as for LDA.

4.3   Distributional Clustering

We use a Skip-Gram Word2Vec model trained with the Russian National Corpus data. We intentionally apply RNC and not a web-trained model, as the goal is to capture established semantic regularities interpretable in terms of general semantic categories, while describing web language peculiarities are represented in the topic models above.

The clustering techniques applied in this task have been compared in [29]. The optimal algorithm used for DSC features is K-means with Euclidean distance, yielding the most homogeneous and precise clusters. Other clustering algorithms and parameters have been applied in preliminary experiments; resulting in various cluster sizes and slightly different cluster contents, different algorithms maintain the basic significant topics unchanged. Function words, numerals and unknown words are out of scope of the semantic model and of the clusters.

4.4   Automatic Label Assignment

In our experiments, we have used the unsupervised graph-based method of automatic topic labelling as described in [27].

For topic models, we generated candidate labels by first querying the top 10 topic words in the Google search engine, then concatenating the titles of the top 30 search results into a text, and applying PageRank [25] in order to evaluate the importance of each term. Next, we constructed a set of syntactically valid key phrases by means of morphological patterns. The key phrases were ranked according to the sums of the individual PageRank scores.

In order to make the procedure applicable for cluster labelling as well, we first ranked terms within each group using Euclidean distance to its centroid, which enabled us to select the top 10 closest words for querying the search engine. We also used Yandex search engine[1] instead of Google in this case, as Google implicitly identified word2vec as the source of the synonymous word lists and suggested word2vec-related pages in most of the cases. The rest of the algorithm remained the same.

5   Author Gender Profiling

5.1   Experiment

Gender profiling of Facebook users is applied as a testbed for topic features. We apply three feature sets: LDA topics, ATM topics, and distributional clusters. Preprocessing consisted of tokenization with *happierfuntokenizer*[2] for social media and morphological normalization with *PyMorphy* [13]. We apply lemma features as a baseline, including all the lemmas used by at least 5% authors. In every experiment we perform feature selection by choosing the most informative

---

[1] https://yandex.ru/.

[2] http://wwbp.org.

features (ANOVA F-value) with p< 0.01, corrected for multiple hypotheses with the Benjamini-Hochberg False-Discovery Rate correction [4].

We apply LinearSVM binary classification with C = 0.5, 10-fold cross-validation. All the experiments are performed using the *sklearn* Python package [31]. The question of the best classification algorithm is not raised in this work; on the contrary, we apply the widely used linear SVM for all our feature sets in order to control for the overfitting-generalizability continuum. The value of the C-parameter was chosen as a trade-off between accuracy and generalizability, whereas lower C indicates lower results which are supposed to be more generalizable to new data, and higher C applies to higher results with a higher chance of overfitting. In our experiments a lower C-value also results in a larger gap between the highest and the lowest results, while a higher C corresponds to more similar performance across the features. However, preliminary experiments using both a different C-value and different classification algorithms have resulted in the same performance patterns across the various feature sets.

5.2   Results

Table 1 contains the results of the classification task in tems of mean accuracy and standard deviation for 10-fold cross-validation. Results representing significant improvement over the lemmas baseline (p < 0.01, two-tailed t-test [12]) are highlighted in bold.

Table 1. Gender classification results

Features	Accuracy	σ
Lemmas	.6372	.0307
LDA	.6456	.0193
ATM	**.6860**	.0400
DSC	.6033	.0333
LDA + lemmas	.6456	.0193
ATM + lemmas	**.6920**	.0403
DSC + lemmas	.6348	.0440
Lemmas + LDA + ATM + DSC	**.6854**	.0384

The best result (Accuracy = **.6920**) is obtained by a combination of baseline and ATM features. LDA features improve the performance insignificantly, while DSC features show no improvement. It is clear that ATM is the best feature set, as it always adds significant improvement to the baseline, both individually and in combination with other features. The best results significantly outperform those reported as state-of-the-art in the English social media domain [2] (.55), but are directly comparable to those reported for Spanish social media [34] (.68); however, direct result comparison might be limited by the different social media platforms employed. Our result in terms of F1-measure (**.7186**) is higher than the

SVM-based Russian-language gender classification result reported by authors of [20] (.66) and comparable to the best learning algorithm result (.74), where both semantic and content-independent features were used; however, in the latter case the data genre was different and depended on a strictly defined communication task given to the respondents.

5.3 Correlation Analysis

For illustration we present four most significant features correlating with each gender in each feature group (see **Tables** 2, 3, 4, 5 for original features, and **Appendix, Tables** 6, 7, 8, 9 for translation into English). The features are ordered by the mean ANOVA **P-value** accross the 10 folds of the experiment. We also show **Spearman's R** between the feature and gender based on the full dataset. Topic and cluster features are represented by the automatically assigned **label**; their **content** is also illustrated with the five most significant words belonging to the topic/cluster.

Table 2. Significant lemmas

Lemma	P	R
Male		
российский	2e-12	.24
россия	7e-11	.28
путин	6e-10	.24
государство	3e-09	.22
Female		
любить	6.e-14	.18
мой	4e-13	.25
мужчина	5e-10	.13
любимый	6-10	.26

It is clear that except for the lemmas and ATM cases, female features are critically under-represented in the list of significant features: the most significant male features score much higher both in terms of classification impact (P-value) and overall correlation (R). ATM is thus a more balanced feature aggregation technique in terms of gender-specific topics.

In terms of the most informative content features in gender classification, politics-related words, topics and clusters in male language clearly stand out, including war, authority figures and international affairs. They cover most of the highly significant features of male language in terms of lemmas, clusters and topics. The highest-scoring female features in clusters and ATM are both related

Table 3. Significant LDA topics

Topic label	P	R	Contents
Male			
ситуация в россии в июле	2e-11	.23	политический россия германия запад практика
геополитика	3e-10	.17	бизнес лидер политик пензенский национальный
кандидаты и доктора	5e-10	.16	учёный америка необходимость доказать обратный
военная история	5e-10	.20	народ чиновник служить власть никита
Female			
мальчики и девочки	1e-05	.05	девочка мальчик самолёт ой лук
поздравление в прозе	4e-04	.14	любимый поздравление дорогой друг намного
поздравление и пожелание в стихах	7e-04	.09	любовь счастие радость любить пусть
афоризмы об искушении и соблазны	1e-03	.06	прекрасный сотрудник верно репутация ева

Table 4. Significant clusters

Cluster label	P	R	Contents
Male			
фашизм	7e-21	.27	империалист фашист большевик фашизм повстанец
горбачев и ельцин	1e-18	.28	горбачев премьер президент путин ельцин
демократия и монархия	5e-16	.26	плюрализм верховенство государственность демократизм демократия
вор и мошенник	2e-14	.23	хулиган обманщик авантюрист мошенник пьяница
Female			
мама и бабушка	3e-13	.23	бабушкин дедушкин женин катин мамин
форумчане в лицах	7e-11	.20	мальчик девочка хорошенький девчонка удаленький
юля и таня в поезде	1e-10	.17	маша катя таня наташа настя
имена для брака	2e-09	.14	ирина мария нина елена татьяна

Table 5. Significant ATM topics

Topic label	P	R	Contents
Male			
государственная культурная политика	4e-17	.27	политический народ президент экономика заявить
удар по авиабазе в сирии	8e-16	.24	сша президент гражданин рф территория
дом в москве	2e-13	.21	москва самый россия место вид
история земли	5e-13	.21	народ мир власть дело жертва
Female			
проблемы в семье на ребенка	1e-10	.23	ребёнок семья девочка родитель дочь
приветы и поздравление для эфира	2e-10	.23	илья любимый прекрасный любовь утро
цитаты и афоризмы эриха марии	2e-08	.14	жить глаз женщина большой думать
прогноз погоды в лабытнанги	1e-06	.15	новый фото праздник город друг

to family members; the other features are different: the clusters represent female names and diminutives, while the LDA and ATM topics are related to admiration and love, festivities, career, and general aphorisms about life. Previous authors find that the most significant topics distinguishing gender in English-speaking social networks are those related to work, home and leisure [2]; specifically for Facebook emotional, psychological and social processes, family, first-person singular pronouns were reported as characteristic of female language, while swear words, object references, sport, war and politics - of male language [37]. Our findings in Russian are totally in line with these results, except for the overwhelming presence of political categories in male language in our data, which appear to leave far behind the male-specific topics reported in previous work in English.

6 Conclusions

We have successfully applied three statistical feature aggregation techniques to author gender classification in Russian-speaking Facebook. To our knowledge, this is the first feature aggregation approach in Russian gender identification, and the first endeavor to compare author-specific and author-independent topic modeling techniques in gender language. Our results (accuracy = **0.69**, F1-measure = **0.72**) mostly overcome state-of-the-art approaches in a different genre in Russian and in other languages in the same genre, although our approach is specifically focused on content features, with no account for any morphological or other content-independent information.

The best feature aggregation technique in our setting is the author-topic model, performing consistently and significantly higher than other models. It also gives balanced results in terms of male- and female-specific topics. Both of these facts indicate that user-specific topic modelling is a suitable and highly interpretable technique for content-based author profiling. The difference between the performance of ATM and LDA in gender profiling can be due to the fact that ATM had access to the authorship information that is essential for the task. At the same time, not only was LDA unaware of authors, but also it had to deal with short-length texts, which is generally challenging for probabilistic models.

Our findings in terms of semantic categories highly indicative of male and female language in Russian are in line with previous research in English. However, there is an important exception in our sample: political issues appear to dominate in male topics, leaving far behind other topics traditionally attributed to male language.

Future research will include application of ATM to other issues in author profiling, including personality assessment.

Acknowledgments. The authors acknowledge Saint-Petersburg State University for a research grant 8.38.351.2015. The reported study is also supported by RFBR grant 16-06-00529.

Appendix

Table 6. Significant lemmas (English translation)

Lemma	P	R
Male		
russian	2e-12	.24
russia	7e-11	.28
putin	6e-10	.24
state	3e-09	.22
Female		
love (*verb*)	6.e-14	.18
my	4e-13	.25
man	5e-10	.13
beloved	6-10	.26

Table 7. Significant LDA topics (English translation)

Topic label	P	R	Contents
Male			
situation in Russia in July	2e-11	.23	political russia germany west practice
geopolitics	3e-10	.17	business leader politician *fromPensa* national
candidates and doctors	5e-10	.16	academic america necessity prove opposite
war history	5e-10	.20	nation officer serve power nikita (*malename*)
Female			
boys and girls	1e-05	.05	girl boy plane ouch look
congratulations in prose	4e-04	.14	beloved congratulation dear friend much
congratulations and wishes in poetry	7e-04	.09	love (*noun*) happiness joy love (*verb*) let
aphorisms about temptation	1e-03	.06	wonderful colleague correct reputation Eve

Table 8. Significant clusters (English translation)

Cluster label	P	R	Contents
Male			
fascism	7e-21	.27	imperialist fascist bolshevik fascism revolter
gorbachev and yeltsin	1e-18	.28	gorbachev prime (*minister*) president putin yeltsin
democracy and monarchy	5e-16	.26	pluralism domination statehood democratism democracy
thief and fraud	2e-14	.23	hooligan deceiver adventurer fraud drunkard
Female			
mom and grandma	3e-13	.23	grandma's grandpa's wife's kate's mom's
chat forum's people	7e-11	.20	boy girl cute chicklet sporty
yulia and tanya in the train	1e-10	.17	masha katya tanya natasha nastya (*diminutive female names*)
names for the marriage	2e-09	.14	irina maria nina elena tatiana (*full female names*)

Table 9. Significant ATM topics (English translation)

Topic label	P	R	Contents
Male			
state	4e-17	.27	political nation president economics declare
cultural policy			
syrian air base attack	8e-16	.24	usa president citizen rf (*RussianFederation*) terrain
moscow property	2e-13	.21	moscow most russia place view
earth history	5e-13	.21	nation world power business sacrifice (*noun*)
Female			
family problems on a child	1e-10	.23	child family girl parent daughter
greetings and congratulations on air	2e-10	.23	ilya (*malename*) beloved wonderful love (*noun*) morning
excerpts and aphorisms by erich maria	2e-08	.14	live eye woman big think
weather forecast in labytnanghi	1e-06	.15	new photo holiday city friend

References

1. Aletras, N., Stevenson, M.: Labelling topics using unsupervised graph-based methods. In: ACL, vol. 2, pp. 631–636 (2014)
2. Álvarez-Carmona, M.A., López-Monroy, A.P., Montes-y-Gómez, M., Villaseñor-Pineda, L., Meza, I.: Evaluating topic-based representations for author profiling in social media. In: Montes-y-Gómez, M., Escalante, H.J., Segura, A., Murillo, J.D. (eds.) IBERAMIA 2016. LNCS (LNAI), vol. 10022, pp. 151–162. Springer, Cham (2016). https://doi.org/10.1007/978-3-319-47955-2_13
3. Amir, S., Coppersmith, G., Carvalho, P., Silva, M.J., Wallace, B.C.: Quantifying mental health from social media with neural user embeddings. arXiv preprint arXiv:1705.00335 (2017)
4. Benjamini, Y., Hochberg, Y.: Controlling the false discovery rate: a practical and powerful approach to multiple testing. J. Roy. Stat. Soc.: Ser. B (Methodol.) **57**(1), 289–300 (1995)
5. Biemann, C.: Chinese whispers: an efficient graph clustering algorithm and its application to natural language processing problems. In: Proceedings of the First Workshop on Graph Based Methods for Natural Language Processing, pp. 73–80. Association for Computational Linguistics (2006)
6. Bird, S., Klein, E., Loper, E.: Natural Language Processing With Python: Analyzing Text With The Natural Language Toolkit. O'Reilly Media Inc, Sebastopol (2009)
7. Blei, D.M., Ng, A.Y., Jordan, M.I.: Latent dirichlet allocation. J. Mach. Learn. Res. **3**, 993–1022 (2003)
8. Bogolyubova, O., Tikhonov, R., Ivanov, V., Panicheva, P., Ledovaya, Y.: Violence exposure, posttraumatic stress, and subjective well-being in a sample of russian adults: a facebook-based study. J. Interpersonal Violence **30**, 1153–1167 (2017). http://journals.sagepub.com/doi/abs/10.1177/0886260517698279
9. Ding, T., Pan, S., Bickel, W.K.: 1todayor2 tomorrow? the answer is in your facebook likes. arXiv preprint arXiv:1703.07726 (2017)
10. Gliozzo, A., Biemann, C., Riedl, M., Coppola, B., Glass, M.R., Hatem, M.: Jobimtext visualizer: a graph-based approach to contextualizing distributional similarity. In: Graph-Based Methods for Natural Language Processing, p. 6 (2013)
11. Hulpus, I., Hayes, C., Karnstedt, M., Greene, D.: Unsupervised graph-based topic labelling using dbpedia. In: Proceedings of the Sixth ACM International Conference on Web Search and Data Mining, pp. 465–474. ACM (2013)

12. Jones, E., Oliphant, T., Peterson, P., et al.: SciPy: open source scientific tools for python (2001). http://www.scipy.org/
13. Korobov, M.: Morphological analyzer and generator for russian and ukrainian languages. In: Khachay, M.Y., Konstantinova, N., Panchenko, A., Ignatov, D.I., Labunets, V.G. (eds.) AIST 2015. CCIS, vol. 542, pp. 320–332. Springer, Cham (2015). https://doi.org/10.1007/978-3-319-26123-2_31
14. Kosinski, M., Matz, S.C., Gosling, S.D., Popov, V., Stillwell, D.: Facebook as a research tool for the social sciences: opportunities, challenges, ethical considerations, and practical guidelines. Am. Psychol. **70**(6), 543 (2015)
15. Kou, W., Li, F., Baldwin, T.: Automatic labelling of topic models using word vectors and letter trigram vectors. In: Zuccon, G., Geva, S., Joho, H., Scholer, F., Sun, A., Zhang, P. (eds.) AIRS 2015. LNCS, vol. 9460, pp. 253–264. Springer, Cham (2015). https://doi.org/10.1007/978-3-319-28940-3_20
16. Kulkarni, V., Kern, M.L., Stillwell, D., Kosinski, M., Matz, S., Ungar, L., Skiena, S., Schwartz, H.A.: Latent human traits in the language of social media: an open-vocabulary approach (2017)
17. Kutuzov, A., Andreev, I.: Texts in, meaning out: neural language models in semantic similarity task for Russian. arXiv preprint arXiv:1504.08183 (2015)
18. Lau, J.H., Grieser, K., Newman, D., Baldwin, T.: Automatic labelling of topic models. In: Proceedings of the 49th Annual Meeting of the Association for Computational Linguistics: Human Language Technologies, vol. 1, pp. 1536–1545. Association for Computational Linguistics (2011)
19. Lau, J.H., Newman, D., Karimi, S., Baldwin, T.: Best topic word selection for topic labelling. In: Proceedings of the 23rd International Conference on Computational Linguistics: Posters, pp. 605–613. Association for Computational Linguistics (2010)
20. Litvinova, T., Seredin, P., Litvinova, O., Zagorovskaya, O., Sboev, A., Gudovskih, D., Moloshnikov, I., Rybka, R.: Gender prediction for authors of Russian texts using regression and classification techniques. In: CDUD 2016–The 3rd International Workshop on Concept Discovery in Unstructured Data, p. 44 (2016). https://cla2016.hse.ru/data/2016/07/24/1119022942/CDUD2016.pdf#page=51
21. Lui, M., Baldwin, T.: Langid. py: an off-the-shelf language identification tool. In: Proceedings of the ACL 2012 System Demonstrations, pp. 25–30. Association for Computational Linguistics (2012)
22. Magatti, D., Calegari, S., Ciucci, D., Stella, F.: Automatic labeling of topics. In: Ninth International Conference on Intelligent Systems Design and Applications ISDA 2009, pp. 1227–1232. IEEE (2009)
23. Mehrotra, R., Sanner, S., Buntine, W., Xie, L.: Improving lda topic models for microblogs via tweet pooling and automatic labeling. In: Proceedings of the 36th International ACM SIGIR Conference on Research and Development in Information Retrieval, pp. 889–892. ACM (2013)
24. Mei, Q., Shen, X., Zhai, C.: Automatic labeling of multinomial topic models. In: Proceedings of the 13th ACM SIGKDD International Conference on Knowledge Discovery and Data Mining, pp. 490–499. ACM (2007)
25. Mihalcea, R., Tarau, P.: Textrank: bringing order into texts. Association for Computational Linguistics (2004)
26. Mikolov, T., Sutskever, I., Chen, K., Corrado, G.S., Dean, J.: Distributed representations of words and phrases and their compositionality. In: Advances in Neural Information Processing Systems, pp. 3111–3119 (2013)
27. Mirzagitova, A., Mitrofanova, O.: Automatic assignment of labels in topic modelling for Russian corpora. In: Proceedings of 7th Tutorial and Research Workshop on Experimental Linguistics, ExLing, pp. 115–118 (2016)

28. Panchenko, A., Loukachevitch, N., Ustalov, D., Paperno, D., Meyer, C., Konstanti-nova, N.: Russe: the first workshop on Russian semantic similarity. In: Computational Linguistics and Intellectual Technologies: Papers from the Annual Conference. Dialogue, vol. 2, pp. 89–105 (2015)
29. Panicheva, P., Ledovaya, Y., Bogoliubova, O.: Revealing interpetable content correlates of the dark triad personality traits. In: Russian Summer School in Information Retrieval (2016)
30. Panicheva, P., Ledovaya, Y., Bogolyubova, O.: Lexical, morphological and semantic correlates of the dark triad personality traits in Russian facebook texts. In: Artificial Intelligence and Natural Language Conference (AINL) IEEE, pp. 1–8. IEEE (2016)
31. Pedregosa, F., Varoquaux, G., Gramfort, A., Michel, V., Thirion, B., Grisel, O., Blondel, M., Prettenhofer, P., Weiss, R., Dubourg, V., et al.: Scikit-learn: machine learning in python. J. Mach. Learn. Res. **12**, 2825–2830 (2011)
32. Pennebaker, J.W., Francis, M.E., Booth, R.J.: Linguistic inquiry and word count: Liwc 2001. Mahway: Lawrence Erlbaum Associates 71 (2001)
33. Prince, S.J.: Computer Vision: Models, Learning and Inference. Cambridge University Press, Cambridge (2012)
34. Rangel, F., Rosso, P., Potthast, M., Trenkmann, M., Stein, B., Verhoeven, B., Daeleman, W., et al.: Overview of the 2nd author profiling task at pan 2014. In: CEUR Workshop Proceedings, vol. 1180, pp. 898–927. CEUR Workshop Proceedings. https://riunet.upv.es/handle/10251/61150
35. Rehurek, R., Sojka, P.: Gensim–python framework for vector space modelling. NLP Centre, Faculty of Informatics, Masaryk University, Brno (2011)
36. Rosen-Zvi, M., Griffiths, T., Steyvers, M., Smyth, P.: The author-topic model for authors and documents. In: Proceedings of the 20th Conference on Uncertainty in Artificial Intelligence, pp. 487–494. AUAI Press (2004)
37. Schwartz, H.A., Eichstaedt, J.C., Kern, M.L., Dziurzynski, L., Ramones, S.M., Agrawal, M., Shah, A., Kosinski, M., Stillwell, D., Seligman, M.E., et al.: Personality, gender, and age in the language of social media: the open-vocabulary approach. PLoS ONE **8**(9), e73791 (2013)
38. Zhang, X., Zhao, J., LeCun, Y.: Character-level convolutional networks for text classification. In: Advances in Neural Information Processing Systems, pp. 649–657 (2015)
39. Zhiqiang, T., Wenting, W.: Dlirec: aspect term extraction and term polarity classification system. In: Proceedings of the 8th International Workshop on Semantic Evaluation (SemEval 2014) (2014)

Using Linguistic Activity in Social Networks to Predict and Interpret Dark Psychological Traits

Arseny Moskvichev[1(✉)], Marina Dubova[1], Sergey Menshov[2],
and Andrey Filchenkov[2]

[1] Saint Petersburg State University, Saint Petersburg, Russia
arseny.moskvichev@gmail.com
[2] ITMO university, Saint Petersburg, Russia

Abstract. Studying the relationships between one's psychological characteristics and linguistic behaviour is a problem of a profound importance in many fields ranging from psychology to marketing, but there are very few works of this kind on Russian-speaking samples. We use Latent Dirichlet Allocation on the Facebook status updates to extract interpretable features that we then use to identify Facebook users with certain negative psychological traits (the so-called Dark Triad: narcissism, psychopathy, and Machiavellianism) and to find the themes that are most important to such individuals.

1 Introduction

The problem of linking individual characteristics and the digital records of one's behaviour has been given much attention in recent literature. Often, the primary goal is to predict individual characteristics based on the user's activity in social networks. This idea was applied to a broad range of target variables, and it was repeatedly demonstrated that it is possible to predict demographic (age, gender, sexual orientation, ethnicity) [7,13,31] and psychological characteristics (agreeableness, neuroticism, happiness) [24–26], as well as political preferences [1, 19]. Another dimension along which one can compare the works in this field is the choice of features. The most common options include user likes, geotags, and wall-posts, but sometimes more original sources of information are used, as in [10], where authors analyzed mobile device logs in order to predict user's personality.

The best predictive performance is usually achieved by combining different sources of information, as it was done, for example, in [14], where the authors improved venue recommendations by combining information from several social networks, or in [11], where the authors described an efficient substance use detection system. A similar approach was applied with considerable success in [21] and in [5] for the problem of predicting psychological variables measured using the Big Five personality model. In these works, a broad set of features was used,

A. Filchenkov et al. (Eds.): AINL 2017, CCIS 789, pp. 16–26, 2018.
https://doi.org/10.1007/978-3-319-71746-3_2

ranging from a number of photos uploaded by user to word forms extracted through linguistic analysis.

The downside of this attitude, however, is that interpretability is often sacrificed for the sake of achieving higher accuracy. Since the primary purpose of our article is to explore the relationship between certain psychological traits and language, we restrict our further analysis to the works that mostly rely on text-based features.

Among the works that utilize texts as the primary source of information, the results are most impressive for the predictions of demographic variables such as gender or age, with the achieved accuracy and R-squared metrics reaching numbers as high as 0.9 and 0.8 for gender and age respectively [27]. There are also works of this kind that focus on Russian-speaking samples, for example, predicting age based on users' wallposts [3].

At the same time, the achieved accuracy values are relatively low, when it comes to predicting psychological characteristics. For example, in one twitter-based study [29], the authors hosted an open competition on Kaggle, with the winning model achieving an AUC of 0.641 for Psychopathy (the results for other psychological traits they used were even worse). Other psychological variables could be even harder to predict, with standard methods giving accuracy values in the sub-0.6 range [2]. This might be due to the fact that the psychological variables themselves are difficult to define and measure, so there is a large amount of noise in the target variable [20].

On the other side of the research spectrum, in the fields of psychology, psychiatry, and sociology, there is a lasting effort to understand how the specific personality traits manifest themselves through behaviour and language. Such studies usually focus on the correlations between psychological traits and specific words or word categories (usually predefined), paying less attention to the predictive performance. The most commonly used predefined word categories include dictionaries like ANEW (Affective Norms fro English Words) that maps words to their emotional values and LIWC (Linguistic Inquiry and Word Count) that provides a number of "psychologically meaningful" word interpretations [22,30]. The problem with this approach is that it lacks flexibility. Not only relevant categories can emerge or disappear from the public discourse with time, it is also difficult to adapt these dictionaries to other languages, since the translations require thorough validation. Therefore, the data-driven approaches to category extraction are becoming more and more popular, and, as shown in [27], they could also lead to superior predictive performance.

In our work, we focus on the following two questions:

1. Are there specific semantic preferences related to the Dark Triad of psychological traits?
2. Can we predict individual's psychological characteristics based on the high-level semantic content of the texts they write?

For English-speaking samples, the answer is "yes", as it can be seen from [16, 27,29]. However, it is unclear, whether the same results can be achieved on the Russian segment of Facebook users. It is especially true for the second question, since while there were studies that study the linguistic correlates of the Dark Triad of psychological traits [23] in Russian samples, the predictive performance was not investigated in that article.

2   Method

2.1   Psychometrics

In order to measure individual psychological traits constituting the psychological Dark Triad, we used the Russian version [12] of the Short Dark Triad questionnaire [18]. We chose the short version to maximize the chances of survey completion.

We also introduced three questions from the classical social desirability scale questionnaire [9] to detect cases when a participant provides dishonest answers in order to seem a "better" person according to social standards.

In addition, one "trap question" was used. It is a simple instruction of the form "please, choose the third option" that is used to check whether the participant is actually paying attention and reading the questions rather than choosing random answers.

2.2   Topic Models

In order to extract high-level topics relevant to the Russian-speaking segment of the Facebook audience, we used the Latent Dirichlet Allocation, which is one of the standard techniques for this task [4].

LDA is based on several assumptions. Each document is assumed to contain text related to several topics and relatedness to a topic is precisely described by containing words related to this topic. More formally, each document is considered to be generated in the following way: given a distribution of its topics and a distribution of words for each topic, a new word in the document is generated by choosing its topic and then choosing the word of that topic. All the choices are independent. Distributions of words and topics are assumed to be Multinomial, while distribution of their parameters is Dirichlet.

2.3   Predictive Models

We used standard classification algorithms, such as Support Vector Machine with a radial basis function kernel, Random Forest ensemble classifier and a Multinomial Naive Bayes classifier [15].

In order to obtain the binary labels from the ordinal measurements of personality, we used the median split on all available data, as it was done in [29]. It should be noted that since there are multiple posts associated with each user, there are different ways to approach this classification problem. One possibility is to train classifiers on single posts entities and to average the predictions on the test phase. In this case, the cross-validation scheme should be chosen appropriately, so as to preclude the event when the posts from one participant are present in both training and test sets. Another option is to average the features for each participant before training the classifier.

Both options were explored and gave almost identical results. Because we use the median split, care should be taken when using the first strategy, in order to account for the slightly changing class imbalances (occurring due to the fact that different participants could have significantly different numbers of posts). Overall, the pre-averaging approach is slightly more natural in this scenario, so we only report the classification results obtained using it.

2.4 Statistical Analysis

Although the methods of statistical inference used in this article are limited to the calculation of Pearson's correlations criterion, it is important to note that in order to account for multiple hypothesis testing, we applied the Benjamini-Hochberg correction (FDR) [6]. By doing this we can ensure that the correlations that we found do indeed reflect the presence of a statistical relationship between two variables rather than being a consequence of excessive hypothesis testing.

3 Experiment

3.1 Data Collection

The data were obtained through a Facebook application that was created for this study. The participants were presented with an option to take part in the study by filling-in the psychological questionnaire and by giving access to their Facebook profile demographic information. No monetary incentives were used to attract participants, with their primary motivation being to receive the feedback on their psychological traits. In order to inform more participants about our study we ran an advertising campaign through Facebook Advertising Services.

For each participant, we collected the following data:

1. The measurements of the participant's individual psychological traits, including the measurements of the so-called psychological "Dark Triad" (Psychopathy, Narcissism, and Machiavellianism), on which we focus in this article.
2. User-generated texts, obtained from the Facebook status updates (wall-posts).
3. Demographic and other information from the user's Facebook profile. This portion of data includes age, gender, location, and likes.

3.2 Data Preprocessing

Initially, this procedure resulted in a sample of 8367 participants, with 56% of the sample being women, 41 person (0.5%) of unidentified gender, and the rest being males. The average age was 46 years, with a standard deviation of 13.46 years, 4% of participants did not provide their age.

During the initial filtering stage, we kept the participants who satisfied the following criteria:

1. They completed the questionnaire
2. They answered correctly to the "trap" question
3. The social desirability scale total is less than 13 points (15 being the maximum)
4. The number of "fast" responses (less than 5 s) is fewer than 36.

This resulted in a sample of 3341 participants. After we additionally filtered out participants with no posts containing the non-empty "message" field, we obtained the final sample with the size of 2852.

3.3   Implementation Details

In order to obtain the user-generated texts, we used the "message" field of the Facebook API post object, as it was done in other studies. Unfortunately, the manual inspection revealed a presence of posts that were automatically generated by Facebook applications and the posts containing copied materials from various sources. Since there is no simple and reliable way of sorting such posts out, and since these posts, while not being written by the user, still do reflect his or her interests and attitudes, we decided to leave them in the dataset.

We used the **word_tokenizer** function from the nltk library to separate message strings into words; we also removed the punctuation symbols and English and Russian stop words (also obtained through the nltk library) in order to make the topics more interpretable. In addition to that, we excluded all words with document frequency less than 10^{-4}.

The next step was to build the bag of words document representation. The Russian language exhibits a rich morphological structure, and in order to reduce this complexity and avoid introducing excessive amounts of variables into the document-word matrix, we extracted the normal form of each word using the pymorphy2 package before building the bag of words representation.

In order to extract topics, we used an LDA implementation from the LDA library for Python[1]. For other machine learning methods, we used the scikit-learn Python library.

Lastly, the statistical analysis was performed using the R programming language.

4   Results

4.1   Prediction

To evaluate the predictive performance of different classifiers, we used a 10-fold cross-validation scheme. Results in Table 1 summarize the algorithm predictive performances for the cases when extracted topics were used as features. It is important to note that the Random Forest classifier repeatedly outperformed all other models in all cases, therefore we only report scores obtained by this model.

---

[1] https://pypi.python.org/pypi/lda.

Table 1. Classification results for topic-based predictions

	Psych.	Mac.	Nar.	Gender
Baseline accuracy	0.52	0.507	0.552	0.531
Random Forest Accuracy	0.558	0.516	0.562	0.691
Random Forest AUC	0.571	0.526	0.558	0.748
Baseline accuracy H/L	0.507	0.531	0.534	-
Random Forest Accuracy H/L	0.572	0.581	0.587	-
Random Forest AUC H/L	0.591	0.576	0.612	-

To make our model comparable to a broader set of works, we also calculated the accuracy for the truncated sample. This truncated sample is obtained by throwing out the cases falling in the interval of $\pm$ one standard deviation from the mean.

It is important to note that by using the raw bag-of-words matrix (instead of 25 topics extracted using LDA), we get the accuracies that do not significantly differ from those listed in the Table 1. Moreover, other methods of dimensionality reduction (such as, for example, PCA or feature selection from the elastic net regression) result in worse prediction performance.

4.2 Statistical Analysis

We calculated the Pearson's correlation between self-reported Dark Triad scores and the estimated presence of each LDA-selected topic (averaged across all posts for each user). In order to account for multiple hypothesis testing, we applied the Benjamini-Hochberg false discovery rate correction (FDR) [6].

Machiavellianism. As it can be seen from the Table 2, we found the following patterns in topics for participants with high Machiavellianism scores:

1. Writing less about God, faith and soul. It is consistent with the idea that Machiavellianism is characterized by cynical disregard for morality [17].
2. Writing more about business and work. It is also consistent with the belief that Machiavellianism is described by concentration on self-interest [17].
3. Writing more posts with patriotic feeling: about Homeland and political situation in Russia. Appeal to patriotic feeling could be an effective method of manipulation of others (the key characteristic of Machiavellianism [17]).

Narcissism. These patterns of Facebook activity turned out to be the indicators of Narcissism:

1. Large diversity of topics among the posts.

2. Writing more posts describing friendship and social relationships. It could a way to brag about happy relationships that is largely consistent with Narcissism [8].
3. Writing more about health, body condition and illnesses. It is consistent with the most well-known characteristic of Narcissism: the concentration on oneself [8].

Psychopathy. Psychopathy is characterized by the following topics activity:

1. Writing more posts on Homeland and political situation: about Russia, Ukraine, USA, Putin, Crimea etc. It could be a form of consistent antisocial behavior (Internet terrorism) related to Psychopathy [28,32].
2. Writing more about daily activity. Small stories describing trivial mundane situations could be related to the selfishness characterizing Psychopathy [28].
3. Writing posts describing parties and celebrations.
4. Writing less about weather, season and time of day.
5. Writing more about working activity, projects, earnings and economical situation. It could also be consistent with selfishness characteristic of Psychopathy [28].

5   Discussion

Fist of all, we did not focus on optimizing the achieved accuracies at all costs (for example we avoided engineering new features and performed only a bare minimum of manual hyperparameter optimization (none for the best performing model)). The reasons to avoid extensive optimizations of this kind were as follows: the primary purpose of this article was to provide the proof of concept, and we deemed it reasonable to start with a simple baseline solution that works "from the box". The other reason is that our dataset is very small, therefore we limited the model evaluation to the cross-validation technique and we did not want to introduce the possibility of our conclusions being contaminated by the cross-validation set overfitting.

Having said that, we should first note that the obtained accuracies are lower than the state of the art predictive models applied to English-speaking segments of social networks [27,29]. At the same time, it is important to mention that the accuracies are generally low for the predictions of psychological variables, and the gap is not very big. Indeed, some studies focusing on predicting the Big Five personality traits report that their standard methods give very similar results, despite using a much larger dataset [2]. Moreover, there are very few works focusing specifically on the Dark Triad prediction, which are particularly difficult to predict, judging by the results of Kaggle competition, described in [29]. Lastly, our study replicates the pattern of differing predictive difficulty found in other articles, with Psychopathy being the most predictable among the Dark Triad psychological traits [16].

Table 2. Semantic correlates of the Dark Personality Traits, $*p < 0.05$, $**p < 0.01$, No signs: $p < 0.06$, FDR-corrected

Machiavellianism		Narcissism		Psychopathy	
Topic	Cor.	Topic	Cor.	Topic	Cor.
Faith** (holy, word, God, church, soul, Christ, faith, pray, sin)	−0.068	Diversity of topics in posts**	0.075	Patriotism** (Russia,nation, Putin, Ukraine, federation, politics, Crimea, citizen, west, USA)	0.068
Business* (money, Russia, work, rouble, company, price, business, project)	0.052	Friendship* (best, good, friend, love, attitude, true)	0.059	Daily Routine* (talk, car, go, think, money, road, phone, decide, do, see, stand, buy)	0.064
Patriotism (Russia, nation, Putin, Ukraine, federation, politics, Crimea, citizen, west, USA)	0.049	Health (water, help, body, doctor, organism, health, energy, illness, treatment)	0.051	Celebration* (celebration, congratulate, Birthday, love, health, greeting)	0.056
				Environment* (morning, summer, good, evening, Moscow, night, weather, autumn, rain)	−0.055
				Business (money, Russia, work, rouble, company, price, business, project)	0.050

There are a few potential explanations for the fact that the achieved performance metrics are not very high. The first and the most obvious is that the amounts of data that we have are smaller by an order of magnitude than the amounts data used in most cases, which may very well be a decisive factor [27]. Another possibility is that the texts that we collected contain too many copied or irrelevant material and are thus more noisy and less reliable. Lastly, there is a chance that the psychometric methods adapted to Russian are less precise in identifying psychological traits.

In order to partially answer to this question, we measured the accuracy of gender prediction (assuming that the self-reported gender is measured with equal precision in Russian and English-speaking samples). The achieved accuracy of (0.69) is very similar that achieved in another study (0.72) [33], where a relatively small dataset and similar prediction techniques were used. At the same time, the studies on larger datasets [27] usually achieve accuracies around 0.9. This observation corroborates the view that the size of the dataset might have been the primary limiting factor.

On the psychological side, we can see that by using topic modeling, we can indeed identify interpretable topics that give insightful information on the ways in which the psychological traits manifest themselves through the linguistic behaviour in social networks.

6   Conclusion

In this paper, we analyzed relationship between Russian-speaking Facebook users' texts and their psychological characteristics. We used topic modeling approach to represent user-generated texts as the mixtures of automatically generated high-level semantic categories. This model was used for two purposes corresponding to the two research questions of this paper.

Firstly, we identified specific semantic preferences related to the Dark Triad of psychological traits, including the following observations:

- Machiavellianists have a tendency to write about business-related and patriotic topics more often, while religious discourse is rare in their texts.
- Narcissistic users have a tendency to write about personal and social aspects of well-being, writing more often about wellness and social acceptance, as well as showing increased diversity in their choice of topics.
- Users with high Psychopathy scores show semantic preferences to business and patriotism topics. They are also more prone to describing the details of their daily routine and actions, while giving less attention to the properties of their surroundings like weather or the time of year.

Secondly, we have shown that it is possible to use these extracted features to predict the psychological characteristics of social network users. Although the accuracies were low in general sense, they were significantly above the chance level, which is a good result, considering the intrinsic noisiness of psychological measurements. Moreover, while not being applicable on practice for individual user profiling, these results could be applied to detect groups of people exhibiting certain negative psychological traits.

We see the main impact of this article in that we have shown that the flexible data-driven methodology previously only applied to English-speaking samples can be successfully adapted to the Russian segment of social networks in order to predict and better understand personal traits based on user-generated texts.

Acknowledgements. The authors acknowledge Saint Petersburg State University for a research grant 8.38.351.2015.

References

1. Agarwal, S., Sureka, A.: Applying social media intelligence for predicting and identifying on-line radicalization and civil unrest oriented threats. arXiv preprint arXiv:1511.06858 (2015)
2. Alam, F., Stepanov, E.A., Riccardi, G.: Personality traits recognition on social network-facebook. In: WCPR (ICWSM-13), Cambridge, MA, USA (2013)
3. Alekseev, A., Nikolenko, S.I.: Predicting the age of social network users from user-generated texts with word embeddings. In: Artificial Intelligence and Natural Language Conference (AINL), IEEE, pp. 1–11. IEEE (2016)
4. Alghamdi, R., Alfalqi, K.: A survey of topic modeling in text mining. Int. J. Adv. Comput. Sci. Appl. (IJACSA) **6**(1) (2015)
5. Bachrach, Y., Kosinski, M., Graepel, T., Kohli, P., Stillwell, D.: Personality and patterns of facebook usage. In: Proceedings of the 4th Annual ACM Web Science Conference, pp. 24–32. ACM (2012)
6. Benjamini, Y., Yekutieli, D.: The control of the false discovery rate in multiple testing under dependency. Ann. Stat. **29**, 1165–1188 (2001)
7. Buraya, K., Farseev, A., Filchenkov, A., Chua, T.-S.: Towards user personality profiling from multiple social networks. In: AAAI, pp. 4909–4910 (2017)
8. Campbell, W.K., Miller, J.D.: The handbook of narcissism and narcissistic personality disorder: theoretical approaches, empirical findings, and treatments. Wiley, Hoboken (2011)
9. Crowne, D.P., Marlowe, D.: A new scale of social desirability independent of psychopathology. J. Consult. Psychol. **24**(4), 349 (1960)
10. de Montjoye, Y.-A., Quoidbach, J., Robic, F., Pentland, A.S.: Predicting personality using novel mobile phone-based metrics. In: Greenberg, A.M., Kennedy, W.G., Bos, N.D. (eds.) SBP 2013. LNCS, vol. 7812, pp. 48–55. Springer, Heidelberg (2013). https://doi.org/10.1007/978-3-642-37210-0_6
11. Ding, T., Bickel, W.K., Pan, S.: Social media-based substance use prediction. arXiv preprint arXiv:1705.05633 (2017)
12. Egorova, M., Sitnikova, M.: Parshikova ov adaptatsiia korotkogo oprosnika temnoi triady [adaptation of the short dark triad]. Psikhologicheskie issledovaniia **8**(43), 1 (2015)
13. Farseev, A., Nie, L., Akbari, M., Chua, T.-S.: Harvesting multiple sources for user profile learning: a big data study. In: Proceedings of the 5th ACM on International Conference on Multimedia Retrieval, pp. 235–242. ACM (2015)
14. Farseev, A., Samborskii, I., Chua, T.-S.: bbridge: A big data platform for social multimedia analytics. In: Proceedings of the 2016 ACM on Multimedia Conference, pp. 759–761. ACM (2016)
15. Friedman, J., Hastie, T., Tibshirani, R.: The elements of statistical learning. Springer series in statistics, vol. 1. Springer, Berlin (2001)
16. Garcia, D., Sikström, S.: The dark side of facebook: Semantic representations of status updates predict the dark triad of personality. Pers. Individ. Differ. **67**, 92–96 (2014)
17. Jakobwitz, S., Egan, V.: The dark triad and normal personality traits. Pers. Individ. Differ. **40**(2), 331–339 (2006)
18. Jones, D.N., Paulhus, D.L.: Introducing the short dark triad (sd3) a brief measure of dark personality traits. Assessment **21**(1), 28–41 (2014)
19. Kosinski, M., Stillwell, D., Graepel, T.: Private traits and attributes are predictable from digital records of human behavior. Proc. Natl. Acad. Sci. **110**(15), 5802–5805 (2013)

20. Lambiotte, R., Kosinski, M.: Tracking the digital footprints of personality. Proc. IEEE **102**(12), 1934–1939 (2014)
21. Markovikj, D., Gievska, S., Kosinski, M., Stillwell, D.J.: Mining facebook data for predictive personality modeling. In: Seventh International AAAI Conference on Weblogs and Social Media (2013)
22. Nielsen, F.Å.: A new anew: evaluation of a word list for sentiment analysis in microblogs. arXiv preprint arXiv:1103.2903 (2011)
23. Panicheva, P., Ledovaya, Y., Bogolyubova, O.: Lexical, morphological and semantic correlates of the dark triad personality traits in russian facebook texts. In: Artificial Intelligence and Natural Language Conference (AINL), IEEE, pp. 1–8. IEEE (2016)
24. Peng, Z., Hu, Q., Dang, J.: Multi-kernel svm based depression recognition using social media data. Int. J. Mach. Learn. Cybern. 1–15 (2017)
25. Preotiuc-Pietro, D., Carpenter, J., Giorgi, S., Ungar, L.: Studying the dark triad of personality through twitter behavior. In: Proceedings of the 25th ACM International on Conference on Information and Knowledge Management, pp. 761–770. ACM (2016)
26. Preoţiuc-Pietro, D., Carpenter, J., Giorgi, S., Ungar, L.: Studying the dark triad of personality using twitter behavior (2016)
27. Schwartz, H.A., Eichstaedt, J.C., Kern, M.L., Dziurzynski, L., Ramones, S.M., Agrawal, M., Shah, A., Kosinski, M., Stillwell, D., Seligman, M.E., et al.: Personality, gender, and age in the language of social media: the open-vocabulary approach. PloS One **8**(9), e73791 (2013)
28. Skeem, J.L., Polaschek, D.L., Patrick, C.J., Lilienfeld, S.O.: Psychopathic personality: bridging the gap between scientific evidence and public policy. Psychol. Sci. Public Interest **12**(3), 95–162 (2011)
29. Sumner, C., Byers, A., Boochever, R., Park, G.J.: Predicting dark triad personality traits from twitter usage and a linguistic analysis of tweets. In: 11th International Conference on Machine Learning and Applications (ICMLA), 2012, vol. 2, pp. 386–393. IEEE (2012)
30. Tausczik, Y.R., Pennebaker, J.W.: The psychological meaning of words: Liwc and computerized text analysis methods. J. Lang. Soc. Psychol. **29**(1), 24–54 (2010)
31. Wang, P., Guo, J., Lan, Y., Xu, J., Cheng, X.: Multi-task representation learning for demographic prediction. In: Ferro, N., Crestani, F., Moens, M.-F., Mothe, J., Silvestri, F., Di Nunzio, G.M., Hauff, C., Silvello, G. (eds.) ECIR 2016. LNCS, vol. 9626, pp. 88–99. Springer, Cham (2016). https://doi.org/10.1007/978-3-319-30671-1_7
32. Williams, K., McAndrew, A., Learn, T., Harms, P., Paulhus, D.L.: The dark triad returns: entertainment preferences and antisocial behavior among narcissists, machiavellians, and psychopaths. In: Poster presented at the 109th Annual Convention of the American Psychological Association, San Francisco, CA (2001)
33. Zhang, C., Zhang, P.: Predicting gender from blog posts. University of Massachusetts Amherst, USA (2010)

Boosting a Rule-Based Chatbot Using Statistics and User Satisfaction Ratings

Octavia Efraim[1], Vladislav Maraev[2(✉)], and João Rodrigues[3]

[1] LIDILE EA3874, University of Rennes 2, Rennes, France
`octavia-edie.efraim@univ-rennes2.fr`
[2] CLASP, University of Gothenburg, Gothenburg, Sweden
`vladislav.maraev@gu.se`
[3] Department of Informatics, Faculty of Sciences,
University of Lisbon, Lisbon, Portugal
`joao.rodrigues@di.fc.ul.pt`

Abstract. Using data from user-chatbot conversations where users have rated the answers as good or bad, we propose a more efficient alternative to a chatbot's keyword-based answer retrieval heuristic. We test two neural network approaches to the near-duplicate question detection task as a first step towards a better answer retrieval method. A convolutional neural network architecture gives promising results on this difficult task.

1 Introduction

A task-oriented conversational agent which returns predefined answers from a fixed set (as opposed to generating responses in real time) can provide a considerable edge over a fully-human answering system, if it handles correctly most of the repetitive queries which require no personalised answer. Indeed, at least in our experience, many of the questions asked by users and their expected answer look like entries in a list of frequently asked questions (FAQ): "What are your opening hours?", "Do you deliver to this area?", etc. An effective conversational agent, or chatbot, can act as a filter, sifting out such questions and only passing on to human agents those it is unable to deal with: those which are too complex (e.g. made up of multiple queries), those for which there simply is no response available, or those which require consulting a client database in order to provide a personalised answer (e.g. the status of a specific order or request). Such questions may occur at the very beginning or at some later point during a conversation between a customer and the automated agent. In the latter case, a well-performing chatbot will at least have saved human effort up to the moment where the difficulty emerged (provided it also hands on to the human a summary of the dialogue).

If the job of such retrieval-based conversational agents may seem easy enough to be successfully handled through a rule-based approach, in reality, questions coming from users exhibit much more variation (be it lexical, spelling-related, or syntactic) that is feasibly built into hand-crafted rules for question parsing.

A. Filchenkov et al. (Eds.): AINL 2017, CCIS 789, pp. 27–41, 2018.
https://doi.org/10.1007/978-3-319-71746-3_3

Approaches based on statistical learning from data may therefore benefit such answer retrieval systems.

Our goal is to improve on an existing closed-domain chatbot which returns answers from a closed set using keywords as a retrieval heuristic and human-defined priority rules to break ties between multiple candidate answers. Assuming a question does have an answer in the closed answer repository, this chatbot may fail to find it because it misunderstands the question (in which case it replies with the wrong answer) or because it is unable to "understand" it (i.e. map it to an available response) altogether (it then asks the user to provide an alternative formulation). This design means that the chatbot's ability to recognise that two distinct questions can be accurately answered by the same reply is very limited. Potential improvements to this system design may target the answer retrieval method, the candidate answer ranking method, and the detection of out-of-domain questions. We choose to address answer retrieval.

This paper is organised as follows: in Sect. 2 we review some tasks and solutions which are potentially relevant to our goal; Sect. 3 gives an overview of the system we set out to improve; Sect. 4 describes the data available to us, and our problem formulation; in Sect. 5 we outline the procedure we applied to our data in order to derive from it a new dataset suited to our chosen task; Sect. 6 gives an account of our proposed systems; in Sect. 7 we sum up and discuss our results; finally, Sect. 8 outlines some directions for follow-up work.

2 Related Work

The ability to predict a candidate answer's fitness to a question is a potentially useful feature in a dialogue system's answer selection module. A low-confidence score for a candidate answer amounts to a problematic turn in a conversation, one that warrants corrective action. Addressing **success/failure prediction in dialogue**, both [28] (human-computer dialogues in the customer relationship domain) and [23] (human-human task-oriented dialogues) distinguish between a predictive task with immediate utility for corrective action in real time, and a post-hoc estimation task for analysis purposes. If the former authors learn a set of classification rules from meta-textual and meta-conversational features only, the latter find that, with an SVM classifier, lexical and syntactic repetition reliably predict the success of a task solved via dialogue.

Answer selection for question answering has recently been addressed using deep learning techniques. In [8], for instance, the task is treated as a binary classification problem over question-answer (QA) pairs: the matching is appropriate or not. The authors propose a language-independent framework based on convolutional neural networks (CNN). The power of 1-dimensional (1D) convolutional-and-pooling architectures in handling language data stems from their sensitivity to local ordering information, which turns them into powerful detectors of informative n-grams [9]. Some of the CNN architectures and similarity metrics tested in [8] on a dataset from the insurance domain achieve good accuracy in selecting one answer from a closed pool of candidates.

The answer selection problem has also been formulated in terms of **information retrieval**. For example, [15] reports on an attempt to answer open-domain questions asked by users on Web forums, by searching the answer in a large but limited set of FAQ QA pairs collected in a previous step. The authors use simple vector-space retrieval models over the user's question treated as a query and the FAQ question, answer, and source document indexed as fields making up the item to be returned. Also taking advantage of the multi-field structure of answers in QA archives, [31] combines a translation-based language model estimated on QA pairs viewed as a parallel corpus, and a query likelihood model with the question field, the answer field, and both combined. A special application of information retrieval, **SMS-based FAQ retrieval** – which was proposed as a shared task at the Forum for Information Retrieval Evaluation in 2011 and 2012 – faces the additional challenge of very short and noisy questions. The authors of [11] break the task down into: question normalisation using rules learnt on several corpora annotated with error corrections; retrieval of a ranked list of answers using a combination of a term overlap metric and two search engines with BM25 as the ranking function, over three indexes (FAQ question, FAQ answer, and both combined); finally, filtering out-of-domain questions using methods specific to each retrieval solution.

Equating new questions to past ones that have already been successfully answered has been proposed as another way of tackling question answering. Such **duplicate question detection** (DQD) approaches fall under near-duplicate detection, and are related to paraphrase identification and other such instances of the broader problem of textual semantic similarity, with particular applications, among others, to community question answering (*cf.* Task 3 at SemEval-2015, 2016, and 2017). In turn, DQD may be cast as an information retrieval problem [4], where the comparison for matching is performed on different entities: new question with or without its detailed explanation if available, old question with or without the answer associated with it; where the task is not to reply to new questions, but rather to organise a QA set, answers have even been compared to each other in order to infer the similarity of their respective questions [14]. Identifying semantically similar questions entails at least two major difficulties: similarity measures targeted at longer documents are not suited to short texts such as regular questions; and word overlap measures (such as Dice's coefficient or the Jaccard similarity coefficient) cannot account for questions which mean the same but use different words. Notwithstanding, word overlap features have been shown to be efficient in certain settings [13, 22]. CNN architectures, which, since their adoption from computer vision, have proved to be very successful feature extractors in text processing [9], have recently started to be applied to the task of DQD. [6] reports impressive results with word-based CNN on data from the StackExchange QA forum. In [25], the authors obtain very good performance on a subset of the AskUbuntu section of StackExchange by combining a similar word-based CNN with an architecture based on [2].

Answer relevancy judgements by human annotators on the output of dialogue systems are a common way of evaluating this technology. The definition of relevancy is tailored to each experimental setup and research goal. In [24] annotators assess whether the answer generated by a system based on statistical machine translation in reply to a Twitter status post is on the same topic as that post and "makes sense" in response to it. More recently—to cite just one example taken from a large body of work on neural response generation—, to evaluate the performance of the neural conversation model in [27], human judges are asked to choose the better of two replies to a given question: the output of the experimental system and a chatbot. The role of human judgements in such settings is nonetheless purely evaluative: the judge assesses post hoc the quality of a small sample of the system output according to some relevancy criterion. In contrast to these experiments, ours is not an unsupervised response generation system, but a **supervised retrieval-based system**, as defined in [19], insofar as it does "explicitly incorporate some supervised signal such as task completion or user satisfaction". Our goal is to take advantage of this feature not only for evaluation, but also for the system's actual design. As far as the evaluation of unsupervised response generation systems goes, this is a challenging area of research in its own right [18,19].

3   Overview of the Rule-Based Chatbot

The chatbot we are aiming at improving is deployed on the website of a French air carrier as a chat interface with an animated avatar. The system was developed by a private company and we had no participation in its conception or implementation. Its purpose is, given a question, to return a suitable predefined answer from a closed set. The French-speaking chatbot has access to a database of 310 responses, each of which is associated unambiguously with one or more keywords and/or skip-keyphrases (phrases which allow for intervening words). An answer is triggered whenever the agent detects in the user's query one of the keywords or keyphrases associated with that answer. A set of generic priority rules is used to break ties between competing candidate answers (which are simultaneously induced by the concurrent presence in the question of their respective keywords).

While this chatbot is closed-domain (air travel), a few responses have been included to handle general conversation (weather, personal questions related to the chatbot, etc.), usually prompting the user to go back on topic. A few other answers are given in default of keywords in the query: the chatbot informs the user that it has not understood the question, and prompts them to rephrase it. Some answers include one or several links either to pages on the company's website or to another answer; in the latter case, a click on the link will trigger a pseudo-question (a query is generated automatically upon the click, and recorded as a new question from the user). By virtue of its design, this system is deterministic: it will always provide the same answer given the same question.

The user interface provides a simple evaluation feature: two buttons (a smiling face and a sad face) enabling users to mark an answer as relevant or irrelevant to the query that prompted it. This evaluation feature is optional and not systematically used by customers. Exchanges with the chatbot usually consist of a single QA pair. There are, however, longer conversations too. Such dialogues can span a few minutes up to many hours, as no limit is imposed on the duration of a period of inactivity (the dialogue box does not close automatically). We improperly denote all input coming from a user as a question: in fact, in longer conversations a message can be phatic, evaluative of the previous answer of the chatbot's performance, it may convey information, or it may be asking a question properly.

4 Raw Data and Task Definition

4.1 Data

Our original data consists of QA pairs from conversations with the chatbot, where users have rated the system's answer using the smiley button. For our purposes, a smiling face rating amounts to a label of "good" and a sad face rating to a label of "bad", assigned to the answer in relation to the question. This binary assessment scheme is far from the complexity of the many multidimensional evaluation frameworks that have been proposed over time to assess the subjective satisfaction or acceptance of users of dialogue systems, chiefly spoken ones [12, 29]. But, while a more nuanced evaluation scale might have been desirable, this simple binary scheme (which is not of our making, but was built into the system) is also lighter on the user. We do not equate the binary judgements with an objective measure of task success, because of their subjective component: many aspects of the user's experience with the system may influence the rating. Therefore we term the "good/bad" ratings in our data "user satisfaction ratings".

We have limited ourselves to one-turn dialogues (which are also the most common), in order to deal with self-contained questions. Our dataset contains 48,114 QA pairs from one-line dialogues. The proportions of classes are 0.28 for "good" and 0.72 for "bad". We conjecture that the predominance of negative ratings is partly a matter of negativity bias [3]: since customers are free, but not required, to evaluate the chatbot's answer, they may choose to do so mostly when they have strong feelings (which are more often negative) about it. Questions are relatively short (13 words and 70 characters on average; median values: 11 words and 57 characters), but there are a few outliers (a maximum of 241 words and 1357 characters).

4.2 Approach Chosen

As mentioned above (Sect. 1), our goal is to improve the chatbot's performance on retrieving answers. We break down the answer retrieval problem into two steps:

1. **Duplicate question detection (DQD).** Given a question, classify it as semantically similar or dissimilar to questions from a set of past questions with known answers.
2. **Answer selection.** Select an answer to the new question based on the DQD output.

In this paper we address the DQD task. We **define semantic similarity** for the task at hand in line with the definition of semantic equivalence in [6], with an additional requirement as per [22]: **we take two questions to be semantically similar if they can be correctly answered by the same answer, whose hypothetical existence suffices, provided that this answer is as specific as possible**. As a point of terminology, "similarity" seems more permissive than "equivalence" as to how far two questions are allowed to diverge from one another: "What time does the flight to New York depart on Monday 12th?" and "When is the departure time for NY on Monday 26th?" may be considered similar because they instantiate the same underlying question ("What is the departure time for New York on Mondays?"), but not strictly equivalent, since the actual details (the dates) differ.

A successful approach to answer retrieval based on DQD addresses our desired improvements to the rule-based chatbot system. It improves the retrieval performance, since it results in more questions being successfully linked to their correct answer. Additionally, the tool can present the user with a set of candidate answers if it is not confident enough to select one.

5   Data Preparation

From the original set of question-answer-label triplets, we produced **a set of question-question (QQ) pairs labelled for semantic similarity**. The transformation we applied to the data is equivalent to interpreting the result of the chatbot's retrieval heuristic in terms of DQD. Thus, all questions answered correctly by a particular answer make up a set of semantically similar questions; all questions answered incorrectly by a particular answer form pairs of semantically dissimilar questions with each of the questions for which that same answer is correct. In line with this interpretation, we generated QQ pairs as described below.

First, we grouped all the questions in our dataset by the answer they received. At this point, each answer is linked to a set of questions for which users have rated it as a good answer (its "positive" group), and to another set of questions for which it has been rated as bad (the answer's "negative" group). Second, we selected a subset of most rated (either as good or as bad) and most informative answers. We discarded very generic answers (e.g. greetings, thanks) and those stating the chatbot's inability to understand the question. An analysis of the distribution of answers in the dataset then revealed that, of the remaining 246 unique answers, the 40 answers with the highest number of total "positive" and "negative" questions made up 79% of the dataset overall, 73% of all "positive" questions, and 81% of all "negative" questions. Those 40 answers were the ones we selected for learning, since they are arguably the most useful ones: they are the most frequently given ones overall, and also comprise both the best-rated answers and the most heavily rejected by users. Next, for each of the 40 answers, we generated exhaustively: pairs of "positive" questions – these are pairs of semantically similar questions (according to our definition of semantic similarity); and pairs made up of one "positive" and one "negative" question – these are pairs of semantically dissimilar questions. Lastly, to keep the data for learning of manageable size, we sampled QQ pairs from the full pairings generated at the previous step. In order to avoid issues related to learning from an imbalanced dataset (there are more dissimilar than similar pairs), we took an equal number of similar and dissimilar pairs, by randomly sampling 10,000 similar pairs and 10,000 dissimilar pairs, which amounts to undersampling the majority class.

6   Experimental Setup

6.1   Data Preprocessing

Questions in our dataset share many features with SMS and with other types of user-generated content, such as social media. The text is riddled with spelling mistakes (e.g. *merci mais ca ne me precise pas le retard de marseille la reunion le 25 09 a 190h et j essai de vous appeler en vains car au bout de 15 mn ca racroche*), but also with the deliberate use of simplifying and expressive devices [26]: repeated punctuation, capitalisation, graphemic stretching, emoticons (e.g. *merciiiiiiiii :) :), NON NON NON!!!!!!!!! J'ai besoin du numero de vol de CDG a JFK qui arrive ce soir*).

For our task, the text of the questions in our QQ dataset underwent a number of cleaning and preprocessing steps. We cleaned up HTML markup and entities, and certain characters. Basic normalisation included lowercasing, removing punctuation, collapsing sequences of more than two repeated characters [1], restoring elided vowels, standardising spelling variations of in-domain terms and proper names and merging those which are multi-word (e.g. *ny, nyc, new york*, and *newyork* all become *newyork*; *AR, aller-retour, <>*, etc. are all replaced by *allerretour*), and grouping sequences that match specific patterns under semantic and formal classes inspired from Bikel et al. [5]: dates, telephone numbers, prices, measurements, other numeric expressions, URLs, e-mail addresses, etc. Given the poor performance and strong disagreement of four language detection packages that we tried on our data, most probably due to the very short size of our questions, we abandoned the idea of automatically filtering out questions in a language other than French. We produced five versions of the text:

1. **Preprocessed** as described above.
2. **Lemmatised** using *MElt* [7] on the preprocessed text. *MElt* is a maximum-entropy Markov-model POS tagger and lemmatiser with a normaliser/corrector wrapper trained on user-generated corpora annotated by hand. Some post-lemmatisation cleaning was needed, mainly for lemma disambiguation.
3. **PoS:** a version of the preprocessed text where tokens were replaced with their part-of-speech tags as output by *MElt*.
4. **Stemmed** using Porter's algorithm on the lemmatised version.

5. **Stemmed after removing accents from lemmas.** Because customers use accents rather haphazardly, it seems reasonable to assume that reducing word forms to stems after stripping accents may decrease considerably the size of the vocabulary.

6.2   Baselines

Our weak baseline is the chatbot in its current form, taken as an (indirect) detector of similar questions. The construction procedure of our QQ dataset means that this baseline has 50% accuracy on our class-balanced data. Indeed, the chatbot correctly identifies all the similar QQ pairs as similar, but it also takes all the dissimilar ones for similar.

For the remainder of systems, including the second baseline, the same train/test split on the data was used, with an 80/20 ratio. We take as strong baseline the Jaccard similarity coefficient, a measure of overlap between sets which is common in information retrieval [21], and which has been used for textual entailment recognition [20] and for near-duplicate detection tasks [30]. For each QQ pair, we compute the Jaccard coefficient between the two questions represented as a set of n-grams (with n running from 1 to 4). The cutoff value is optimised on the training set, and evaluated on the test set.

6.3   Proposed Systems

The systems we are testing are two CNN architectures developed specifically for DQD, which performed very well on a dataset in English from the AskUbuntu forum [6]. CNN architectures have shown great success at a number of natural language processing tasks, such as classifying sentences [16] or modelling sentence pairs [32].

CNN. Our system is based on the CNN architecture for DQD proposed in [6]. First, the CNN obtains vector representations of the words, also known as word embeddings, from the two input segments. Next, a convolution layer constructs a vector representation for each of the two segments. Finally, the two representations are compared using cosine similarity. If the value of this metric exceeds an empirically estimated threshold, the two segments are classified as duplicate. The same feature maps (for word embedding and the convolution layer) are used to produce the representation of both questions.

Our CNN architecture is also inspired from [17]. The authors of that paper use the concatenation of several convolution filters with multiple feature widths. We improve the architecture proposed in [6] by changing the convolution layer to a set of convolution filters with multiple feature widths (*cf.* diagram in Fig. 1).

The vector representation uses an embedding layer of 200 randomly initiated neurons which are trainable. Each convolution layer uses 100 neurons for the output of the filters, and the widths of the filters are 2, 3, and 5. The optimisation algorithm used for the network is stochastic gradient descent (SGD) with a learning rate of 0.005.

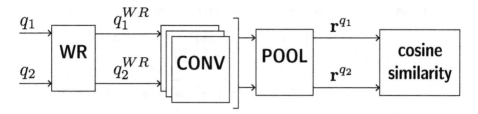

Fig. 1. CNN architecture, with layers: word representation (WR) for a pair of questions (q_n; output q_n^{WR}); concatenated set of convolution filters (CONV); max pooling (POOL); question representation ($\mathbf{r}^{q_n}$); and cosine similarity measurement.

Hybrid Deep CNN (CNN-Deep). The second system we tested is described in detail in [25]. It combines a CNN similar to our first proposed system with a deep neural network with three hidden, fully-connected, layers, based on the architecture described in [2]. A diagram of the system is shown in Fig. 2.

The vector representation uses an embedding layer of 300 randomly initiated neurons which are trainable. The convolution layer uses 300 neurons for the output of filters with a kernel size of 15 units, and each deep layer has 50 neurons. The optimisation algorithm used for the network is SGD with a learning rate of 0.01.

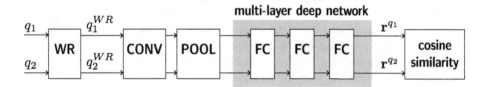

Fig. 2. CNN-Deep architecture: as the CNN, with the addition of fully-connected layers (FC).

7   Results and Discussion

7.1   System Performance

From Table 1, it is immediately obvious that two of the three systems (Jaccard and CNN-Deep) perform barely better than a random classifier (such as our weak baseline, which has accuracy of 50%), while CNN is at the order of 20% points above both on four of the five text versions.

The poor accuracy of the Jaccard similarity baseline goes to confirm that, for our task, word overlap is not a reliable indicator of semantic similarity. For example, the questions in pair 1 in Table 2 are similar (according to our definition) despite sharing almost no words (they share fewer words in French than in our English translation). On the other hand, CNN-Deep scoring barely better

Table 1. Accuracy for the DQD task on the five versions of the data.

	Jaccard	CNN	CNN-Deep
Preprocessed	52.8	**74.9**	55.8
Lemmatised	55.2	72.5	53.0
PoS	51.6	59.7	55.7
Stemmed	54.3	72.4	56.0
Stemmed unaccented lemmas	54.2	72.1	55.3

than the Jaccard baseline is consistent with the results reported in [25] for a general-domain corpus. It is striking that, on this data and this task, a tool of this level of sophistication is on par with a very simple overlap measure. The complexity of CNN-Deep's architecture might be ill-suited to the needs of the task at hand. Conversely, the simpler CNN architecture performs better.

Although for each system the differences in accuracy when applied to the different versions of the text are generally small, each system seems to perform best on a specific version. Nonetheless, the three methods do not agree on which level of preprocessing is the most efficient: Jaccard seems to prefer lemmas over stems, while CNN-Deep does the worst on lemmas and the best on stems; and the performance of the CNN deteriorates with any additional processing on top of the initial preprocessing. Surprisingly enough, although stems from unaccented lemmas are more powerful in collapsing the vocabulary, they do not lead to improved performance compared to stems over lemmas as such. Representing the text exclusively as parts of speech has a negative impact on Jaccard and results in an even more marked drop in accuracy for CNN, but does not seem to affect CNN-Deep. Overall, it is hard to assess the benefit of the different types of text preprocessing. Depending on the tool and on the task, the effects may differ.

7.2 Difficulty of the Task

The task we set out to tackle is hard. Two human annotators asked to label independently as semantically similar or not a random sample of 100 QQs pairs from our data have achieved a Cohen's kappa as low as 0.332. The annotators were given our definition of semantic similarity and a few examples (including a reply which is so general that it could arguably answer any query), and were instructed to decide whether the two questions in each pair are similar according to the definition. The agreement is very low not only between the annotators, but also between each of them and the ground truth. Very low correlation between raters has been reported in the literature for hard tasks. For instance, on a task that consisted in rating three aspects related to user satisfaction with the dialogue turns of an automated or human interlocutor, [10] reports near-zero Spearman's rank correlation between two raters, including on the easiest of the three aspects, which is deciding if the interlocutor is a good listener or not. Such

Table 2. Example QQ pairs from our dataset (preprocessed version). The errors in the French are the users' (*cf.* Sect. 6.1), and the English translation mimics the French.

	Question 1	Question 2
1	*comment puis je choisir ma place dans le avion* "how can i choose my seat in the plane"	*je souhaiterai savoir comment faire pour réserver un siège en ligne* "i would like to know how to book a seat online"
2	*je ai dépassé la date pour réservé un siège car je pars dans NUMBER jours comment je peus faire* "i have missed the deadline for booking a seat because i leave in NUMBER days what can i do"	*bonsoir puis je réservé mon siege pour cancun* "hello can i book my seat for cancun"
3	*est il possible de payer par chèque* "do you accept cheques"	*peut on payer plusieurs mensualite* "do you accept instalment payments"
4	*quels sont les moyens de paiement* "what are the payment options"	*est il possible de payer par paypal* "can i pay with paypal"
5	*je souhaiterai savoir pourquoi vous ne avez pas de autres dates de disponible pour debut septembre NUMBER* "i would like to know why there are no other dates available for early september NUMBER"	*bonjour je peut pas reserver pour avril* "hello i cannot book for april"
6	*a partir de quand puis je choisir mon siège* "when will i be able to choose my seat"	*j ai reserve et je voudrais savoir ou je suis assise ou si je dois choisir ma place* "i have booked a ticket and i would like to know where i am seated or if i need to choose my seat"
7	*poids* "weight"	*pour un deuxieme bagage vers les dom on a droit a combien de kg* "what is the maximum weight for a second piece of luggage for the overseas departments"
8	*bonjour je suis à la réunion* "i am in réunion"	*cherche vol reunion charles de gaule* "looking for a reunion charles de gaule flight"
9	*je peux prendre le bagage sup sur le vol retour ajout excédents* "can i take the extra baggage on the flight back extra baggage"	*bonjour concernant le bagage supplémentaire quel est le tarif vol cdg neew york jfk classe éco* "hello about the extra baggage what are the fees cdg neew york jfk economy flight"

low agreement may suggest that the task is very hard for a human to solve, that the data may be too noisy for any patterns to be discoverable, or even that there may be no patterns to learn in the data in the first place. We believe the first hypothesis to be plausible in our case. The fact that our best system (CNN) achieves 60% accuracy on the same sample – which, while not as high as the performance on our test sets, is a considerable improvement over random label assignment – points to there being some actual patterns to learn from the data, even if they may not be easily discernible to a human judge.

Specification of the User's Information Need. We believe the difficulty of deciding whether two questions are semantically similar according to our definition may stem from the complexity of correctly inferring the user's real information need from the question they ask. The potential discrepancy between a user's actual information need and what may be inferred from its expression in a textual query is a pervasive problem in information retrieval [21]. As an example, to assess how well suited to a question the answers retrieved by their system were, the authors of [15] had raters "back-generate" a possible information need behind each question before judging the quality of the answers provided by the system. Those researchers point out that for some questions the assessors were unable to reconstruct the original information need, which means they were unable to judge the quality of the answers. Some of the questions in our dataset exhibit an underspecification of the information need (e.g. question 1 in example QQ pair 7), while others are extremely specific (e.g. question 2 in example QQ pair 7); further details are needed about the first one to decide whether the same answer could fit them both. Some questions are incomplete, as question 1 in our example pair 8; if we assume an information need (perhaps the most likely one, or perhaps Paris is the only destination reachable from the origin stated by the user), this question may be viewed as similar to question 2 in pair 8.

Annotators' Knowledge of the Domain and Context. Assessing the quality of answers in the domain at hand does not require any technical knowledge, so the "expert/novice" annotator distinction in [19] does not apply here sensu stricto; still, the level of familiarity with the domain (air carrier's products) may affect an annotator's perception of an answer's relevancy. Our annotators were not familiar with the domain, which complicates their assessment of whether the answer is specific enough to satisfy the query. Example QQ pair 2 in Table 2 shows two questions which may very well be acceptably answered by a reply providing comprehensive details about the company's seat reservation policy; however, the first user may expect a reply dealing specifically and exclusively with seat reservation when the deadline has expired. That goes for example pairs 3 and 4: as long as the generic answer is, in fact, exhaustive, it is perfectly valid for any question whose specific answer is included in the generic one. Human raters, however, will find it difficult to decide on the semantic similarity of two questions without some knowledge of the context and the domain. To decide whether the questions in example QQ pair 5 may be similar, one would need to know if both questions were asked a certain number of months earlier than the desired travel date, and if the company does have a policy for handling early bookings, in which case a common answer may satisfy both queries. Likewise, the semantic similarity of example pair 6 may hinge on the actual availability of a seat choice option; if there is none, this will be the valid answer to both questions. QQ pair 9 may be a case of similarity if the company's excess baggage policy is the same regardless of the route.

8   Conclusion and Future Work

Deciding whether two questions are semantically similar or not is a hard task for humans. Notwithstanding, one of the systems tested in this paper, the CNN, achieved good accuracy on a QQ set derived from user-chatbot exchanges labelled for user satisfaction, outperforming the rule-based chatbot on this task. By simply learning from user-labelled data collected over time, a chatbot can thus improve significantly its ability to detect similar questions in the course of time.

But ultimately, our goal is to assess the usefulness of DQD as part of an answer-retrieving chatbot. Therefore, our next step will be to test our system on Step 2 (*cf.* Sect. 4), i.e. the actual retrieval of an answer using the output of Step 1 (DQD). To evaluate the performance of our proposed system on this task against the existing system as a baseline, we are preparing a set of questions labelled for their correct answer. Another issue to tackle will be an optimal way of performing fast and efficiently the comparisons between the incoming question and the ones in the reference set as that set grows over time.

In this experimental setup we have restricted ourselves to one-line dialogues, but conversations offer a good ground for yet another application of DQD: detecting the rephrasing of a question during a dialogue, which may be indicative of a problem that requires attention. It would also be interesting to assess the impact of more advanced spelling normalisation and correction on our best system's performance. In addition, new experiments could take account of the known correct answer to a past question when assessing its similarity with a new question. Last but not least, it will be interesting to validate the results reported here on a similar corpus coming from a different chatbot in a different domain.

Acknowledgements. This research is partly funded by the Regional Council of Brittany through an ARED grant. The present research was also partly supported by the CLARIN and ANI/3279/2016 grants. We are grateful to Telsi for providing the data.

References

1. Accorsi, P., Patel, N., Lopez, C., Panckhurst, R., Roche, M.: Seek & hide: anonymising a french sms corpus using natural language processing techniques. Lingvisticæ Investigationes **35**(2), 163–180 (2012)
2. Afzal, N., Wang, Y., Liu, H.: MayoNLP at SemEval-2016 Task 1: semantic textual similarity based on lexical semantic net and deep learning semantic model. In: Proceedings of the 10th International Workshop on Semantic Evaluation, SemEval@NAACL-HLT 2016, San Diego, CA, USA, 16–17 June 2016, pp. 674–679 (2016)

3. Baumeister, R.F., Bratslavsky, E., Finkenauer, C., Vohs, K.D.: Bad is stronger than good. Rev. Gen. Psychol. **5**(4), 323 (2001)
4. Bernhard, D., Gurevych, I.: Answering learners' questions by retrieving question paraphrases from social Q&A sites. In: Proceedings of the Third Workshop on Innovative Use of NLP for Building Educational Applications, pp. 44–52. ACL (2008)
5. Bikel, D.M., Schwartz, R., Weischedel, R.M.: An algorithm that learns what's in a name. Mach. Learn. **34**(1), 211–231 (1999)
6. Bogdanova, D., dos Santos, C.N., Barbosa, L., Zadrozny, B.: Detecting semantically equivalent questions in online user forums. In: Proceedings of the 19th Conference on Computational Natural Language Learning, CoNLL 2015, Beijing, China, 30–31 July 2015, pp. 123–131 (2015)
7. Denis, P., Sagot, B.: Coupling an annotated corpus and a lexicon for state-of-the-art pos tagging. Lang. Resour. Evaluation **46**(4), 721–736 (2012)
8. Feng, M., Xiang, B., Glass, M.R., Wang, L., Zhou, B.: Applying deep learning to answer selection: a study and an open task. In: 2015 IEEE Workshop on Automatic Speech Recognition and Understanding, ASRU 2015, Scottsdale, AZ, USA, 13–17 December 2015, pp. 813–820 (2015)
9. Goldberg, Y.: Neural Network Methods for Natural Language Processing. Morgan & Claypool, San Rafael (2017)
10. Higashinaka, R., Minami, Y., Dohsaka, K., Meguro, T.: Issues in predicting user satisfaction transitions in dialogues: individual differences, evaluation criteria, and prediction models. In: Lee, G.G., Mariani, J., Minker, W., Nakamura, S. (eds.) IWSDS 2010. LNCS (LNAI), vol. 6392, pp. 48–60. Springer, Heidelberg (2010). https://doi.org/10.1007/978-3-642-16202-2_5
11. Hogan, D., Leveling, J., Wang, H., Ferguson, P., Gurrin, C.: Dcu@fire 2011: SMS-based FAQ retrieval. In: 3rd Workshop of the Forum for Information Retrieval Evaluation, FIRE, pp. 2–4 (2011)
12. Hone, K.S., Graham, R.: Subjective assessment of speech-system interface usability. In: INTERSPEECH, pp. 2083–2086 (2001)
13. Jalbert, N., Weimer, W.: Automated duplicate detection for bug tracking systems. In: IEEE International Conference on Dependable Systems and Networks with FTCS and DCC (DSN 2008), pp. 52–61. IEEE (2008)
14. Jeon, J., Croft, W.B., Lee, J.H.: Finding semantically similar questions based on their answers. In: Proceedings of the 28th Annual International ACM SIGIR Conference on Research and Development in Information Retrieval, pp. 617–618. ACM (2005)
15. Jijkoun, V., de Rijke, M.: Retrieving answers from frequently asked questions pages on the web. In: Proceedings of the 14th ACM International Conference on Information and Knowledge Management, pp. 76–83. ACM (2005)
16. Kim, Y.: Convolutional neural networks for sentence classification. arXiv preprint arXiv:1408.5882 (2014)
17. Kim, Y.: Convolutional neural networks for sentence classification. In: EMNLP, pp. 1746–1751. ACL (2014)
18. Liu, C.W., Lowe, R., Serban, I.V., Noseworthy, M., Charlin, L., Pineau, J.: How not to evaluate your dialogue system: an empirical study of unsupervised evaluation metrics for dialogue response generation. arXiv preprint arXiv:1603.08023 (2016)
19. Lowe, R., Serban, I.V., Noseworthy, M., Charlin, L., Pineau, J.: On the evaluation of dialogue systems with next utterance classification. arXiv preprint arXiv:1605.05414 (2016)

20. Malakasiotis, P., Androutsopoulos, I.: Learning textual entailment using SVMs and string similarity measures. In: Proceedings of the ACL-PASCAL Workshop on Textual Entailment and Paraphrasing, pp. 42–47. ACL (2007)

21. Manning, C.D., Raghavan, P., Schütze, H., et al.: Introduction to Information Retrieval, vol. 1. Cambridge University Press, Cambridge (2008)

22. Muthmann, K., Petrova, A.: An automatic approach for identifying topical near-duplicate relations between questions from social media Q/A sites. In: Proceeding of WSDM 2014 Workshop: Web-Scale Classification: Classifying Big Data from the Web (2014)

23. Reitter, D., Moore, J.D.: Predicting success in dialogue. In: Proceedings of the 45th Annual Meeting of the ACL, ACL 2007, 23–30 June 2007, Prague, Czech Republic (2007)

24. Ritter, A., Cherry, C., Dolan, W.B.: Data-driven response generation in social media. In: Proceedings of the Conference on Empirical Methods in Natural Language Processing, pp. 583–593. ACL (2011)

25. Rodrigues, J.A., Saedi, C., Maraev, V., Silva, J., Branco, A.: Ways of asking and replying in duplicate question detection. In: Ide, N., Herbelot, A., Màrquez, L. (eds.) Proceedings of the 6th Joint Conference on Lexical and Computational Semantics, *SEM @ACM 2017, Vancouver, Canada, 3–4 August 2017, pp. 262–270. Association for Computational Linguistics (2017). https://doi.org/10.18653/v1/S17-1030

26. Seddah, D., Sagot, B., Candito, M., Mouilleron, V., Combet, V.: The French Social Media Bank: a treebank of noisy user generated content. In: 24th International Conference on Computational Linguistics, COLING 2012 (2012)

27. Vinyals, O., Le, Q.: A neural conversational model. arXiv preprint arXiv:1506.05869 (2015)

28. Walker, M., Langkilde, I., Wright, J., Gorin, A., Litman, D.: Learning to predict problematic situations in a spoken dialogue system: experiments with how may I help you? In: Proceedings of the 1st North American Chapter of the Association for Computational Linguistics Conference, pp. 210–217. Association for Computational Linguistics (2000)

29. Walker, M.A., Litman, D.J., Kamm, C.A., Abella, A.: PARADISE: A framework for evaluating spoken dialogue agents. In: Proceedings of the Eighth Conference on European Chapter of the Association for Computational Linguistics, pp. 271–280. ACL (1997)

30. Wu, Y., Zhang, Q., Huang, X.: Efficient near-duplicate detection for Q&A forum. In: Fifth International Joint Conference on Natural Language Processing, IJCNLP 2011, Chiang Mai, Thailand, 8–13 November 2011, pp. 1001–1009 (2011)

31. Xue, X., Jeon, J., Croft, W.B.: Retrieval models for question and answer archives. In: Proceedings of the 31st Annual International ACM SIGIR Conference on Research and Development in Information Retrieval, pp. 475–482. ACM (2008)

32. Yin, W., Schütze, H., Xiang, B., Zhou, B.: ABCNN: attention-based convolutional neural network for modeling sentence pairs. arXiv preprint arXiv:1512.05193 (2015)

Speech Processing

Deep Learning for Acoustic Addressee Detection in Spoken Dialogue Systems

Aleksei Pugachev[1,2(✉)], Oleg Akhtiamov[1,3], Alexey Karpov[1,2], and Wolfgang Minker[3]

[1] ITMO University, Saint-Petersburg, Russia
terixoid@gmail.com
[2] SPIIRAS Institute, Saint-Petersburg, Russia
[3] Ulm University, Ulm, Germany

Abstract. The addressee detection problem arises in real spoken dialogue systems (SDSs) which are supposed to distinguish the speech addressed to them from the speech addressed to real humans. In this work, several modalities were analyzed, and acoustic data has been chosen as the main modality by reason of the most flexible usability in modern SDSs. To resolve the problem of addressee detection, deep learning methods such as fully-connected neural networks and Long Short-Term Memory were applied in the present study. The developed models were improved by using different optimization methods, activation functions and a learning rate optimization method. Also the models were optimized by using a recursive feature elimination method and multiple initialization to increase the training speed. A fully-connected neural network reaches an average recall of 0.78, a Long Short-Term Memory neural network shows an average recall of 0.65. Advantages and disadvantages of both architectures are provided for the particular task.

Keywords: Off-talk · Multiparty conversation · LSTM
Fully-connected neural network · Speech processing · Speaking style

1 Introduction

Human-human-computer interaction is a common phenomenon in real spoken dialogue systems (SDSs). Handling this kind of interaction, the system is supposed to distinguish the speech addressed to it (On-talk) from the speech addressed to another human (Off-talk). There are a lot of approaches to solving this problem such as text classification for semantic analysis, acoustic analysis – to detect speech anomalies intrinsic to the speech addressed to an automatic speech recognition system (ASR) (long pauses between words, precise pronunciation, etc.), gaze detection – to detect in which direction the talking human looks.

All of these methods possess advantages and disadvantages:

1. Text classification:

- Best choice for a dialogue system in a specific domain
- Domain dependent - Language dependent

© Springer International Publishing AG 2018
A. Filchenkov et al. (Eds.): AINL 2017, CCIS 789, pp. 45–53, 2018.
https://doi.org/10.1007/978-3-319-71746-3_4

2. Acoustic analysis:

- Still valid by reason of ASR imperfection
- Domain and language independent
- Harder to train than a text classifier

3. Gaze detection:

- Easy to train
- Valid only in those systems where a human wants to see an answer on the system screen

Acoustic data was chosen as the primary modality for this research by reason of the flexibility of its usability: nowadays, people have to change their manner of speech in a way that will be easier recognized by ASR. Also the acoustic modality is universal: there is no dependency on language and domain.

This paper presents results of acoustic classification with helps of different deep learning architectures (DNN [1] and LSTM [2]). The reason why we choose deep learning methods as primary classifiers is the flexibility of hyperparameters that gives us many capabilities to build strong models.

2　Related Work

In [3, 4] four corpora were used: two for human-human interaction and two for human-computer interaction. The first way to solve this problem was an application of linguistic models based on n-grams. To solve the problem with out-of-vocabulary words, the authors decided to replace rare words by their part-of-speech tags. Two linguistic models were proposed: P(w|H) – for human-human and P(w|C) – for human-computer interaction. Then the system returns a score based on the lengthnormalized likelihood ratio of two classes:

$$\frac{1}{|w|} log \frac{P(w|c)}{P(w|H),}$$

where |w| is the number of words in the recognition output w for an utterance. P(w|C) and P(w|H) are obtained from class-specific language models. As a metric, they used equal error rate (EER) and achieved a value of 14.625%.

In [5] five modalities were used:

1. Acoustic
2. Visual
3. System
4. Beam forming
5. ASR

In total, 117 features were extracted. Classification was done by using adaboost with tree stumps. They achieved EER of 9.84%.

In [6], experiments on the Smart Video Corpus [7] were done. Three modalities were used in this work:

1. Acoustic
2. Text
3. Visual

A linear discriminant classifier (LDC) was used for audio feature extraction: 13 metafeatures from 100 prosodic features. The same classifier was used for text feature extraction: 18 meta-features from 30 part-of-speech features. Wavelet transformation was used to extract nine visual meta-features. Finally, all meta features were used as input attributes for a higher level LDC to perform the final classification. The results of this work were an average recall of 0.681 for two classes and an average recall of 0.530 for four classes.

3   Experimental Data

As data, we used the Smart Video Corpus which contains 3.5 h of German speech in three modalities – text, audio and video. The labeling of Off-Talk and On-Talk classes was provided only for the text modality for each word. For audio, time labeling was given only for utterances. In order to create the labeling for audio, we counted all labels in one utterance, and the class with the maximum number of word labels was assigned as the utterance label. Finally, we have got the following number of data points in the Off-talk class – 1115 and in the On-talk class – 1078.

4   Features

In this research, we experimented with two types of features:

1. Features extracted with the OpenSmile [8] toolkit by using the configuration of INTERSPEECH 2013 Computational Paralinguistic Challenge [9]. The features were extracted from an entire utterance: for each utterance we have got a 6373 feature vector. Result feature vector contains such features as:
   a. Chroma features;
   b. Mel-frequency Cepstral Coefficients (MFCC) (25 ms window, 13 coefficients, 13 delta coefficients, 13 acceleration coefficients) with different filtering and normalization;
   c. Preceptual linear prediction (PLP) features (six coefficients, six delta coefficients, six acceleration coefficients) with different processing of zero coefficients and normalization techniques;
   d. Prosodic features;
   e. Features for emotion recognition (voicing probability computed from the autocorrelation function, the fundamental frequency computed from cepstrum, statistical contour features, skewness, kurtosis)

2. The combination of pitch and root mean square energy (RMSE) extracted within a 200 ms window with ¾ overlapping.

5   Models

As a baseline system, we implemented a support vector machine (SVM) [10] in conjunction with the OpenSmile features. This system provides the minimum threshold for our metric: deep learning methods which gave us lower results were marked as not valid for this task or requiring modifications.

For the OpenSmile features, we created a deep fully-connected neural network with two hidden layers (Fig. 1).

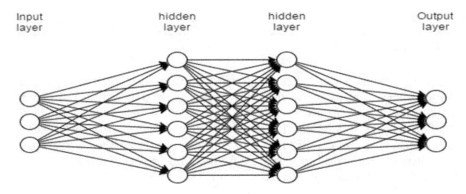

Fig. 1. Fully-connected DNN architecture

The dimensionality of the input layer was 1000 neurons, of hidden layers – 3000 neurons in each, and of the output layer – two neurons. The activation function of hidden layers was exponential linear unit (ELU) [11], of the output layer – sigmoid. As an optimizer, we applied stochastic gradient descent (SGD) [12] with momentum (initial learning rate – 0.2, mass – 0.6) with 200 minibatch length and 50 epochs. Error function – Cross-Entropy.

Also we used a learning rate optimization method: we popped 10% of each class from the training data into a new cross-validation set. On this set, we computed accuracy after each epoch. If accuracy gets lower more than by a threshold – we reduced the learning rate in half. The threshold for the learning rate optimization was 0.01. This method gives us a much better error function behavior (comparison of Figs. 2 and 3)

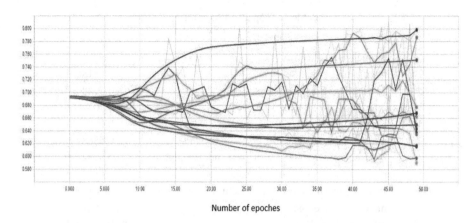

Fig. 2. Error functions for 14-fold cross-validation without learning rate optimization

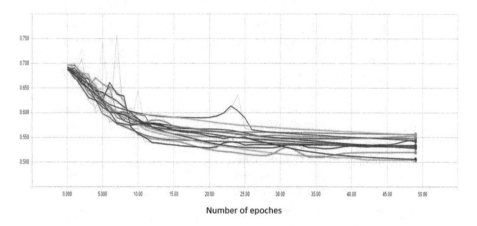

Fig. 3. Error functions for 14-fold cross-validation with learning rate optimization

To increase the speed of the training process, we used two methods:

1. Recursive feature elimination [13] – we took SVM weights, and the features with the highest weights were sent to the fully-connected DNN input layer. With 1000 features we achieved the same results as with 6373, moreover, we did it much faster and avoided the curse of dimensionality. SVM weights were obtained by Rapid-Miner toolkit which calculates the relevance of the attributes by computing for each attribute of the input example set the weight with respect to the class attribute. Top 29 features are listed in Table 1.

Table 1. Top 29 SVM features

Top	Name	Weight
1	F0final_sma_risetime	1
2	mfcc_sma[1]_minPos	0,7618435
3	voicingFinalUnclipped_sma_lpc4	0,7148017
4	mfcc_sma_de[1]_minPos	0,7072656
5	audSpec_Rfilt_sma_de[15]_maxPos	0,6654168
6	mfcc_sma_de[5]_maxPos	0,6368439
7	audspecRasta_lengthL1norm_sma_minRangeRel	0,5982509
8	pcm_RMSenergy_sma_peakRangeRel	0,5865465
9	voicingFinalUnclipped_sma_risetime	0,5819561
10	mfcc_sma[5]_qregc3	0,5801319
11	mfcc_sma[10]_linregc1	0,5718525
12	pcm_fftMag_spectralRollOff25.0_sma_risetime	0,5526515
13	pcm_zcr_sma_minRangeRel	0,5522197
14	pcm_fftMag_spectralRollOff25.0_sma_de_quartile2	0,5481618
15	pcm_fftMag_spectralEntropy_sma_de_lpc4	0,5437438
16	pcm_fftMag_spectralRollOff50.0_sma_risetime	0,5428366
17	mfcc_sma_de[3]_leftctime	0,5415289
18	mfcc_sma[8]_minPos	0,5208094
19	mfcc_sma_de[1]_maxPos	0,4987886
20	mfcc_sma_de[5]_quartile2	0,4985048
21	audSpec_Rfilt_sma[1]_leftctime	0,4971357
22	audSpec_Rfilt_sma_de[8]_skewness	0,4956926
23	mfcc_sma_de[6]_upleveltime90	0,4949792
24	mfcc_sma[8]_stddevRisingSlope	0,4903852
25	audspec_lengthL1norm_sma_peakRangeRel	0,4874503
26	audSpec_Rfilt_sma[2]_centroid	0,48
27	pcm_fftMag_spectralVariance_sma_minPos	0,4697651
28	mfcc_sma_de[6]_peakMeanRel	0,4691406
29	mfcc_sma_de[13]_minPos	0,4650589

2. Multiple initialization – in this method we initialized weights for five models and trained them on five epochs. After five epochs we checked the accuracy of each model and continued learning the model with best result on 45 epochs. This method did not give us a significant improvement in metric, however, due to this approach we avoided the initialization in a local minimum. Without this method, we needed 250 epochs to check the quality of five initializations, while with this method we required only 70 epochs.

In Table 2 presented summary of parameters in our experiments with deep neural network architecture.

Table 2. Deep neural network parameters

Parameters	DNN1	DNN2	DNN3
Input dimension	6373	6373	1000
Number of hidden layers	2	2	2
Hidden layer dimension	1000	3000	3000
Minibatch length	100	200	200
Optimizer	SGD	Momentum	Momentum
Initial learning rate	0.2	0.2	0.2
Mass	–	0.6	0.6
Number of epochs	50	50	50
Learning rate optimization threshold	–	0.01	0.01
Multiple initialization	–	–	+

During these experiments we used different dimensionality of layers, optimizers with different sets of parameters, includes speed up techniques, tried different activation functions. Short summary of deep neural network experiment results is listed in Table 3.

Table 3. Deep neural networks results

	Recall	F1	Accuracy
DNN1	0.69	0.58	0.61
DNN2	0.78	0.78	0.68
DNN3	0.78	0.78	0.69

For the combination of pitch and RMSE, we applied a Bidirectional Long Short-Term Memory (BLSTM) [14] model with the input layer consisting of two neurons, two LSTM (Fig. 4) layers with 50 neurons in each, one fully-connected layer after BLSTM with 50 neurons and the output layer with two neurons.

As an optimizer, we used RMSProp [16] with a mass of 0.5, learning rate of 0.005, decay of 0.5 and minibatch length of 100. We trained this model on 20 epochs. Unfortunately, we determined that pitch information gives us very fast overfitting (on the fourth

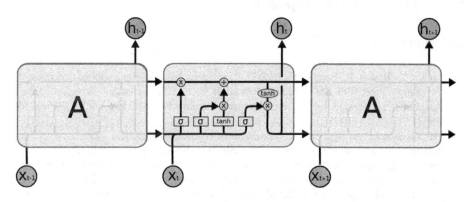

Fig. 4. LSTM architecture [15]

epoch). Therefore, we decided to exclude pitch from consideration and change the number of neurons in the input layer to one.

6   Results

In Table 4 results of models are presented with the average recall metric:

Table 4. Results of different models

	SVM	DNN	BLSTM
Recall	0.7	0.78	0.65

Why does the simple DNN method give better results than BLSTM? In a real SDS, a human who talks to a computer may change the direction of his or her speech within one utterance. By this factor, models based on RNN [17] may get confused, therefore, to avoid this problem we need to get labeling for each word in the audio modality. Another possible reason of such a poor result is the fact that we probably do not have enough data for such complex architectures.

7   Conclusions

In this research, we applied the fully-connected deep neural network which gives us high results especially for the acoustic modality. Also the BLSTM model was developed, and to improve the results of this model, we need to get time labeling for each word in each utterance. In order to do this, we require ASR. Also we need to obtain more data and try data augmentation methods.

In our future work, we will experiment with multi-modal fusion and transfer learning [18] between text and audio modalities to obtain stronger domain and language independent models.

Acknowledgments. This work is partially supported by the grant of the President of Russia (No. MD-254.2017.8) and by the RFBR (project No. 16-37-60100).

References

1. LeCun, Y., Bottou, L., Bengio, Y., Haffner, P.: Gradient-based learning applied to document recognition. Proc. IEEE **86**(11), 2278–2324 (1998)
2. Hochreiter, S., Schmidhuber, J.: Long short-term memory. Neural Comput. **9**(8), 1735–1780 (1997)
3. Lee, H., Stolcke, A., Shriberg, E.: Using out-of-domain data for lexical addressee detection in human-human-computer dialog. In: Proceedings of NAACL, pp. 221–229 (2013)
4. Shriberg, E., Stolcke, A., Ravuri, S.: Addressee detection for dialog systems using temporal and spectral dimensions of speaking style. In: Proceedings of Interspeech (2013)
5. Tsai, T.J., Stolcke, A., Slaney, M.: Multimodel addressee detection in multiparty dialogue systems. In: Proceedings of ICASSP, pp. 2314–2318, April 2015

6. SmartWeb Handled Corpus. http://catalog.elra.info/product_info.php?products_id=1068. Accessed 27 Apr 2017
7. Batliner, A., Hacker, C., Nöth, E.: J Multimodal User Interfaces **2**, 171 (2008). https://doi.org/10.1007/s12193-009-0016-6
8. OpenSmile IS13 configuration. http://www.audeering.com/research-and-opensource/files/openSMILE-book-latest.pdf. Accessed 27 Apr 2017
9. INTERSPEECH 2013 Computational Paralinguistics Challenge. http://emotionresearch.net/sigs/speech-sig/is13-compare. Accessed 21 June 2017
10. Cortes, C., Vapnik, V.: Support-vector networks. Mach. Learn. **20**(3), 273–297 (1995). https://doi.org/10.1007/BF00994018
11. Clevert, D.A., Unterthiner, T., Hochreiter, S.: Fast and accurate deep network learning by exponential linear units (ELUs). In: ICLR (2016)
12. Bottou, L.: Large-scale machine learning with stochastic gradient descent. In: International Conference on Computational Statistics, pp. 177–187 (2010)
13. Zeng, X., Chen, Y.W.: Feature selection using recursive feature elimination for handwritten digit recognition. In: Proceedings of Fifth International Conference on Intelligent Information Hiding and Multimedia Signal Processing, pp. 1205–1208 (2009)
14. Ray, A., Rajeswar, S., Chaudhury, S.: Text recognition using deep blstm network. In: Proceedings of the International Conference on Advances of Pattern Recognition (2015)
15. Understanding LSTM Networks – colah's blog. http://colah.github.io/posts/2015-08Understanding-LSTMs/. Accessed 21 June 2017
16. An overview of gradient descent optimization algorithms. http://sebastianruder.com/optimizing-gradient-descent/index.html#rmsprop. Accessed 21 June 2017
17. Cho, K.: Learning phrase representations using RNN encoder-decoder for statistical machine translation. arXiv:1406.1078 (2014)
18. Weiss, K., Khoshgoftaar, T.M., Wang, D.D.: A survey of transfer learning. J. Big Data **3**(1), 1–40 (2016)

Deep Neural Networks in Russian Speech Recognition

Nikita Markovnikov[1,2(✉)], Irina Kipyatkova[2], Alexey Karpov[2],
and Andrey Filchenkov[1]

[1] ITMO University, Saint-Petersburg, Russia
niklemark@gmail.com
[2] SPIIRAS Institute, Saint-Petersburg, Russia

Abstract. Hybrid speech recognition systems incorporating deep neural networks (DNNs) with Hidden Markov Models/Gaussian Mixture Models have achieved good results. We propose applying various DNNs in automatic recognition of Russian continuous speech. We used different neural network models such as Convolutional Neural Networks (CNNs), modifications of Long short-term memory (LSTM), Residual Networks and Recurrent Convolutional Networks (RCNNs). The presented model achieved 7.5% reducing of word error rate (WER) compared with Kaldi baseline. Experiments are performed with extra-large vocabulary (more than 30 h) of Russian speech.

Keywords: Deep learning · Russian speech · Speech recognition
Acoustic models

1 Introduction

Automatic speech recognition (ASR) is a process of converting speech to text. It can be performed using both acoustic model (AM) and language model (LM) as shown in [1]. In this paper, we consider building and learning of acoustic models only.

Acoustic models are traditionally built using hidden Markov models (HMM) with the Gaussian mixture model (GMM). However, hybrid deep neural networks with Hidden Markov Models (DNN-HMM) models recently showed better results and reduced error of speech recognition [2].

DNN models for languages with strict word order (e.g. English) suit well, but as for the Russian language, these models are not such efficient. Our motivation is to find neural network architecture that would accomplish an improvement of our Kaldi baseline.

Recently, there some promising models were proposed. For instance, recurrent neural networks such as the long short-term memory have achieved significant results in speech recognition. However, LSTMs are easy to overfit. Convolutional neural networks (CNNs) is a popular class of deep neural networks, but it has

© Springer International Publishing AG 2018
A. Filchenkov et al. (Eds.): AINL 2017, CCIS 789, pp. 54–67, 2018.
https://doi.org/10.1007/978-3-319-71746-3_5

not achieved large reduction of recognition error. Our goal is to construct and apply various deep neural networks to the problem of automatic recognition of continuous Russian speech.

To study an acoustic model, we need a large corpus of the Russian speech. In this work, neural networks were constructed using extra-large vocabulary with more than 25 h of the Russian continuous speech that will be described below.

The performance of our ASR systems was evaluated in term of word error rate (WER). This metric is computed using the Levenstein distance between the recognized sequence and the truth sequence and it is expressed in percentage as follows:

$$WER = \frac{D + S + I}{N} \cdot 100\%$$

where N denotes the total number of words in the truth sequence, D is the number of deletions, S is the number of substitutions and I denotes the number of insertions.

The rest of the paper is organized as follows. In Sect. 2, we survey related works. In Sect. 3, we describe architectures of DNNs that we used for the constructing AMs. In Sect. 4, we discuss datasets for a training and testing AMs and our language model. In Sect. 6, we describe our experimental setup and present configurations of neural networks and the results. Finally, we conclude in Sect. 7.

2 Related Work

We give a brief overview of Kaldi [3]. Kaldi is a toolkit written in C++, integrated with OpenFST toolkit for a support of finite state transducers. Also, it uses BLAS and LAPACK libraries for a support of linear algebra operations. Kaldi purpose is to have a modern and flexible code, since it is easy to be extended and modify. Kaldi is an open-source toolkit and it is available for modifications. Kaldi provides two realizations of neural network training. The first one is Kerel's implementation [4] that supports pretraining using deep belief networks and training using GPU. The second realization is Dan's implementation [5] that does not use pretraining, but provides parallel training using several CPUs.

A speech recognition system for the Italian speech on CHILDIT corpus was suggested in [6] using Kaldi toolkit. The best result was shown by a hybrid system with deep neural networks. Kaldi demonstrated the effectiveness of easily usage of DNN in order to reduce recognition error comparing with other toolkits for automatic speech recognition. Kaldi provides a baseline for speech recognition.

A system for recognition Serbian speech was described in [7]. Serbian is in the same language group as Russian, thus it is interesting for us. The system was written using CUDA. System performance was examined using Kaldi. WER for HMM/GMM was 63.39%, while for a hybrid system with deep neural network it was 48.5% resulting into improvement by 15–22% in dependence on testing data.

Also, system for Russian speech recognition was described in [8]. Modeling was performed using deep neural network and studying was provided with GPU.

There were described two types of recognition. The first model used features that were got in the bottleneck and the second model used a hybrid approach with neural network. Baseline was 31.5% and the best result was 25.1%.

A research on hybrid models for Russian speech recognition was presented in [9]. Various configurations of neural networks were learned with various numbers of layers, their dimensions as well as with various activation functions including hyperbolic tangent and p-norm. For a constructing of acoustic models and testing Kaldi toolkit was used. Experiments were performed using acoustic models built with GMM and hybrid models with DNNs. Baseline was 25.32% and the best result was 20.3%, so a reduction of an error was approximately 20%.

3   Architectures of Neural Networks for Acoustic Modeling

In this section we will shortly describe architectures of neural networks that we used for the experiments.

3.1   LSTM

Standard LSTM. LSTM network [10] consists of several LSTM-units which are chained consequentially. LSTM-units have several gates that control data flow for saving and removing from the unit. One of such gates, forget gate layer is

$$f_t = \sigma(W_f \cdot [h_{t-1}, x_t] + b_f),$$

where h_{t-1}, x_t are inputs, σ is the logistic activation function and f_t is an output value between 0 and 1.

Then, LSTM-unit uses layer that filters data for saving. It consists of two parts. The first one is an input gate layer:

$$i_t = \sigma(W_i \cdot [h_{t-1}, x_t] + b_i).$$

And the second one is

$$\tilde{C}_t = \tanh(W_C \cdot [h_{t-1}, x_t] + b_C).$$

Thus, a value of the new state is computed in the following way:

$$C_t = f_t \circ C_{t-1} + i_t \circ \tilde{C}_t.$$

Peephole LSTM. Peephole LSTM is a modification of the standard LSTM. Its gates are allowed to look at a state of a LSTM-unit. So, gates are:

$$f_t = \sigma(W_f \cdot [C_{t-1}, h_{t-1}, x_t] + b_f);$$
$$i_t = \sigma(W_i \cdot [C_{t-1}, h_{t-1}, x_t] + b_i);$$
$$o_t = \sigma(W_o \cdot [C_t, h_{t-1}, x_t] + b_o).$$

BLSTM. One disadvantage of the standard LSTMs is an opportunity to use only previous context. Looking at the future context may be useful in speech recognition of a language with complex grammar as Russian. This aspect is included in bidirectional recurrent neural network (BRNN). They can be described as follows:

$$\overrightarrow{h}_t = \sigma(W_{x\overrightarrow{h}}x_t + W_{\overrightarrow{h}\overrightarrow{h}}\overrightarrow{h}_{t-1} + b_{\overrightarrow{h}});$$

$$\overleftarrow{h}_t = \sigma(W_{x\overleftarrow{h}}x_t + W_{\overleftarrow{h}\overleftarrow{h}}\overleftarrow{h}_{t-1} + b_{\overleftarrow{h}});$$

$$y_t = W_{\overrightarrow{h}y}\overrightarrow{h}_t + W_{\overleftarrow{h}y}\overleftarrow{h}_t + b_y.$$

A combination of BRNN and LSTM gives bidirectional LSTM (BLSTM).

3.2   CNN

The next wide class of neural networks is convolutional neural networks (CNN) [11]. These models consist of layers of three types: convolutional layers, subsampling layers and fully-connected layers. The main idea of CNNs is increasing the density of uncorrelated sections of the features.

Discrete convolution operation is used for building convolutional layers and written as

$$(f * g)[n] = \sum_m f[m]g[n-m],$$

where f is a feature matrix and g is a convolution kernel.

The output of neurons can be represented as

$$h^l = f(h^{l-1} * k^l + b^l),$$

where h^l is an output vector of lth layer, f is an activation function, b is a bias and k is convolution kernel. The result is called feature map.

Subsampling layers reduce dimension of input feature maps. They are divided into several types such as max-pooling, average-pooling, etc.

3.3   ResNet

A deep convolutional residual network was presented for image recognition in [12]. Its main component is a residual unit:

$$y_l = h(x_l) + F(x_l, W_l), x_{l+1} = f(y_l),$$

where x_l and x_{l+1} are an input and an output of the lth unit, F is a residual function.

In that paper, $h(x_l) = x_l$ and f was ReLU. The main idea is to learn the residual function F. F can use some activation function, convolutional layers, etc. Residual unit is presented in Fig. 1.

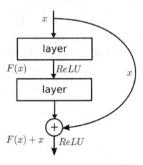

Fig. 1. Residual unit

3.4   RCNN

In papers [13,14], a combination of RNN and CNN was proposed. It was called recurrent convolutional neural network (RCNN). That model was used for object recognition and scene labeling. The main unit is a recurrent convolutional layer (RCL):

$$
h_t(i,j) = \sigma \left(\sum_{i'=-s}^{s} \sum_{j'=-s}^{s} w_k^f \left(i',j'\right) x_t \left(i-i', j-j'\right) \right.
$$

$$
\left. + \sum_{i'=-s}^{s} \sum_{j'=-s}^{s} w_k^r \left(i',j'\right) h_{t-1} \left(i-i', j-j'\right) + b \right),
$$

where w_k^f and w_k^r are kernels.

$\sigma(x) = f(g(x))$ is a superposition of two functions. $g(x)$ can be the sigmoid function or ReLU. $f(\cdot)$ is a normalization function. Batch-normalization [15] can be used as a normalization function to speed up the learning process. There are T time steps. Network depth grows up with growth of T. Also, it can be extended with max-pooling and other layers. Schema of RCNN is presented in Fig. 2.

4   Datasets

4.1   Dataset for the Acoustic Models

In this work, we use the training speech corpus collected at SPIIRAS as in [9] and combined using three databases:

- recordings of 50 native Russian speakers, $16,350$ utterances. Each speaker pronounced a set of 327 phrases;
- recordings of 55 native Russian speakers where each speaker pronounced 105 phrases;

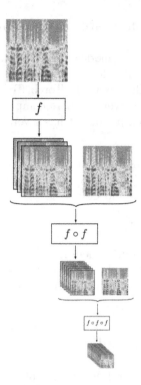

Fig. 2. Recurrent convolutional network ($T = 3$), where f is a convolutional layer

- the third part is an audio part of the audio-visual speech corpus HAVRUS [16]. 20 native Russian speakers (10 male and 10 female speakers) with no language or hearing problems participated in the recordings. Each of them pronounced 200 Russian phrases.

The total duration of the entire speech corpus is more than 30 h.

To test the system, we use a speech database of 500 phrases pronounced by 5 speakers. The phrases were taken from the materials of Russian online newspaper "Fontanka.ru"[1] that was not used in the training data.

4.2   Dataset for the Language Model

Language model is an important part of the recognition system. Our language model is learned using data from a Russian news sites [17]. Dataset for the training of language model contains approximately 300 millions of collocations. As a language model n-gram ($n = 2$) model with KneserNey smoothing [18] is used.

---

[1] http://www.fontanka.ru/.

5 Speech Recognition System Implementation

For building and testing acoustic models, Kaldi toolkit [3] was used.

We need to choose a toolkit for a studying and configuring neural networks. Popular toolkits are Theano, Caffe, Torch, TensorFlow, CNTK, MXNET, Deeplearing4j, etc. There are several papers comparing these toolkits: [19–22]. As a result, CNTK was chosen because it has several advantages in comparison with TensorFlow:

- clear and simple network description using BrainScript or NDL,
- simple realization of combining with Kaldi (lesser number of code lines),
- short description of LSTMs and CNNs,
- a lot of examples,
- wide support of using GPUs.

We used BrainScript for configuring neural networks. Kaldi's features were read with CNTK's Kaldi2Reader module. SGD with learning rate per minibatch 0.1 was used with size of minibatch 256. All experiments were provided using NVIDIA GeForce GT 730M.

6 Experiments and Results

In this section, we describe neural networks configurations and experiments on using them for continuous Russian speech recognition.

6.1 Baseline

Baseline is implemented using standard Kaldi steps as in [9]. Firstly, we extract features (13 mel-frequently cepstral coefficients [23]) from the training and the testing speech datasets. Then, we learn and tune monophone acoustic models. After that, we learn a triphone model using previous models. Finally, LDA, MLLT [24], SAT [25] and fMLLR [26] is applied.

The final step was studying hybrid DNN-HMM model. It takes 440 input features after LDA application. Neural network is a multi-layer perceptron (MLP) that consists of four hidden layers with tanh activation function ending with soft-max layer. Also, weight matrix initialization using DBN [27] is applied.

The baseline achieves 23.96% of WER.

Also, we compared our models' results with model from [8]. It used *nnet3* Kaldi's configuration that applied BLSTMs for speech recognition. The following configuration of the network was applied: three forward and three backward layers, 1024 cell and hidden dimensions, 128 recurrent and nonrecurrent projection dimensions. An initial learning rate was 0.0003 and final learning rate was 0.00003. This model achieves 22.8% of WER.

Table 1. Results for MLPs

	1 model	2 model	3 model	4 model
Layers	3	6	6	6
Dimensions	450×3	2048×6	512×6	512×6
Activation function	sigmoid	sigmoid	tanh	p-norm $(p = 2)$
Iterations	20	20	18	18
WER	25.54%	25.32%	24.96%	24.26%

6.2 MLP

Firstly, experiments on MPLs with various activation functions were provided using CNTK. We test four configurations presented in Table 1 together with their results. The fourth model showed the best result and it used p-norm activation function as in [9]:

$$y = \|x\|_p = \left(\sum_i |x_i|^p \right)^{1/p}.$$

We test models with $p = 1$ and $p = 3$, but the best result was made with $p = 2$.

6.3 LSTM

Configurations and results of applying LSTMs are presented in Table 2. The best result is shown by BLSTM. This model surpassed result of the baseline. However, a disadvantage of LSTM is that it can easily get overfitted and a lot of computational time is required to tune proper parameters.

Table 2. Results for LSTMs

	LSTM	PLSTM	BLSTM
Layers	6	3	3
Dimensions	512×6	512×3	512×3
Iterations	16	14	10
WER	23.32%	24.12%	23.08%

6.4 CNN

To learn CNN, we transformed features into tensors of dimension 40×11, where the first dimension is the time and the second one is the frequency.

The first model is a standard CNN. In the beginning, it has a convolutional layer with 64 output channels, 3×3 kernel, no padding and ReLU activation function. Then, it has a max-pooling layer with 2×2 kernel, with padding and $2 : 2$ stride. Then, the same convolutional layer as the first one is applied, but with 128 output channels. After that, a max-pooling layer is used. Finally, two MLPs (dimension is 4096) with ReLU activation function are used.

WER achieved by CNN is 24.96%.

Standard CNN does not show a good result because of a degradation of the network. So, in [12] this problem was solved using residual units.

6.5 ResNet

We use ResNet architecture presented in Fig. 3 there is an architecture of ResNet that was used. After four iterations it receives WER = 22.17%. This result improved the baseline by 7.5%.

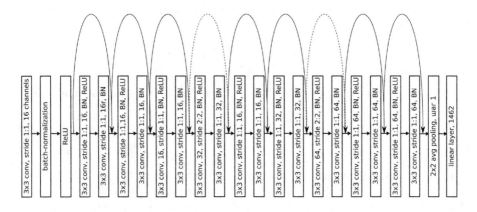

Fig. 3. ResNet

6.6 RCNN

Firstly, features are transformed into tensors of a dimension 40×11 and are sent to RCL stack input. RCL stack has the depth $T = 3$. The first convolution transforms data into 64 channels with padding, $2 : 1$ stride and 10×2 kernel. Then, the batch-normalization and ReLU are applied. The next convolution transforms the result of ReLU application into 4096 channels with padding and 3×1 kernel. After that, the batch-normalization and ReLU are applied to the sum of the previous and the current convolutions. Then, the same RCL-unit is applied. Convolutional layer with ReLU, 16×2 kernel and $1 : 1$ stride is applied to RCL stack output. Finally, three hidden sigmoid layers of 128 dimensions are applied with the batch-normalization.

After 12 iterations, we achieve WER = 22.56%. This result improves the baseline by 5.8%.

6.7   Comparing of Models

The best results shown by BLSTM, ResNet and RCNN are presented in Table 3. The speed of a decoding and a training is presented in Table 4. ResNet has shown the best result, but it was the slowest model. RCNN was faster, but it had a higher error of recognition.

Table 3. The best results of all models

	BLSTM	ResNet	RCNN
WER	23.08%	22.17%	22.56%

Table 4. Average speed of a training (features per second) and a decoding (utterances per second)

Model	Train	Decode
BLSTM	450.7	0.211
ResNet	121.4	0.105
RCNN	325.1	0.162

6.8   New Model

ResNet has shown the best result, but it was the slowest model. RCNN was faster, but it had a higher error of a recognition. LSTMs are difficult to be learned. Since, we can increase the density of uncorrelated sections of the features, simplify input features for the next studying using LSTMs. But CNNs show degradation, so we can use RCNNs and residual units.

Features were transformed into tensors of a dimension 40×11 and were sent to RCNN with $T = 3$. Then, there was a unit that was consist of two convolutional layers with a batch-normalization and ReLU, 3×3 kernel with padding and $1 : 1$ stride. Then, convolutional layer with 2×2 and $1 : 1$ stride. Finally, BLSTM's stack (three layers with 512 units in each layer) was applied.

That model gave WER = 22.34%. Also, other variations were applied. So, with $T > 3$ recognition error was growing up. A batch-normalization increased an error slightly, but it improved the training speed. So, the decoding speed was 0.134 utterances per second and the training speed 227.6 features per second. Also, a replacing of the last convolutional layer by max-pooling decreased error and it became 22.28%. But after an adding a residual unit we got a result WER = 22.07%, this result improved the baseline by 7.8%. The model is shown in Fig. 4.

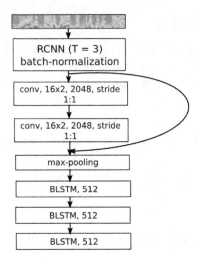

Fig. 4. RCNN and BLSTM union

Table 5. Results

Model	WER
Kaldi baseline	26.62%
Kaldi + DBN baseline	23.96%
Kaldi nnet3	22.80%
MLP-3-sigmoid	25.54%
MLP-6-sigmoid	25.32%
MLP-6-tanh	24.96%
MLP-6-p-norm	24.26%
LSTM	23.32%
PLSTM	24.12%
BLSTM	23.08%
CNN	24.92%
RCNN	22.56%
ResNet	**22.17%**
RCNN + CL + BLSTM	22.34%
RCNN + max-pooling + BLSTM	22.28%
RCNN + residual unit + max-pooling + BLSTM	**22.07%**

6.9 Summarization

The results of all discussed models that we used are shown in Table 5. So, ResNet and RCNN showed good results. A reduction of a recognition error was approximately 7.5%.

7   Conclusion

In this work, we consider the task of Russian speech recognition using hybrid DNN-HMM acoustic models. We used Kaldi and CNTK toolkits.

We used various neural network architectures: multilayer perceptron, LSTMs and theirs modifications, convolutional networks, residual convolutional networks and recurrent convolutional networks. The best result was shown by residual convolutional networks. After four iterations WER was 22.17%.

In the future we will provide experiments on using residual units, union with other models like BLSTMs using score fusion, applying an augmentation of the data. Also, we can use models that we've got for other languages (e.g. English). Moreover, we are interested in applying end-to-end systems for Russian speech.

Acknowledgments. This research is partially supported by the Council for Grants of the President of the Russian Federation (project No. MK-1000.2017.8) and by the Russian Foundation for Basic Research (project No. 15-07-04322).

References

1. Benesty, J., Sondhi, M.M., Huang, Y.: Introduction to speech processing. In: Benesty, J., Sondhi, M.M., Huang, Y.A. (eds.) Springer Handbook of Speech Processing. Springer Handbooks, pp. 1–4. Springer, Heidelberg (2008). https://doi.org/10.1007/978-3-540-49127-9_1
2. Hinton, G., Deng, L., Yu, D., Dahl, G.E., Mohamed, A.R., Jaitly, N., Kingsbury, B.: Deep neural networks for acoustic modeling in speech recognition: the shared views of four research groups. IEEE Sig. Process. Mag. **29**(6), 82–97 (2012)
3. Povey, D., Ghoshal, A., Boulianne, G., Burget, L., Glembek, O., Goel, N., Silovsky, J.: The Kaldi speech recognition toolkit. In: IEEE 2011 Workshop on Automatic Speech Recognition and Understanding (No. EPFL-CONF-192584). IEEE Signal Processing Society (2011)
4. Vesel K., Ghoshal, A., Burget, L., Povey, D.: Sequence-discriminative training of deep neural networks. In: Interspeech, pp. 2345–2349 (2013)
5. Povey, D., Zhang, X., Khudanpur, S.: Parallel training of DNNs with natural gradient and parameter averaging (2014). arXiv preprint arXiv:1410.7455
6. Cosi, P.: A KALDI-DNN-based ASR system for Italian. In: 2015 International Joint Conference on Neural Networks (IJCNN), pp. 1–5. IEEE (2015)
7. Popović, B., Ostrogonac, S., Pakoci, E., Jakovljević, N., Delić, V.: Deep neural network based continuous speech recognition for Serbian using the Kaldi toolkit. In: Ronzhin, A., Potapova, R., Fakotakis, N. (eds.) SPECOM 2015. LNCS, vol. 9319, pp. 186–192. Springer, Cham (2015). https://doi.org/10.1007/978-3-319-23132-7_23
8. Prudnikov, A., Medennikov, I., Mendelev, V., Korenevsky, M., Khokhlov, Y.: Improving acoustic models for Russian spontaneous speech recognition. In: Ronzhin, A., Potapova, R., Fakotakis, N. (eds.) SPECOM 2015. LNCS, vol. 9319, pp. 234–242. Springer, Cham (2015). https://doi.org/10.1007/978-3-319-23132-7_29

9. Kipyatkova, I., Karpov, A.: DNN-based acoustic modeling for Russian speech recognition using Kaldi. In: Ronzhin, A., Potapova, R., Németh, G. (eds.) SPECOM 2016. LNCS, vol. 9811, pp. 246–253. Springer, Cham (2016). https://doi.org/10.1007/978-3-319-43958-7_29

10. Hochreiter, S., Schmidhuber, J.: Long short-term memory. Neural Comput. **9**(8), 1735–1780 (1997)

11. LeCun, Y., Bengio, Y.: Convolutional networks for images, speech, and time series. In: The Handbook of Brain Theory and Neural Networks, vol. 3361, no. 10. The MIT Press, Cambridge (1995)

12. He, K., Zhang, X., Ren, S., Sun, J.: Deep residual learning for image recognition. In: Proceedings of the IEEE Conference on Computer Vision and Pattern Recognition, pp. 770–778 (2016)

13. Liang, M., Hu, X.: Recurrent convolutional neural network for object recognition. In: Proceedings of the IEEE Conference on Computer Vision and Pattern Recognition, pp. 3367–3375 (2015)

14. Liang, M., Hu, X., Zhang, B.: Convolutional neural networks with intra-layer recurrent connections for scene labeling. In: Advances in Neural Information Processing Systems, pp. 937–945. Morgan Kaufmann Publishers Inc., San Francisco (2015)

15. Ioffe, S., Szegedy, C.: Batch normalization: accelerating deep network training by reducing internal covariate shift (2015). arXiv preprint arXiv:1502.03167

16. Verkhodanova, V., Ronzhin, A., Kipyatkova, I., Ivanko, D., Karpov, A., Železný, M.: HAVRUS corpus: high-speed recordings of audio-visual Russian speech. In: Ronzhin, A., Potapova, R., Németh, G. (eds.) SPECOM 2016. LNCS, vol. 9811, pp. 338–345. Springer, Cham (2016). https://doi.org/10.1007/978-3-319-43958-7_40

17. Kipyatkova, I.S., Karpov, A.A.: Automatic processing and statistic analysis of a news text corpus for a language model of a Russian language speech recognition system. Inf. Upravl. Sist. **4**(47), 28 (2010)

18. Chen, S.F., Goodman, J.: An empirical study of smoothing techniques for language modeling. In: Proceedings of the 34th Annual Meeting on Association for Computational Linguistics, pp. 310–318. Association for Computational Linguistics (1996)

19. Fox, J., Zou, Y., Qiu, J.: Software frameworks for deep learning at scale. Internal Indiana University Technical report (2016)

20. Kovalev, V., Kalinovsky, A., Kovalev, S.: Deep Learning with Theano, Torch, Caffe, Tensorflow, and Deeplearning4J: Which One is the Best in Speed and Accuracy? (2016)

21. Chen, T., Li, M., Li, Y., Lin, M., Wang, N., Wang, M., Zhang, Z.: Mxnet: a flexible and efficient machine learning library for heterogeneous distributed systems (2015). arXiv preprint arXiv:1512.01274

22. Bahrampour, S., Ramakrishnan, N., Schott, L., Shah, M.: Comparative study of deep learning software frameworks (2015). arXiv preprint arXiv:1511.06435

23. Ganchev, T., Fakotakis, N., Kokkinakis, G.: Comparative evaluation of various MFCC implementations on the speaker verification task. In: Proceedings of the SPECOM, vol. 1, pp. 191–194 (2005)

24. Gopinath, R.A.: Maximum likelihood modeling with Gaussian distributions for classification. In: Proceedings of the 1998 IEEE International Conference on Acoustics, Speech and Signal Processing, vol. 2, pp. 661–664. IEEE (1998)

25. Anastasakos, T., McDonough, J., Makhoul, J.: Speaker adaptive training: a maximum likelihood approach to speaker normalization. In: 1997 IEEE International Conference on Acoustics, Speech, and Signal Processing, ICASSP 1997, vol. 2, pp. 1043–1046. IEEE (1997)
26. Gales, M.J.: Maximum likelihood linear transformations for HMM-based speech recognition. Comput. Speech Lang. **12**(2), 75–98 (1998)
27. Ackley, D.H., Hinton, G.E., Sejnowski, T.J.: A learning algorithm for Boltzmann machines. Cogn. Sci. **9**(1), 147–169 (1985)

Combined Feature Representation for Emotion Classification from Russian Speech

Oxana Verkholyak[1,2(✉)] and Alexey Karpov[1,2]

[1] ITMO University, St. Petersburg, Russia
overkholyak@gmail.com
[2] SPIIRAS Institute, St. Petersburg, Russia
karpov@iias.spb.su

Abstract. Acoustic feature extraction for emotion classification is possible on different levels. Frame-level features provide low-level description characteristics that preserve temporal structure of the utterance. On the other hand, utterance-level features represent functionals applied to the low-level descriptors and contain important information about speaker emotional state. Utterance-level features are particularly useful for determining emotion intensity, however, they lose information about temporal changes of the signal. Another drawback includes often insufficient number of feature vectors for complex classification tasks. One solution to overcome these problems is to combine the frame-level features and utterance-level features to take advantage of both methods. This paper proposes to obtain low-level feature representation feeding frame-level descriptor sequences to a Long Short-Term Memory (LSTM) network, combine the outcome with the Principal Component Analysis (PCA) representation of utterance-level features, and make the final prediction with a logistic regression classifier.

Keywords: Emotion classification · Long Short-Term Memory
Logistic regression · Principal Component Analysis

1 Introduction and Related Work

Emotion recognition is an important aspect of human-computer communication, which allows for more natural interaction between people and machines. Although a significant amount of research has been devoted to this topic and several approaches have been proposed to solve the task, the problem of automatic recognition of emotions remains an open issue in the field.

Much attention in the field of emotion recognition was focused on feature extraction, selection and representation methods for robust classification. Because it is important to model temporal structure of the data, many experiments were conducted to extract acoustic features on segments of different length, including fixed-size frame level, phoneme level, turn-level and utterance level [1]. There is a trade-off between the ability to capture temporal changes and the big picture of emotions: the smaller the segments, the more detailed information about temporal changes, but a poorer representation of the emotional content of the whole utterance. Utterance-level features, such as overall

A. Filchenkov et al. (Eds.): AINL 2017, CCIS 789, pp. 68–73, 2018.
https://doi.org/10.1007/978-3-319-71746-3_6

loudness or rate of speech, contain relevant information about emotional color of the whole utterance, nevertheless, they fail to account for the temporal changes of the voice signal, and therefore lose important information about temporal structure of the data. Moreover, because there is only one such feature vector per utterance, the total number of feature vectors may not be enough to apply complex classification methods [2]. On the other hand, extracting low-level descriptors (LLDs) on the frame level alone is not enough to model emotional states; however, they allow taking advantage of the temporal structure of the data. Therefore, the focus of this paper is to combine the two different types of features, utterance-level and frame level, to take advantage of the both approaches simultaneously.

Recently research has demonstrated that deep neural networks can be effectively used to generate new representations of original feature sets. These representations provide discriminative features that approximate non-linear dependencies between features in the original set [3]. A type of recurrent neural network with LSTM [4] in particular has become a popular choice among authors because of the ability to model arbitrarily long time dependencies of input data. The proposed method was inspired by [3], who experimented with Deep Belief Network models to generate features for audio-visual emotion recognition, and [5], who combined frame level and turn-level information for recognition of emotions within speech.

2   Proposed Method

The proposed method can be split into 4 stages. At the first stage, frame-level LLD sequences and utterance-level functionals are extracted from the audio files. At the second stage, a new feature representation is obtained separately for LLD features and functionals. At the third stage, the results of the previous stage are combined to form a single feature vector. At the last stage, the predictions are obtained via logistic regression classifier. The whole procedure is presented in Fig. 1.

2.1   Feature Extraction

The feature extraction was performed via commonly used open source tool-kit open-SMILE [6]. The feature set was chosen to be the one used in INTERSPEECH 2010 paralinguistic challenge [7], because it has previously shown a better performance in comparison to other sets on the RUSLANA database [8] and other emotional speech databases, for example [9]. The feature set contains 38 LLDs with regression coefficients and 21 functionals. The summary of the features and the functionals are given for convenience in Table 1.

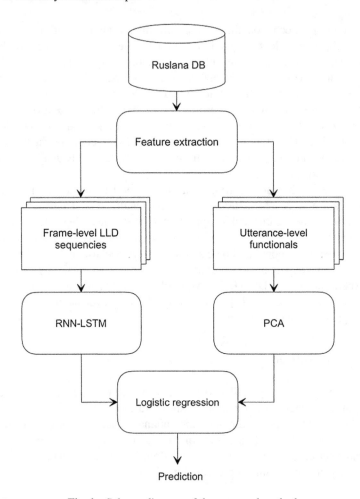

Fig. 1. Scheme diagram of the proposed method

Table 1. INTERSPEECH 2010 openSMILE paralinguistic challenge feature set

Low-level descriptors	Functionals
PCM loudness	Position maximum/minimum
MFCC [0–14]	Arithmetic mean, Standard deviation
Log Mel Freq. Band [0–7]	Skewness, Kurtosis
LSP Frequency [0–7]	Linear regression coefficients 1/2
F0 by sub—Harmonic sum	Linear regression error Q/A
F0 envelop	Quartile 1/2/3
Voicing probability	Quartile range 2-1/3-2/3-1
Jitter local	Percentile 1/99
Jitter DDP	Percentile range 99-1
Shimmer local	Up-level time 75/90

The utterance-level features were derived by applying 21 functionals to LLD features depicted above. The frame-level features were obtained by extracting 38 LLDs at the rate of 100 frames per second. They come pre-smoothed by simple moving average low-pass filtering. Hence, each utterance is represented by one utterance-level feature vector and several LLD sequences. After extraction all the raw features were normalized to have a zero mean and unit variance.

2.2 Feature Representation

Because the dimensionality of the extracted features is very high it is necessary to find a feature representation that will allow reducing the dimensionality and decorrelating the features. Two separate feature representation techniques for frame level and utterance-level features are proposed.

Utterance-Level. To obtain feature representation of the utterance-level feature vectors we applied PCA [10]. PCA is a linear feature learning approach that allows minimum information loss while reducing the number of features. The number of principal components was fixed and equal to the optimal number of components established in our previous research [8], which was found to be 300.

Frame-Level. The LLD feature representation was obtained via Recurrent Neural Network (RNN) with LSTM architecture and 2 hidden layers. At the time t, the utterance-level feature vector x_t was presented at the input of the network. The l-th hidden layer was calculated based on the given input x_t and the activation from the previous time step $h^{(l)}(x_{t-1})$:

$$h^{(l)}(x_t) = f(W^{(l)}h^{(l-1)}(x_t) + b^{(l)} + U^{(l)}h^{(l)}(x_{t-1})) \tag{1}$$

Where W and U are weight matrices, b – bias vector, f(·) – tanh activation function. The output layer was calculated as following:

$$\hat{y}_t = W^{(3)}h^{(2)}(h^{(1)}(x_t)) + b \tag{2}$$

The output from the second hidden layer was used as the new feature representation:

$$x'. \overset{\text{def}}{=} W^{(2)}h^{(1)}(x_t) + b^{(2)} + U^{(2)}h^{(2)}(x_{t-1}) \tag{3}$$

2.3 Classification

For the final step of classification, the two feature representations obtained earlier were concatenated and the final predictions were made via logistic regression. The choice of classification method is based on the previous research that showed that logistic regression outperforms other classifiers in the task of emotion recognition based on RUSLANA corpus [8, 11]. For comparison of the results we used a baseline method established in our previous work [8]. It includes processing of utterance level features with PCA dimensionality reduction and following logistic regression classification. The choice of the baseline method is based on the results of a comparison study of classifiers applied to the RUSLANA database. All the experiments are 60 repetitions of 60-fold cross-validation, i.e. the training was implemented on 60 speakers and the testing on the remaining speaker to achieve speaker-independent properties of the system.

3   Experimental Settings and Results

3.1   RUSLANA Database

RUSsian LANguage Affective speech database (RUSLANA) [12] is a database containing emotional utterances from acted speech. 61 university students (49 female and 12 male) in the age of 16–28 years old were asked to pronounce 10 phonetically representative sentences portraying the following 6 emotional states: Neutral (N), Anger (A), Happiness (H), Surprise (S), Fear (F) and Sadness (D). Each sentence represented one of the 10 syntactic types, corresponding to distinct intonation contours intrinsic to Russian language. Therefore, the database contains overall $61 \times 10 \times 6 = 3660$ utterances ranging from 2 to 5 s long.

3.2   Experimental Results

We used classification accuracy as the measure of effectiveness of classification because the RUSLANA database is well-balanced and it is easier to compare the results with the work of other authors working with the same database. The baseline method showed maximum classification accuracy of 47.2%, similar to that achieved in the state-of-the-art system by Sidorov in [11], who obtained 47.3% with the boosted logistic regression, and 53.5% using additional speaker adaptive information. Implementing the newly proposed method allowed us to obtain classification accuracy of 49.5% – an improvement of 2.3%. The results are summarized in Table 2.

Table 2.   Comparison of classification accuracy

	Baseline method	Proposed method
Proposed method	47.2%	49.5%
State-of-the-art [11]	47.3%	53.5%

4   Conclusion

We have proposed a new method of combining two feature representations for emotion classification from speech: a frame-level representation of low-level descriptors and an utterance level representation of LLD functionals. The proposed method showed increased classification accuracy compared to the baseline method, although it was not possible to overcome the performance of the state-of-the-art approach. One possible reason includes different scale of the two feature representations. Therefore, the direction of future research will be to investigate possible ways of post-processing of the obtained feature representations, as well as fine-tuning the parameters of the system to guarantee best possible performance.

Acknowledgments. This work was financially supported by the Ministry of Education and Science of the Russian Federation (contract 14.575.21.0132, ID RFMEFI57517X0132), as well as by the Council for grants of the President of the Russian Federation (project № MD–254.2017.8) and by the RFBR (project № 16-3760100).

References

1. Metallinou, A., Wollmer, M., Katsamanis, A., Eyben, F., Schuller, B., Narayanan, S.: Context-sensitive learning for enhanced audiovisual emotion classification. IEEE Trans. Affect. Comput. **3**(2), 184–198 (2012)
2. Vlasenko, B., Schuller, B., Wendemuth, A., Rigoll, G.: Frame vs. turn-level: emotion recognition from speech considering static and dynamic processing. In: Paiva, A.C.R., Prada, R., Picard, R.W. (eds.) ACII 2007. LNCS, vol. 4738, pp. 139–147. Springer, Heidelberg (2007). https://doi.org/10.1007/978-3-540-74889-2_13
3. Kim, Y., Honglak, L., Provost, E.M.: Deep learning for robust feature generation in audiovisual emotion recognition. In: Proceedings of the International Conference on Acoustics, Speech and Signal Processing ICASSP-2013, pp. 3687–3691 (2013)
4. Hochreiter, S., Jürgen, S.: Long short-term memory. Neural Comput. **9**(8), 1735–1780 (1997)
5. Vlasenko, B., Schuller, B., Wendemuth, A., Rigoll, G.: Combining frame and turn-level information for robust recognition of emotions within speech. In: Proceedings of 8th International Conference INTERSPEECH-2007, Antwerp, Belgium, pp. 2249–2252 (2007)
6. Eyben, F., Wöllmer, M., Schuller, B.: openSMILE – the Munich versatile and fast open-source audio feature extractor. In: Proceedings of 18th ACM International Conference on Multimedia, Florence, Italy, pp. 1459–1462 (2010)
7. Schuller, B., Steidl, S., Batliner, A., Burkhardt, F., Devillers, L., Müller, C.A., Narayanan, S.S.: The INTERSPEECH 2010 paralinguistic challenge. In: Proceedings 11th International Conference INTERSPEECH-2010, Makuhari, Japan, pp. 2795–2798 (2010)
8. Verkholyak, O.: Research on methods of automatic emotion recognition in Russian speech. Ms. dissertation, ITMO University, St. Petersburg, Russia (2017)
9. Kaya, H., Karpov, A.A., Salah, A.A.: Robust acoustic emotion recognition based on cascaded normalization and extreme learning machines. In: Cheng, L., Liu, Q., Ronzhin, A. (eds.) ISNN 2016. LNCS, vol. 9719, pp. 115–123. Springer, Cham (2016). https://doi.org/10.1007/978-3-319-40663-3_14
10. Jolliffe, I.: Principal Component Analysis. Wiley, Indianapolis (2002)
11. Sidorov, M.: Automatic recognition of paralinguistic information. Ph.D. dissertation, Ulm University, Ulm, Germany (2016)
12. Makarova, V., Petrushin, V.A.: RUSLANA: a database of Russian emotional utterances. In: Proceedings of 7th International Conference on Spoken Language Processing ICSLP-2002, Denver, Colorado, USA, pp. 2041–2044 (2002)

Information Extraction

Active Learning with Adaptive Density Weighted Sampling for Information Extraction from Scientific Papers

Roman Suvorov[✉], Artem Shelmanov, and Ivan Smirnov

Federal Research Center "Computer Science and Control"
of the Russian Academy of Sciences, Moscow, Russia
{rsuvorov,shelmanov,ivs}@isa.ru

Abstract. The paper addresses the task of information extraction from scientific literature with machine learning methods. In particular, the tasks of definition and result extraction from scientific publications in Russian are considered. We note that annotation of scientific texts for creation of training dataset is very labor insensitive and expensive process. To tackle this problem, we propose methods and tools based on active learning. We describe and evaluate a novel adaptive density-weighted sampling (ADWeS) meta-strategy for active learning. The experiments demonstrate that active learning can be a very efficient technique for scientific text mining, and the proposed meta-strategy can be beneficial for corpus annotation with strongly skewed class distribution. We also investigate informative task-independent features for information extraction from scientific texts and present an openly available tool for corpus annotation, which is equipped with ADWeS and compatible with well-known sampling strategies.

Keywords: Information extraction · Deep linguistic analysis
Active machine learning · Scientific texts analysis

1 Introduction

Scientific publications are the main sources of information about a research. The automatic information extraction from these texts could be very beneficial for various fields including research itself. For example, NLP techniques can be especially useful for analysis of publications in biomedicine, since this field produces a lot of works with experimental results that are hard or expensive to replicate. Having elaborated tools for information extraction from such texts can facilitate the development of scientific search engines or summarization systems and eventually save time and expenses for field review and excess experimental investigations.

Our work aims at various types of scientific text processing and information extraction. In this paper, we consider the tasks of definition and result extraction

© Springer International Publishing AG 2018
A. Filchenkov et al. (Eds.): AINL 2017, CCIS 789, pp. 77–90, 2018.
https://doi.org/10.1007/978-3-319-71746-3_7

(extraction of fragments that express the achievement of a result) from scientific publications in Russian. However, our goal is not just to solve the particular problems for a particular language but rather create more general methods and tools that can be adapted for different tasks of information extraction from scientific publications and for different languages.

The majority of state-of-the-art information extraction systems rely either on supervised machine learning (ML) or hand-crafted rules and dictionaries (or their combination). Both approaches require solving very laborious tasks. For supervised machine learning, creation of large labeled datasets is required. For rule-based systems, the development of dictionaries and grammars is needed, which requires rare and expensive combination of skills: deep domain knowledge and programming experience. There are a number of approaches that can help to reduce the amount of labor needed to build an information extraction system: bootstrapping, unsupervised/semi-supervised/distant learning, construction of rules for hint generation, and active learning. Each of these techniques has its own scope of applicability. In this paper, we focus on active learning because it suits information extraction from scientific papers well.

Active learning (AL) is an interactive approach to simultaneously building a labeled dataset and training a machine learning model. The general algorithm of the AL procedure is the following. First, a relatively large corpus of unlabeled texts is gathered. Then, a domain expert labels a few positive samples in the corpus. After that, an interactive computer-aided annotation begins: a classifier is trained on labeled samples; then, it is applied to the rest of the corpus; the expert is asked to label only samples that are most "useful" (e.g., increase classification performance).

The crucial component of an AL system is a strategy that guides the process of object sampling from unannotated part of the corpus and gives objects to a human-expert for annotation. There is a number of different strategies for AL [1], e.g., one of the most well-known strategies is uncertainty sampling [2]. In this work, we propose a novel meta-strategy – the adaptive density weighted sampling. This meta-strategy leverages the analysis of object density in the feature space for effective exploration of different parts of the dataset and avoiding outliers. We use it to construct two new AL strategies and apply them to the tasks of information extraction from scientific texts.

The main contributions of this paper are the following:

1. The openly available dataset for definition extraction from scientific papers in Russian.
2. New adaptive density weighted sampling (ADWeS) meta-strategy for AL.
3. The openly available tool for corpus annotation equipped with ADWeS and compatible with well-known sampling strategies.

The rest of the paper is structured as follows. Section 2 outlines the related work. Section 3 discusses the active learning sampling strategies used in our work and presents ADWeS meta-strategy. Section 4 presents the feature set and the classification pipeline. The annotation tool is described in Sect. 5.

The experiments and their results are described in Sect. 6. The Sect. 7 concludes and presents the future work.

2   Related Work

Information extraction from scientific texts is an emerging research field. One of the recent efforts in this field – ScienceIE SemEval 2017 shared task [3] was devoted to extraction of keyphrases and their relations from scientific documents. These pieces of information are considered crucial for understanding described processes, tasks, and materials. The problem of definition extraction from English texts has had a lot of attention from the research community. Definition extraction was applied to some high-level tasks such as ontology learning, relation extraction, and question answering. It is commonly solved by a combination of different information extraction techniques including various types of machine learning and manual constructed rules [4,5]. We should note that performance of most methods in this field is very far from ideal: as summarized in [5], researchers achieve only up to 60–70% of F_1 score when focus on specific domains and patterns. This is due to high variance of definition patterns. We note that processing Russian scientific texts is even harder since free word order and flectiveness of the language generate even more variants of definition patterns. For Russian, the most elaborated works related to definition extraction propose hand-crafted rules built with a specific pattern construction language, e.g., [6].

The scarce usage of supervised machine learning in this area for Russian is obviously related to the little amount of annotated corpora. This problem is encountered by researchers and engineers in different domains and languages on the regular basis. It is one of the reasons (however, not the most important) why many industry systems rely primarily on hand-crafted rule sets, while research community is mostly embraced by machine learning [7]. The well-researched supervised learning methods in many cases do not solve the final problem, since creating large annotated corpus for model training is also a laborious and tedious task. For successful practical usage, we need techniques for labor reduction.

There are four main approaches to reducing the amount of labor required for corpora annotation: bootstrapping, annotation with hand-crafted rules, semi-supervised/unsupervised/distant learning, active learning, and construction of rules for hint generation.

Bootstrapping is a general term that encompasses strategies of iterative classification model construction: first, a small set of positive examples is acquired, which is used for training the basic classifier on the consecutive steps. The classifier is used for finding new training examples that are subsequently used to improve the model, which in its turn helps to find another portion of training examples. The steps of finding new examples and training the model alternate each other until the quality of the model stops improving. The bootstrapping technique is used in many applications to increase the dictionary coverage and for creating information extraction rules [8].

The unsupervised/semi-supervised/distant learning are the common approaches in open information extraction. The methods for knowledge base construction and extension are often rely on these techniques [9]. They provide the biggest labor reduction in cases when facts (entities and relations) appear many times in texts. However, they cannot be used when the mentions of the facts are scarce.

Active learning and construction of rules for hint generation are interactive and iterative approaches. They take into account the feedback from an expert during the annotation process and classifier construction. In AL, the feedback from an expert usually consists in annotating a small amount of examples specifically chosen in a way to increase the quality of the model. There are also publications that use multi-modal approach, in which features are annotated as well. The experiments show that this technique speeds up and simplifies the process of corpus annotation [10]. Another developing approach of AL lies in adopting external knowledge resources that do not require any human annotation, e.g., word embeddings [11] or ontologies that are used for choosing more adequate examples for annotation [12]. Automatic construction of rules for hint generation may be treated as a kind of feature annotation that fits rule-based classifiers best [13].

In this work, we focus on AL since we see the most potential in it for high-level tasks of information extraction from scientific publications. Annotation of scientific papers can be very expensive since relatively cheap crowdsourcing is inapplicable. In most cases, an annotator must have a high specific qualification (e.g., PhD grade in particular sciences). Another problem is sparsity: a scientific paper may contain only a couple of positive samples, but an annotator needs to read it through anyway. Using distant learning for such tasks as result extraction from scientific texts is not preferable since formulations of scientific results are very different, scarce, and are not covered by any resources like knowledge bases. These problems can be tackled with AL.

3   Sampling Strategies for Active Learning

In this paper, we consider and experiment with the following standard AL strategies: random strategy, uncertainty sampling (US) [2], choosing examples by maximum probability (MPERR). We also present a novel meta-strategy adaptive density weighted sampling (ADWeS).

The *random strategy* simply draws samples uniformly from the pool of all unannotated examples. This strategy in fact corresponds to completely manual annotation without active learning in that sense that the choice of the next example for annotation is not directed. However, they are not exactly the same, since during the standard annotation process an expert reads the document from the beginning to the end, therefore, examples are chosen nonuniformly.

The *uncertainty sampling* tends to select samples that lie closely to the discriminative surface (and thus have minimum margin). For binary classification, when a classifier is able to predict probabilities of classes (e.g., logistic regression model), the closeness of class probability to 0.5 is used as an estimation of example importance. This strategy aims to clarify the position of the surface by relatively small changes. If the initial approximation of the surface is far away from optimal, the strategy could require many steps to converge. More over, this strategy does not take into account the similarities between objects and thus may select outliers, which are hard to label and almost useless for the model construction.

The *maximum probability strategy* selects unannotated samples that are considered positive by a classifier with high confidence. In such an approach, the training dataset is constructed from both annotated and unannotated examples. The samples labeled as positive so far constitute the positive class. All other samples (including unlabeled) are treated as negative ones. The classifier is fitted to the dataset and is applied to unlabeled samples. The samples, which have the highest scores of being positive are given to experts, who decide whether they are actually true positives or not.

Adaptive density weighted sampling is a meta-strategy that aims on taking into account joint distribution of feature values. In this approach, the examples are ranked according to the score, which is calculated as a harmonic mean of two values: model-based interest estimate (e.g., uncertainty or predicted probability) for a sample $MI(x)$ and the sample object centrality score $Cen(x)$.

$$Score(x) = \frac{2MI(x)Cen(x)}{MI(x) + Cen(x)}$$

The centrality shows how many other objects are located near the given one. The maximal centrality corresponds to centroids (or modes) of clusters of training samples, the minimal centrality corresponds to noise examples.

To calculate initial centralities, we use average cosine similarity of the given example and its closest neighbors:

$$Cen_0(x) = \frac{1}{k} \sum_{w \in ClosestNeighbors(x)} Cosine(x, w)$$

For similarity search, we use tree-based approximate nearest neighbor algorithm implemented in Annoy package[1]. We should note that there are many accurate and mathematically grounded methods for centrality calculation, e.g., PageRank [14]. However, they are much more computationally complex. We suggest that the proposed simple similarity search-based algorithm works satisfactory in practice.

The strategy is called "adaptive" because the centrality scores are changing as more and more samples become labeled. When an example is annotated, its centrality value is set to zero, and the centrality values of its neighbors are decreased proportionally to their similarity to the annotated example. More formally, given a labeled sample x, neighbor sample w, and two hyper parameters – centrality update rate $rate \in [0, 1]$ and distance non-linearity power $pow \in (0, 1)$, centrality update rule is the following:

---

[1] https://github.com/spotify/annoy.

$$Cen_i(w) = Cen_{i-1}(w) \left(1 - rate \cdot \left(1 - \frac{1 - Cosine(x, w)}{max_{q,t}[1 - Cosine(q, t)]} \right)^{pow} \right)$$

The ADWeS marks the well-explored regions of a feature space of a training set and tries to avoid them again in the future iterations. It also tends to avoid outliers, since it configured to choose more "central" objects. ADWeS meta-strategy can be used in conjunction with many AL strategies. In this work, we combined it with uncertainty sampling and choosing by maximum probable error, which results in new strategies referred as ADWeS-US and ADWeS-MPERR, correspondingly.

4   Task-Independent Features and Classification Pipeline

The goal of our work consists in creating generalizable methods and tools that can be adapted for different tasks of information extraction from scientific publications. Therefore, we do not rely on task specific or domain specific features but rather on generic ones.

The tasks of definition and result extraction were treated as sentence classification tasks. We find that most of the definitions and scientific results are well expressed by a single sentence. However, the features considered in this section are not limited to sentence classification and can be used for classification of different types of text fragments as well as for sequence labeling.

In the current work, we consider four groups of features:

- n-grams of word lemmas and POS tags;
- syntax phrases;
- predicate-argument structures;
- context linear rules.

The n-grams of word lemmas and POS tags are very primitive features, they require only morphological analyzer. We use AOT.ru framework for POS tagging and lemmatization[2]. We use distinct lowercased lemmas as features, as well as bigrams and trigrams of lemmas and POS tags.

Phrases of syntactically linked words should provide better generalization than bigrams because syntax relations are more abstract and are not dependent on the word order. We use MaltParser framework [15] trained on SynTagRus corpus [16] for parsing.

Predicate-argument structures is the next level of abstraction. They are constructed using results of semantic role labeling (SRL) [17]. This type of semantic parsing usually provides high-level information about phrases, takes into account global semantic structure by respecting linguistic constraints on semantic arguments and roles, and leverages additional information from various semantic resources (corpora, lexicons, thesauri). Therefore, such structures can be informative generic features in many information extraction tasks. In this work, predicate-argument features were constructed as pairs of a semantic role and

---

[2] http://www.aot.ru/docs/sokirko/sokirko-candid-eng.html.

a predicate lemma. To perform SRL, we use the semantic parser for Russian developed in FRC CSC RAS [18].

Linear context rules (LCRs) aim at capturing many generic linear templates. LCR takes into account morphological features of words (grammar case, plurality, animacy, etc.) and word lemmas. We use the following pairs and triples as LCRs:

- <main word lemma>$\Longrightarrow$ <morphological features of word to the right of main>
- <morphological features of word to the left of the main>$\Longrightarrow$ <main word lemma>
- <left word morphological features>$\Longrightarrow$ <main word lemma>$\Longrightarrow$ <right word morphological features>

Iteratively, every word of a sentence is considered as "main" and is used for generation of the aforementioned templates. During the template construction, words in a window around the main word are used to fill in slots of rules. Such templates are more general than the simple n-grams since they are less dependent on the specific word positions.

Since we use task-independent features, during the data preprocessing, the excessive number of features is generated. In the classification pipeline, we use linear SVM with L1 regularization to prune the majority of features, leaving less than a percent of the original feature space.

For classification itself, for the sake of simplicity, we use logistic regression model with L2 regularization and gradient boosting algorithm implemented in LightGBM[3]. For experiments with active learning, we use only logistic regression model, since it trains very fast and can be efficiently used in interactive mode in practice.

5   Annotation Tool

For active learning to be efficiently applied in practice, easy-to-use yet flexible tools are needed. In this paper, we propose a simple tool for active learning. The target audience of this tool are data scientists, who need to create ML-based classifiers and collaborate with domain experts. The tool is a widget written in Python for the popular integrated development environment Jupyter[4] (Fig. 1). Python has the most developed data science infrastructure, so it may be useful for a broad spectrum of applications including annotation of texts and images.

The widget can be configured with various active learning strategies (including ADWeS-US and ADWeS-MPERR) and visualizators of objects to annotate. In Fig. 1, the tool is configured to display the sentence and other information via table. If we were solving the task of image recognition, the examples could

---

[3] https://github.com/Microsoft/LightGBM.
[4] http://jupyter.org/.

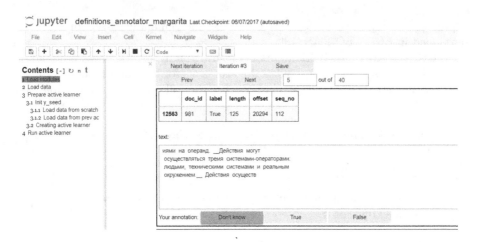

Fig. 1. The preview of the active learning annotation tool

be presented as images. After tool object is invoked, the Jupyter IDE displays the widget and interactive labeling begins.

The examples to annotate are organized in pages. A human annotator sees their visual representation and selects proper class using buttons below. A mini-batch of examples (e.g., 40) correspond to an iteration of active learning algorithm. A user can annotate all or just a part of them and invoke the next iteration of active learning algorithm with "Next iteration" button. After that, the machine learning model is retrained on the updated dataset and the sampling strategy chooses new examples to annotate. The user can save the answers on disk after every iteration. All the changes are automatically synchronized with a Pandas DataFrame[5], which stores objects and labels.

This tool might be useful for rapid annotation in small to medium projects. However, we note that it lacks many useful features, e.g., the ability to work with multiple users at once. The code of the tools is openly available[6].

6   Experiments

6.1   Data

For definition extraction task, we prepared two corpora of scientific texts in Russian. The corpora consist of publications on different topics from scientific journals and proceedings of conferences. They contain annotations of spans that are related to definitions and terms that are explicitly defined in texts. Examples (in English): "X is named as Y", "X – is ...", "We define X as ...", etc.

The first corpus is a gold-standard that was annotated and verified by multiple experts. It contains 36 texts of volume of more than 180,000 tokens with

---

[5] http://pandas.pydata.org/.

[6] http://nlp.isa.ru/adwes_tools/.

439 definitions. After extraction of sentences from the corpus, we got a dataset containing 408 positive and 10,256 negative examples.

The second corpus is also manually annotated, however, it was not verified and contains erroneous examples, it also lacks many annotations. This corpus contains 207 texts with more than 500,000 tokens. There are 767 annotations related to definitions. After extraction of sentences from the corpus we got 560 positive and 55,413 negative examples.

The latter corpus also contains annotations of results. The annotations are also not verified, therefore, the resulting dataset contains some errors and does not cover many positive cases. There are 721 positive and 56,369 negative examples.

The given statistics about the corpora shows that the distribution of classes is very skewed, which complicates the task of manual annotation. However, this problem could be tackled by active learning.

6.2   Evaluation Without Active Learning

The goal of the first experiment is to assess the quality of features and the models on the small gold-standard corpus without active learning. It reveals the best possible performance of models.

In this experiment, instead of standard cross-validation we used multiple shuffle splits with folds grouped by documents. We grouped folds by documents to mitigate the influence of document lexis on the final results. The classifiers could overfit to specific lexis and due to this the standard cross-validation can produce biased estimations. Since the length and a number of annotations vary from document to document, we use random shuffle split for 50 iterations. For evaluation, we use four metrics: precision, recall, F_1-score, and ROC AUC. The results of the evaluation are presented in Table 1.

Table 1. The performance of definition extraction models

Classifier	Precision,%	Recall,%	F_1,%	ROC AUC, %
Log. reg.	64.6 ± 8.6	38.7 ± 8.3	48.0 ± 8.1	89.6 ± 3.1
LightGBM	59.3 ± 9.2	52.6 ± 9.1	55.4 ± 8.2	93.5 ± 1.7

The best performance achieved is $F_1 = 55.4\%$, which is relatively high for Russian considering very unbalanced distribution of classes, the small size of the training dataset, big number of definition patterns in Russian, and usage of only task-independent features. Although, it is definitely possible to perform elaborated feature engineering to decrease the bias, however, we note that it is not the goal of the current work, since active learning annotation tools should mostly rely on features of the general type.

High deviation appears due to the limited size of the corpus and variance of numbers of examples in the splits: there are too many different types of definitions that cannot be distributed uniformly due to the small size of the corpus.

The feature importance evaluation revealed that the most significant features are n-grams of lemmas. We notice several predicate argument structures among the highly significant features that correspond to the frequent patterns of term definitions. The models also reveal many human interpretable context rules relevant to the task of definition extraction. Surprisingly, the features that are based on syntax relations did not appear very informative. We note that most of them were filtered as insignificant in favor of n-grams.

6.3   Evaluation with Active Learning

The goal of the second experiment is to assess how much of labor input can be saved when active learning is used for corpus annotation and model training. In this experiment, we run 10 simulations of active learning on gold-standard dataset with annotated definitions using different selection strategies. In each simulation, 200 iterations of annotate-and-retrain were executed. In each iteration, "the expert" labeled 10 sentences. Before the iterative procedure started, the whole corpus was randomly split into train and test subsets as 70/30. In the train subset, initially, only two randomly chosen positive samples were labeled. We measured precision, recall, and F_1-score both on training and test subsets (Figs. 2, 3 and 5).

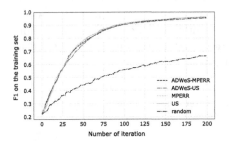

Fig. 2. Dynamics of F_1-score on the training subset

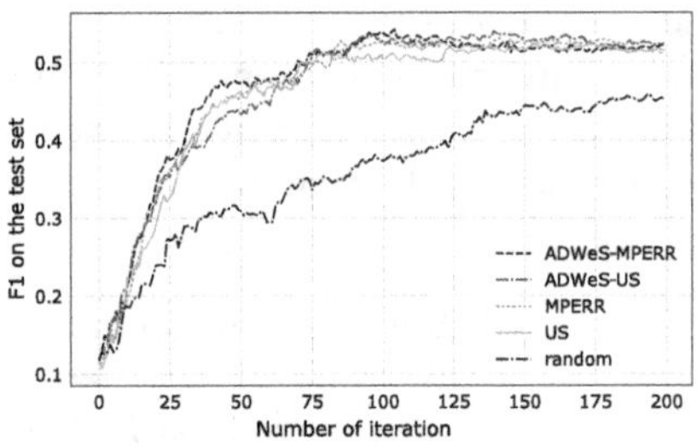

Fig. 3. Dynamics of F_1-score on the test subset

Figure 2 shows that the dataset in the chosen feature space is well-separable with a linear classifier. All strategies except random lead to approximately the same learning curve. Main differences between the strategies lie in the way they explore the area of positive samples.

The charts in Fig. 4 show the ratio of positive samples labeled during the active learning annotation among all positive labels present in the dataset. As expected, the MPERR and ADWeS-MPERR strategies discover positive samples faster than other strategies.

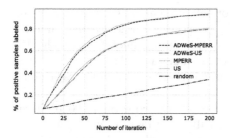

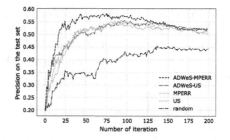

Fig. 4. The dynamics of labeled positive examples ratio discovered by AL strategies

Fig. 5. Dynamics of precision on the test subset

Figure 3 shows that the maximum performance of the models on the test subset is achieved ($F_1 \approx 51.0\%$) after labeling only 900–1000 sentences: all strategies (except random) need 90–100 iterations until F_1-score stabilizes.

ADWeS strategies only slightly reduce the number of iterations needed to achieve the best F_1-score. However, the charts that reflect the precision dynamics (Fig. 5) and the speed of positive examples discovery (Fig. 4) demonstrate that ADWeS strategies behave differently compared to the standard ones. Therefore, each sampling strategy has its own scope of applicability, e.g., ADWeS-MPERR better suits rapid corpus construction with strongly skewed class distribution, since it provides the fastest way to discover dense areas of positive examples, while ADWeS-US optimizes model generalization.

Overall, experiments show that in the task of definition extraction, active learning provides a very high labor reduction. The 1000 examples needed to achieve the best performance is approximately just 14% from all examples in the training subset. Thus, active learning allows to reduce the amount of labor needed to create a training corpus by factor of 7. To sum up, we conclude that active learning is a very efficient technique for scientific text mining.

6.4   Corpus Improvement with Active Learning

The third type of experiment consists in correcting erroneous datasets using active learning. As mentioned in Sect. 6.1, there is a dataset annotated with definitions and results, which contains errors and lacks some annotations. The erroneous dataset has a substantial size, therefore, fixing it in a standard manual way would require significant amount of time and human labor. At the same time, the bigger dataset can be used to improve the diversity of training examples for ML models, which can improve their performance and decrease variance. We do not need a perfect dataset for that purpose, however, big amount of errors can lead to model degradation.

To improve the quality of the bigger corpus with definition and result annotation, we run ADWeS-MPERR to increase the recall of the positive classes and asked an expert to perform 50 iterations. For definition annotation, we also used AL to increase precision of corpus. For this purpose, we used standard uncertainty sampling but consider all positive examples as unannotated. We should note that the model is trained using the whole annotated dataset (with fixed labels so far) on every iteration. This procedure is similar to finding noisy examples in a dataset.

After the AL annotation, we got the corpus with more than 900 positive examples of definitions, which is one and a half times more than in the original dataset. The number of positive annotations of results is also substantially increased to more than 1100.

For assessment of the contribution of the active learning, we performed evaluations of two models (log.reg. and LightGBM) trained on the original corpus and on the improved corpus.

For evaluation of definition extraction models, we used gold-standard corpus as a hold-out. The results of the evaluation are presented in Table 2.

Table 2. The performance of definition extraction before and after the AL annotation

Corpus	Classifier	Precision,%	Recall,%	F_1,%	ROC AUC, %
Before AL	Log. reg.	73.0	6.6	12.1	85.6
	LightGBM	54.2	41.2	46.8	91.1
After AL	Log. reg.	84.5	20.1	32.5	87.9
	LightGBM	65.0	45.6	53.6	91.8

The results show that the AL annotation improved the corpus drastically. Some examples became linear separable, the F_1-score of linear classifier increased by more than 20%. The LightGBM model also works significantly better, the F_1-score increases by almost 7%.

For result extraction, we do not have gold-standard corpus for evaluation on a hold out, therefore, we used the evaluation technique with many random splits described in Subsect. 6.2. Table 3 presents the performance of the models before and after the annotation with AL.

Table 3. The performance of result extraction before and after the AL annotation

Corpus	Classifier	Precision,%	Recall,%	F_1,%	ROC AUC, %
Before AL	Log. reg.	35.8 ± 8.2	10.9 ± 3.0	16.6 ± 4.2	75.1 ± 3.3
	LightGBM	29.4 ± 6.2	18.4 ± 4.3	22.3 ± 4.4	79.8 ± 2.2
After AL	Log. reg.	77.9 ± 6.0	39.4 ± 4.8	52.0 ± 3.9	88.5 ± 2.3
	LightGBM	62.4 ± 5.1	50.1 ± 5.8	55.2 ± 3.4	89.2 ± 1.7

Just like in the task of definition extraction, the application of AL for a few iterations, drastically improved the corpus and models for the task of result extraction.

7   Conclusion

In this paper, we considered the task of information extraction from scientific literature, provided results of experiments with machine learning methods for definition and result extraction from scientific publications in Russian, investigated informative task-independent features and the usefulness of the active learning techniques in this area. A novel adaptive density-weighted sampling (ADWeS) meta-strategy for active learning was proposed and evaluated. Experiments showed that while strategies adjusted by ADWeS perform just slightly better in terms of F_1-score compared to standard sampling strategies, there is significant difference in how positive and negative examples are explored, which could be leveraged in different applications. It was demonstrated that active learning can be a very efficient technique for scientific text mining. For the task of definition extraction, according to the conducted experiments, active learning can reduce efforts needed to build a model with the best performance by a factor of 7. It was also shown that it is possible to use active learning not only to create classifiers and annotate corpora, but also to improve existing linguistic resources. We present an openly available tool for corpus annotation equipped with ADWeS and compatible with well-known sampling strategies.

Main directions of the future work include developing more elaborated task-independent feature sets and deeper investigation of adaptive sampling strategies. We are going to apply the developed methods and the tool to broad spectrum of tasks of information extraction from biomedical texts.

Acknowledgments. The project is supported by the Russian Foundation for Basic Research, project number: 16-29-07210 "ofi_m".

References

1. Settles, B.: Active learning literature survey. University of Wisconsin, Madison, 52(55–66), 11 (2010)
2. Lewis, D., Gale, W.: Training text classifiers by uncertainty sampling (1994)
3. Augenstein, I., Das, M., Riedel, S., Vikraman, L., McCallum, A.: SemEval 2017 task 10: ScienceIE - extracting keyphrases and relations from scientific publications. In: Proceedings of the 11th International Workshop on Semantic Evaluation (SemEval-2017), pp. 546–555 (2017)
4. Del Gaudio, R.: Automatic extraction of definitions. PhD thesis, University of Lisbon (2014)
5. Navigli, R., Velardi, P.: Learning word-class lattices for definition and hypernym extraction. In: Proceedings of the 48th Annual Meeting of the Association for Computational Linguistics, pp. 1318–1327 (2010)

6. Bolshakova, E., Efremova, N., Noskov, A.: LSPL-patterns as a tool for information extraction from natural language texts. In: New Trends in Classification and Data Mining, pp. 110–118 (2010)

7. Chiticariu, L., Li, Y., Reiss, F.R.: Rule-based information extraction is dead! Long live rule-based information extraction systems! In: Proceedings of the 2013 Conference on Empirical Methods in Natural Language Processing, pp. 827–832 (2013)

8. Gupta, S., Manning, C.: SPIED: Stanford pattern based information extraction and diagnostics. In: Proceedings of the Workshop on Interactive Language Learning, Visualization, and Interfaces, pp. 38–44 (2014)

9. Augenstein, I., Maynard, D., Ciravegna, F.: Relation extraction from the web using distant supervision. In: Janowicz, K., Schlobach, S., Lambrix, P., Hyvönen, E. (eds.) EKAW 2014. LNCS (LNAI), vol. 8876, pp. 26–41. Springer, Cham (2014). https://doi.org/10.1007/978-3-319-13704-9_3

10. Jun, K.S., Zhu, J., Settles, B., Rogers, T.: Learning from human-generated lists. In: International Conference on Machine Learning, pp. 181–189 (2013)

11. Kholghi, M., De Vine, L., Sitbon, L., Zuccon, G., Nguyen, A.: The benefits of word embeddings features for active learning in clinical information extraction. arXiv preprint arXiv:1607.02810 (2016)

12. Kholghi, M., Sitbon, L., Zuccon, G., Nguyen, A.: External knowledge and query strategies in active learning: a study in clinical information extraction. In: Proceedings of the 24th ACM International on Conference on Information and Knowledge Management, pp. 143–152. ACM (2015)

13. Dalvi, B., Bhakthavatsalam, S., Clark, C., Clark, P., Etzioni, O., Fader, A., Groeneveld, D.: IKE-an interactive tool for knowledge extraction. In: Proceedings of the 5th Workshop on Automated Knowledge Base Construction, AKBC@ NAACL-HLT, pp. 12–17 (2016)

14. Page, L., Brin, S., Motwani, R., Winograd, T.: The PageRank citation ranking: Bringing order to the web. Technical report, Stanford InfoLab (1999)

15. Nivre, J., Hall, J., Nilsson, J., Chanev, A., Eryigit, G., Kübler, S., Marinov, S., Marsi, E.: MaltParser: A language-independent system for data-driven dependency parsing. Nat. Lang. Eng. **13**(2), 95–135 (2007)

16. Nivre, J., Boguslavsky, I.M., Iomdin, L.L.: Parsing the SynTagRus treebank of Russian. In: Proceedings of the 22nd International Conference on Computational Linguistics, pp. 641–648 (2008)

17. Gildea, D., Jurafsky, D.: Automatic labeling of semantic roles. Comput. Linguist. **28**(3), 245–288 (2002)

18. Shelmanov, A.O., Smirnov, I.V.: Methods for semantic role labeling of Russian texts. In: Computational Linguistics and Intellectual Technologies, Papers from the Annual International Conference "Dialogue-2014", vol. 13, pp. 607–620 (2014)

Application of a Hybrid Bi-LSTM-CRF Model to the Task of Russian Named Entity Recognition

The Anh Le[1,2], Mikhail Y. Arkhipov[1(✉)], and Mikhail S. Burtsev[1]

[1] Neural Networks and Deep Learning Lab,
Moscow Institute of Physics and Technology, Dolgoprudny, Russia
{arkhipov.mu,burtcev.ms}@mipt.ru
[2] Faculty of Information Technology, Vietnam Maritime University,
Haiphong, Viet Nam
anhlt@vimaru.edu.vn

Abstract. Named Entity Recognition (NER) is one of the most common tasks of the natural language processing. The purpose of NER is to find and classify tokens in text documents into predefined categories called tags, such as person names, quantity expressions, percentage expressions, names of locations, organizations, as well as expression of time, currency and others. Although there is a number of approaches have been proposed for this task in Russian language, it still has a substantial potential for the better solutions. In this work, we studied several deep neural network models starting from vanilla Bi-directional Long Short Term Memory (Bi-LSTM) then supplementing it with Conditional Random Fields (CRF) as well as highway networks and finally adding external word embeddings. All models were evaluated across three datasets Gareev's, Person-1000 and FactRuEval 2016. We found that extension of Bi-LSTM model with CRF significantly increased the quality of predictions. Encoding input tokens with external word embeddings reduced training time and allowed to achieve state of the art for the Russian NER task.

Keywords: NER · Bi-LSTM · CRF

1 Introduction

There are two main approaches to address the named entity recognition (NER) problem [1]. The first one is based on handcrafted rules, and the other relies on statistical learning. The rule based methods are primarily focused on engineering a grammar and syntactic extraction of patterns related to the structure of language. In this case, a laborious tagging of a large amount of examples is not required. The downsides of fixed rules are poor ability to generalize and inability to learn from examples. As a result, this type of NER systems is costly to develop and maintain. Learning based systems automatically extract patterns relevant to the NER task from the training set of examples, so they don't require

A. Filchenkov et al. (Eds.): AINL 2017, CCIS 789, pp. 91–103, 2018.
https://doi.org/10.1007/978-3-319-71746-3_8

deep language specific knowledge. This makes possible to apply the same NER system to different languages without significant changes in architecture.

NER task can be considered as a sequence labeling problem. At the moment one of the most common methods to address problems with sequential structure is Recurrent Neural Networks (RNNs) due their ability to store in memory and relate to each other different parts of a sequence. Thus, RNNs is a natural choice to deal with the NER problem. Up to now, a series of neural models were suggested for NER. To our knowledge on the moment of writing this article the best results for a number of languages such as English, German, Dutch and Spanish were achieved with a hybrid model combining bi-directional long short-term memory network with conditional random fields (Bi-LSTM + CRF) [2]. In our study we extended the original work by applying Bi-LSTM + CRF model to NER task in Russian language. We also implemented and experimented with a series of extensions of the NeuroNER model [3]. NeuroNER is a different implementation of the same Bi-LSTM + CRF model. However, the realizations of the models might differ in such details as initialization and LSTM cell structure. To reduce training time and improve results, we used the FastText[1] model trained on Lenta corpus[2] to obtain external word embeddings. We studied the following models:

- Bi-LSTM (char and word);
- Bi-LSTM (char and word) + CRF;
- Bi-LSTM (char and word) + CRF + external word embeddings;
- Default NeuroNER + char level highway network;
- Default NeuroNER + word level highway Bi-LSTM;
- Default NeuroNER + char level highway network + word level highway Bi-LSTM.

To test all models we used three datasets:

- Gareev's dataset [4];
- FactRuEval 2016[3];
- Persons-1000 [18].

Our study shows that Bi-LSTM + CRF + external word embeddings model achieves state-of-the-art results for Russian NER task.

2   Neuronal NER Models

In this section we briefly outline fundamental concepts of recurrent neural networks such as LSTM and Bi-LSTM models. We also describe a hybrid architecture which combines Bi-LSTM with a CRF layer for NER task as well as some extensions of this baseline architecture.

---

[1] An open-source library for learning text representations and text classifiers. URL: https://fasttext.cc/.

[2] A Russian public corpus for some tasks of natural language processing. URL: https://github.com/yutkin/lenta.ru-news-dataset.

[3] The dataset for NER and Fact Extraction task given at The International Conference on Computational Linguistics and Intellectual Technologies - Moscow 2016.

2.1 Long Short-Term Memory Recurrent Neural Networks

Recurrent neural networks have been employed to tackle a variety of tasks including natural language processing problems due to its ability to use the previous information from a sequence for calculation of current output. However, it was found [10] that in spite theoretical possibility to learn a long-term dependency in practice RNN models don't perform as expected and suffer from gradient descent issues. For this reason, a special architecture of RNN called Long Short-Term Memory (LSTM) has been developed to deal with the vanishing gradient problem [11]. LSTM replaces hidden units in RNN architecture with units called memory blocks which contain 4 components: input gate, output gate, forget gate and memory cell. Formulas for these components are listed below:

$$i_t = \sigma(W_{ix}x_t + W_{ih}h_{t-1} + b_i), \tag{1}$$

$$f_t = \sigma(W_{fx}x_t + W_{fh}h_{t-1} + b_f), \tag{2}$$

$$c_n = g(W_{cx}x_t + W_{ch}h_{t-1} + b_c), \tag{3}$$

$$c_t = f_t \circ c_{t-1} + i_t \circ c_n, \tag{4}$$

$$h_t = o_t \circ g(c_t), \tag{5}$$

$$o_t = \sigma(W_{ox}x_t + W_{oh}h_{t-1} + b_o), \tag{6}$$

where σ, g denote the sigmoid and $tanh$ functions, respectively; $\circ$ is an element-wise product; W terms denotes weight matrices; b are bias vectors; and i, f, o, c denote input gate, forget gate, output gate and cell activation vectors, respectively.

2.2 Bi-LSTM

Correct recognition of named entity in a sentence depends on the context of the word. Both preceding and following words matter to predict a tag. Bi-directional recurrent neuronal networks [12] were designed to encode every element in a sequence taking into account left and right contexts which makes it one of the best choices for NER task. Bi-directional model calculation consists of two steps: (1) the forward layer computes representation of the left context, and (2) the backward layer computes representation of the right context. Outputs of these steps are then concatenated to produce a complete representation of an element of the input sequence. Bi-directional LSTM encoders have been demonstrated to be useful in many NLP tasks such as machine translation, question answering, and especially for NER problem.

2.3 CRF Model for NER Task

Conditional Random Field is a probabilistic model for structured prediction which has been successfully applied in variety of fields, such as computer vision, bioinformatics, natural language processing. CRF can be used independently to solve NER task ([13,15]).

The CRF model is trained to predict a vector $\mathbf{y} = \{y_0, y_1, .., y_T\}$ of tags given a sentence $\mathbf{x} = \{x_0, x_1, .., x_T\}$. To do this, a conditional probability is computed:

$$p(\mathbf{y}|\mathbf{x}) = \frac{e^{Score(\mathbf{x},\mathbf{y})}}{\sum_{\mathbf{y}'} e^{Score(\mathbf{x},\mathbf{y}')}}, \tag{7}$$

where $Score$ is computed by the formula below [2]:

$$Score(\mathbf{x}, \mathbf{y}) = \sum_{i=0}^{T} A_{y_i, y_{i+1}} + \sum_{i=1}^{T} P_{i, y_i}, \tag{8}$$

where $A_{y_i, y_{i+1}}$ denotes the emission probability which represents the score of transition from tag i to tag j, $P_{i,j}$ is transition probability which represents the score of the jth tag of the word i^{th}.

In the training stage, log probability of correct tag sequence $log(p(\mathbf{y}|\mathbf{x}))$ is maximized.

2.4   Combined Bi-LSTM and CRF Model

Russian is a morphologically and grammatically rich language. Thus, we expected that a combination of CRF model with a Bi-LSTM neural network encoding [2] should increase the accuracy of the tagging decisions. The architecture of the model is presented on the Fig. 1.

In the combined model characters of each word in a sentence are fed into a Bi-LSTM network in order to capture character-level features of words. Then these character-level vector representations are concatenated with word embedding vectors and fed into another Bi-LSTM network. This network calculates a sequence of scores that represent likelihoods of tags for each word in the sentence. To improve accuracy of the prediction a CRF layer is trained to enforce constraints dependent on the order of tags. For example, in the IOB scheme (I – Inside, O – Other, B – Begin) tag I never appears at the beginning of a sentence, or "O I B O" is an invalid sequence of tags.

Full set of parameters for this model consists of parameters of Bi-LSTM layers (weight matrices, biases, word embedding matrix) and transition matrix of CRF layer. All these parameters are tuned during training stage by back propagation algorithm with stochastic gradient descent. Dropout is applied to avoid over-fitting and improve the system performance.

2.5   Neuro NER Extensions

NeuroNER is an open-source software package for solving NER tasks. The neural network architecture of NeuroNER is similar to the architecture proposed in the previous section.

Inspired by success of character aware networks approach [20] we extended NeuroNER model with a highway layer on top of the Bi-LSTM character embedding layer. This extension is depicted on Fig. 2. Dense layer makes character

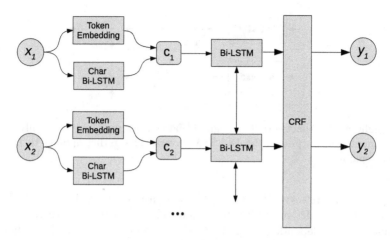

Fig. 1. The architecture of Bi-LSTM neural network for solving NER task. Here, x_i is a representation of word in a sequence. It is fed into character and word level embedding blocks. Then character and word level representations are concatenated into c_i. Bi-LSTM performs conditioning of the concatenated representations on the left and right contexts. Finally CRF layers provide output tag predictions y_i.

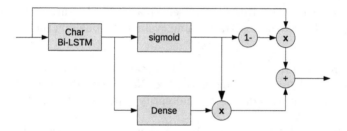

Fig. 2. Highway network on top of the character embedding network. The Dense layer serves for compute higher level representations of the input. The sigmoid layer computes gate values. This values are used to dynamically balance between high and low level representations. Block (1-) subtracts input from 1, and block (x) perform multiplication of the inputs.

embedding network deeper. The carry gate presented by sigmoid layer provides a possibility to choose between dense and shortcut connections dynamically. A highway network can be described by the following equation:

$$y = H(\mathbf{x}, \mathbf{W_H}) \cdot G(\mathbf{x}, \mathbf{W_G}) + \mathbf{x} \cdot (1 - G(\mathbf{x}, \mathbf{W_G})) \tag{9}$$

where $\mathbf{x}$ is the input of the network, $H(\mathbf{x}, \mathbf{W_H})$ is the processing function, $G(\mathbf{x}, \mathbf{W_G})$ is the gating function. The dimensionality of $\mathbf{x}$, $\mathbf{y}$, $G(\mathbf{x}, \mathbf{W_G})$, and $H(\mathbf{x}, \mathbf{W_H})$ must be the same.

Another extension of NeuroNER we implemented is a Bi-LSTM highway network [21]. The architecture of this network is quite similar to the character-aware

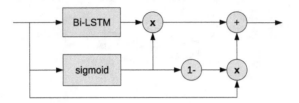

Fig. 3. Highway LSTM network. Here sigmoid gate layer is used to dynamically balance between input and output of the Bi-LSTM layers. The gating applied to the each direction separately.

highway network. However, the carry gate is conditioned on the input of the LSTM cell. The gate provides an ability to dynamically balance between raw embeddings and context dependent LSTM representation of the input. The scheme of our implementation of the highway LSTM is depicted in Fig. 3.

3   Experiments

3.1   Datasets

Currently, there are a few Russian datasets created for the purpose of developing and testing NER systems. We trained and evaluated models on the three Russian datasets:

- Dataset received from Gareev et al. [4] contains 97 documents collected from ten top cited "Business" feeds in Yandex "News" web directory. IOB tagging scheme is used in this data sets, and entity types are Person, Organization, Other.
- The FactRuEval 2016 corpus [16] contains news and analytical texts in Russian. Sources of the dataset are Private Correspondent[4] web site and Wikinews[5]. Topics of the texts are social and political. Tagging scheme is IOB.
- Person-1000 [18] is a Russian news corpus with marked up person named entities. This corpus contains materials from the Russian on line news services.

Statistics on these datasets are provided in the Table 1.

3.2   External Word Embedding

News and *Lenta* are two external word embeddings we used to initialize lookup table for the training step.

News[6] is a Russian word embeddings introduced by Kutuzov et al. [14]. Corpus for this word embedding is a set of Russian news (from September 2013 until November 2016). Here are more details about *news*:

---

[4] http://www.chaskor.ru/.

[5] http://ru.wikinews.org.

[6] Word embeddings, which are available to download from http://rusvectores.org.

Table 1. Statistics on datasets.

Corpus	Tokens	Words and numbers	Persons	Organizations	Locations
FactRuEval 2016	90322	73807	2087	1181	2686
Gareev's Corpus	44326	35116	486	1317	-
Persons-1000	284221	224446	10600	-	-

- Corpus size: near 5 billion words
- Vocabulary size: 194058
- Frequency threshold: 200
- Algorithm: Continuous Bag of Words
- Vector size: 300

Lenta is a publicly available corpus of unannotated Russian news. This corpus consists of 635000 news from Russian online news resource lenta.ru. The size of the corpus is around 46 million words. The corpus spans vocabulary of size 376000 words.

To train embeddings on this corpus, we use skip-gram algorithm enriched with subword information [17]. Parameters of the algorithm were the following:

- Vector size: 100
- Minimal length of char n-gram: 3
- Maximal length of char n-gram: 6
- Frequency threshold: 10

3.3   Results

The purpose of the first experiment was to compare tagging accuracy of three implementations: Bi-LSTM, Bi-LSTM + CRF, Bi-LSTM + CRF + external word embedding *news*. To do this, we evaluated these implementations on the *Gareev's dataset*. Parameters of the dataset and hyper-parameters of the models are listed below:

- Word embedding dimension: 100
- Char embedding dimension: 25
- Dimension of hidden layer: 100 (for each LSTM: forward layer and backward layer)
- Learning method: SGD, learning rate: 0.005
- Dropout: 0.5
- Number of sentences: 2136 (for training/ validation/ testing: 1282/ 427/ 427)
- Number of words: 25372 (unique words: 7876). 7208 words (account for 91.52% of unique words) was initialized with pre-trained embedding *Lenta*
- epochs: 100

We used ConllEval[7] to calculate metrics of performance.

The result is shown in the Table 2. One can see that adding CRF layer significantly improved prediction. Besides that, using external word embeddings also reduced training time and increased tagging accuracy. Due to absence of lemmatization in our text processing pipeline *news* embeddings matched only about 15% of words in the corpus, embeddings for other words were just initialized randomly. Therefore, the improvement was not really significant and prediction for *Organization* type was even lower with *news* embeddings. To deal with this problem, in the second experiment we decided to use FastText trained on *Lenta* corpus in order to build an external word embedding. After that, we used this embedding to train on Gareev's dataset one more time using the same configuration with the previous experiment.

Table 2. Tagging results of baseline models on Gareev's dataset

Model	Person			Organization			Overall		
	P	R	F	P	R	F	P	R	F
Bi-LSTM	67.11	78.46	72.34	76.56	72.59	74.52	73.04	74.50	73.76
Bi-LSTM CRF	92.93	86.79	89.76	85.24	**81.91**	83.54	87.30	83.25	85.22
Bi-LSTM CRF + *news* word emb.	95.05	90.57	92.75	85.13	81.21	83.12	87.84	83.76	85.75
Bi-LSTM CRF + *Lenta* word emb.	**95.60**	**94.57**	**95.08**	**87.40**	81.62	**84.41**	**89.57**	**84.89**	**87.17**

Table 3 shows the confusion matrix on the test set. We also experimented on two other datasets: Persons-1000, FactRuEval 2016. The summary of experiments on these datasets are shown in the Table 4.

Table 3. The confusion matrix on the test set of Gareev's dataset

Named entity	Total	O	I-ORG	B-ORG	B-PER	I-PER	Percent
O	7688	7647	19	22	0	0	99.467
I-ORG	308	36	268	3	1	0	87.013
B-ORG	272	38	2	229	2	1	84.191
B-PER	92	3	1	0	88	0	95.652
I-PER	69	2	5	0	0	62	89.855

We compare Bi-LSTM + CRF + *Lenta* model and other published results as well as NeuroNER and its extensions on three datasets mentioned in the

---

[7] A Perl script was used to evaluate result of processing CoNLL-2000 shared task: http://www.cnts.ua.ac.be/conll2000/chunking/conlleval.txt.

Table 4. Tagging results of Bi-LSTM + CRF + *Lenta* word embedding on three datasets: Gareev's dataset, FactRuEval 2016, Persons-1000

Datasets	Dev. set			Test set		
	P	R	F	P	R	F
FactRuEval 2016	84.39	81.11	82.72	83.88	80.40	82.10
Gareev's dataset	90.99	86.94	88.92	89.57	84.89	87.17
Persons-1000	98.97	98.20	98.58	99.43	99.09	99.26

Subsect. 3.1. Results are presented in the Table 5. Bi-LSTM + CRF + *Lenta* model significantly outperforms other approaches on Gareev's dataset and Persons-1000. However, the result on FactRuEval 2016 dataset is not as high as we expected.

Table 5. Performance of different models across datasets

Models	Gareev's dataset			Persons-1000			FactRuEval 2016		
	P	R	F	P	R	F	P	R	F
Gareev et al. [4]	67.98	75.05	84.11	-	-	-	-	-	-
Malykh et al. [9]	59.65	65.70	62.49	-	-	-	-	-	-
Trofimov [5]	-	-	-	97.26	93.92	95.57	-	-	-
Rubaylo et al. [19]	-	-	-	-	-	-	77.70	78.50	78.13
Sysoev et al. [8]	-	-	-	-	-	-	**88.19**	64.75	74.67
Ivanitsky et al. [7]	-	-	-	-	-	-	-	-	**87.88**
Mozharova et al. [6]	-	-	-	-	-	97.21	-	-	-
NeuroNER	88.19	82.73	85.37	96.38	96.83	96.60	80.49	79.23	79.86
NeuroNER + Highway char	85.75	**88.40**	87.06	96.56	97.11	96.83	80.59	80.72	80.66
NeuroNER + Highway LSTM	84.35	81.96	83.14	96.49	97.19	96.84	81.09	79.31	80.19
NeuroNER + Highway char + Highway LSTM	83.33	85.05	84.18	96.74	96.83	96.78	79.13	78.76	78.95
Bi-LSTM + CRF + *Lenta*	**89.57**	84.89	**87.17**	**99.43**	**99.09**	**99.26**	83.88	**80.84**	82.10

4 Discussion

Traditional approaches to Russian NER heavily relied on hand-crafted rules and external resources. Thus regular expressions and dictionaries were used in [5] to solve the task. The next step was application of statistical learning methods such as conditional random fields (CRF) and support vector machines (SVM) for entity classification. CRF on top of linguistic features considered as a baseline in the study of [4]. Mozharova and Loukachevitch [6] proposed two-stage CRF algorithm. Here, an input for the CRF of the first stage was a set of hand-crafted linguistic features. Then on the second stage the same input features were combined with a global statistics calculated on the first stage and fed into CRF. Ivanitskiy et al. [7]

applied SVM classifier to the distributed representations of words and phrases. These representations were obtained by extensive unsupervised pre-training on different news corpora. Simultaneous use of dictionary based features and distributed word representations was presented in [8]. Dictionary features were retrieved from Wikidata and word representations were pre-trained on Wikipedia. Then these features were used for classification with SVM.

At the moment deep learning methods are seen as the most promising choice for NER. Malykh and Ozerin [9] proposed character aware deep LSTM network for solving Russian NER task. A distinctive feature of this work is coupling of language modeling task with named entity classification.

In our study we applied current state of the art neural network based model for English NER to known Russian NER datasets. The model consists of three main components such as bi-directional LSTM, CRF and external word embeddings. Our experiments demonstrated that Bi-LSTM alone was slightly worse than CRF based model of [4]. Addition of CRF as a next processing step on top of Bi-LSTM layer significantly improves model's performance and allow to outperform the model presented in [4]. The difference of Bi-LSTM + CRF model from the model presented in [4] is trainable feature representations. Combined training of Bi-LSTM network on the levels of words and characters gave better results then manual feature engineering in [4].

Distributed word representations are becoming a standard tool in the field of natural language processing. Such representations are able to capture semantic features of words and significantly improve results for different tasks. When we encoded words with *news* or *Lenta* embeddings results were consistently better for all three datasets. Up to now, the prediction accuracy of the Bi-LSTM + CRF + *Lenta* model outperforms published models on Gareev's dataset and Persons-1000. However, the results of both Bi-LSTM + CRF + *Lenta* and NeuroNER models on the FactRuEval dataset were better then results reported in [8,19] but not as good as SVM based model reported in [7].

In spite the fact that both models we tested have the same structure, performance of NeuroNER [3] is a bit lower than Bi-LSTM+CRF model [2]. This issue can be explained by different strategies for initialization of parameters.

Our extension of the baseline model with a highway network for character embedding provides moderate performance growth in nearly all cases. Implementation of the Bi-LSTM highway network for tokens resulted in a slight increase of performance for Persons-100 and FactRuEval 2016 datasets and a decrease of performance for Gareev's dataset. Simultaneous extension of the NeuroNER with character and token Bi-LSTM highway networks results in the drop of performance in the most of the cases.

We think that results of LSTM highway network can be improved by different bias initialization and deeper architectures. In the current work the highway gate bias was initialized with 0 vector. However, bias could be initialized to some negative value. This initialization will force the network to prefer processed path to the raw path. Furthermore, stacking highway LSTM layers might improve results allowing a network dynamically adjust complexity of the processing.

Alternatively, character embedding network can be built using convolutional neural networks (CNN) instead of LSTM. A number of authors [20,22] reported promising results with a character level CNN. Another promising extension of presented architecture is an attention mechanism [23]. For NER task this mechanism can be used to selectively attend to the different parts of the context for each word giving additional information for the tagging decision.

5 Conclusions

Named Entity Recognition is an important stage in information extraction tasks. Today, neural network methods for solving NER task in English demonstrate the highest potential. For Russian language there are still a few papers describing application of neural networks to NER. We studied a series of neural models starting from vanilla bi-directional LSTM then supplementing it with conditional random fields, highway networks and finally adding external word embeddings. For the first time in the literature evaluation of models were performed across three Russian NER datasets. Our results demonstrated that (1) basic Bi-LSTM model is not sufficient to outperform existing state of the art NER solutions, (2) addition of CRF layer to the Bi-LSTM model significantly increases it's quality, (3) pre-processing the word level input of the model with external word embeddings allowed to improve performance further and achieve state-of-the-art for the Russian NER.

Acknowledgments. The statement of author contributions. AL conducted initial literature review, selected a baseline (Bi-LSTM + CRF) model, prepared datasets and run experiments under supervision of MB. AM implemented and studied extensions of the NeuroNER model. AL drafted the first version of the paper. AM added a review of works related to the Russian NER and materials related to the NeuroNER modifications. MB, AL and AM edited and extended the manuscript.

This work was supported by National Technology Initiative and PAO Sberbank project ID 0000000007417F630002.

References

1. Patawar, M.L., Potey, M.A.: Approaches to named entity recognition: a survey. Int. J. Innov. Res. Comput. Commun. Eng. **3**(12), 12201–12208 (2015)
2. Lample, G., Ballesteros, M., Subramanian, S., Kawakami, K., Dyer, C.: Neural architectures for named entity recognition. ArXiv preprint arXiv: 1603.01360 (2016)
3. Dernoncourt F., Lee, J.Y., Szolovits P.: NeuroNER: an easy-to-use program for named-entity recognition based on neural networks. ArXiv preprint arXiv:1705.05487 (2017)
4. Gareev, R., Tkachenko, M., Solovyev, V., Simanovsky, A., Ivanov, V.: Introducing baselines for russian named entity recognition. In: Gelbukh, A. (ed.) CICLing 2013. LNCS, vol. 7816, pp. 329–342. Springer, Heidelberg (2013). https://doi.org/10.1007/978-3-642-37247-6_27

5. Trofimov, I.V.: Person name recognition in news articles based on the persons-1000/1111-F collections. In: 16th All-Russian Scientific Conference Digital Libraries: Advanced Methods and Technologies, Digital Collection, RCDL 2014, pp. 217–221 (2014)

6. Mozharova V., Loukachevitch N.: Two-stage approach in Russian named entity recognition. In: 2016 International FRUCT Conference on Intelligence, Social Media and Web (ISMW FRUCT), pp. 1–6 (2016)

7. Ivanitskiy, R., Alexander, S., Liubov, K.: Russian named entities recognition and classification using distributed word and phrase representations. In: SIMBig, pp. 150–156 (2016)

8. Sysoev, A.A., Andrianov, I.A.: Named entity recognition in Russian: the power of Wiki-based approach. dialog-21.ru (2016)

9. Malykh, V., Ozerin, A.: Reproducing Russian NER baseline quality without additional data. In: Proceedings of the 3rd International Workshop on Concept Discovery in Unstructured Data, Moscow, Russia, pp. 54–59 (2016)

10. Bengio, Y., Simard, P., Frasconi, P.: Learning long-term dependencies with gradient descent is difficult. IEEE Trans. Neural Netw. **5**(2), 157–166 (1994)

11. Hochreiter, S., Schmidhuber, J.: Long short-term memory. Neural Comput. **9**(8), 1735–1780 (1997). MIT Press

12. Schuster, M., Paliwal, K.K.: Bidirectional recurrent neural networks. IEEE Trans. Signal Process. **45**(11), 2673–2681 (1997)

13. Chen, W., Zhang, Y., Isahara, H.: Chinese named entity recognition with conditional random fields. In: Proceedings of the Fifth SIGHAN Workshop on Chinese Language Processing, pp. 118–121 (2006)

14. Kutuzov, A., Kuzmenko, E.: WebVectors: a toolkit for building web interfaces for vector semantic models. In: Ignatov, D.I., Khachay, M.Y., Labunets, V.G., Loukachevitch, N., Nikolenko, S.I., Panchenko, A., Savchenko, A.V., Vorontsov, K. (eds.) AIST 2016. CCIS, vol. 661, pp. 155–161. Springer, Cham (2017). https://doi.org/10.1007/978-3-319-52920-2_15

15. Ekbal, A., Haque, R., Bandyopadhyay, S.: Named entity recognition in Bengali: a conditional random field approach. In: IJCNLP Conference, pp. 589–594 (2008)

16. Starostin, A.S., Bocharov, V.V., Alexeeva, S.V., Bodrova, A., Chuchunkov, A.S., Dzhumaev, S.S., Nikolaeva, M.A.: FactRuEval 2016: evaluation of named entity recognition and fact extraction systems for Russian. In: Proceedings of the Annual International Conference Dialogue on Computational Linguistics and Intellectual Technologies, no. 15, pp. 702–720 (2016)

17. Bojanowski, P., Grave, E., Joulin, A., Mikolov, T.: Enriching word vectors with subword information. ArXiv preprint arXiv:1607.04606 (2016)

18. Vlasova, N.A., Suleymanova, E.A., Trofimov, I.V: Report on Russian corpus for personal name retrieval. In: Proceedings of Computational and Cognitive Linguistics, TEL 2014, Kazan, Russia, pp. 36–40 (2014)

19. Rubaylo, A.V., Kosenko, M.Y.: Software utilities for natural language information retrievial. Alm. Mod. Sci. Educ. **12**(114), 87–92 (2016)

20. Kim, Y., Jernite, Y., Sontag, D., Rush, A.M.: Character-aware neural language models. In: Proceedings of the Thirtieth AAAI Conference on Artificial Intelligence, pp. 2741–2749 (2016)

21. Pundak, G., Sainath, T.N.: Highway-LSTM and recurrent highway networks for speech recognition. In: Proceedings of Interspeech 2017, ISCA (2017)

22. Tran, P.-N., Ta, V.-D., Truong, Q.-T., Duong, Q.-V., Nguyen, T.-T., Phan, X.-H.: Named entity recognition for vietnamese spoken texts and its application in smart mobile voice interaction. In: Nguyen, N.T., Trawiński, B., Fujita, H., Hong, T.-P. (eds.) ACIIDS 2016. LNCS (LNAI), vol. 9621, pp. 170–180. Springer, Heidelberg (2016). https://doi.org/10.1007/978-3-662-49381-6_17

23. Bahdanau, D., Cho, K., Bengio, Y.: Neural machine translation by jointly learning to align and translate. arXiv preprint arXiv:1409.0473 (2014)

Web-Scale Data Processing

Employing Wikipedia Data for Coreference Resolution in Russian

Ilya Azerkovich[(✉)]

Higher School of Economics, Moscow, Russia
ilazerkovich@edu.hse.ru

Abstract. Semantic information has been deemed a valuable resource for solving the task of coreference resolution by many researchers. Unfortunately, not much has been done in the direction of using this data when working with Russian data. This work describes the first step of a research, attempting to create a coreference resolution system for Russian based on semantic data, concerned with using Wikipedia information for the task. The obtained results are comparable to ones for English data, which gives reasons to expect their improvement in further steps of the research.

Keywords: Natural language processing · Coreference resolution · Information extraction

1 Introduction

Coreference resolution is a very important part of many natural language processing (NLP) tasks, which generally requires information from several language layers. As a rule, morphological and syntactical information is used, but as of late many researchers have been pointing out the importance of integrating semantical information in the process of solving this task. Use of freely available knowledge sources, such as Wikipedia, which are widely used for such tasks as Named Entity Extraction [3], has also been suggested for coreference resolution, see e.g. [7, 9].

While certain work has already been done on creating systems of automated coreference resolution, the room for improvement still exists (see results, presented in [15]). Because of many linguistic differences between English and Russian, e.g. lack of explicitly expressed definiteness category, and more variation in morphology, approaches used in NLP of one language are not always suitable for the other. Apart from that, Russian segment of Wikipedia is smaller (1,398,000 articles compared to 5,419,000 as of June 8th) and less structured than the English one. Nevertheless, the idea of applying semantic features from Wikipedia to coreference resolution in Russian seemed promising, because this resource is still one of the largest and most developed, compared to other open-source solutions available.

This work presents one stage of the research, dedicated to adapting existing machine learning algorithms, used for coreference resolution in Russian, to include semantic features for analysis. It describes the results that were achieved by training a classifier

© Springer International Publishing AG 2018
A. Filchenkov et al. (Eds.): AINL 2017, CCIS 789, pp. 107–112, 2018.
https://doi.org/10.1007/978-3-319-71746-3_9

algorithm for coreference resolution on a small corpus of news texts, assembled for this task, with additional semantic features, obtained using Wikipedia as the source. While many processes were simplified due to constraints in time and processing power, the achieved results show certain improvement. Future work based on this research includes testing effect of semantic features on other algorithms as well as using other sources of semantic information.

2   Related Work on Topic

Importance of semantical information for various tasks of natural language processing has since long come to the attention of researchers, and open sources of suck knowledge are widely employed for this purpose. One of most popular such sources is Wikipedia, information from which, among other purposes, has been used for Name Entities Extraction [3], or calculating semantic relatedness [13]. Use of Wikipedia data per se as a source of features for coreference resolution has been first suggested in [7], in conjunction with features, obtained from other knowledge sources, such as WordNet. Similar approach, involving combined features from Wikipedia and WordNet, is taken in other works, e.g. [9], where data from YAGO ontology is used, which combines information from these two sources.

In Russian NLP Wikipedia data is mostly used for Named Entity Extraction and Entity Disambiguation, and, as a rule, is not seen as a source of features for coreference resolution. Existing systems have been based on existing proprietary ontologies [15] or not relying on semantic information, but this tendency is gradually changing, e.g. in [14] named entity recognition is used to provide additional features for analysis. This work is an attempt to encourage these changes by showing that using semantic features can be advantageous for analysis.

3   Using Semantic Features from Wikipedia Data to Improve Results of Coreference Resolution

3.1   Text Preprocessing and Feature Extraction

The research was conducted on the base of a corpus, consisting of approximately 1,000 texts, obtained from the site of INTERFAX news network[1]. They were tokenized and morphologically tagged using the TreeTagger system [11], and then processed with a simple noun phrase (NP) extractor, based on recognizing proper names and genitive and adjectival dependencies, based on the tags of the tokens. As a result, the list of coreference candidates – pronouns and noun phrases – was obtained from all the texts. In each noun phrase, tokens comprising its head were additionally marked. Finally, candidates no more than 12 extracted groups apart were joined in pairs and marked according to the features set.

---

[1] http://www.interfax.ru/.

Coreference resolution for the purposes of this research was considered a binary classification task: a classifier is given a pair of NPs and decides, whether they corefer, or not. The SVM-based algorithm was used as the main classifier, as an algorithm that has already been applied to anaphora resolution for Russian with satisfying results [4, 5].

The features used for classifier training were mostly chosen based on the articles [1, 7]. Features used in this research can also be divided into several classes: string, morphological, relative location. The full set is listed below in Table 1.

Table 1. Set of features used for classifier training

Category	Features
String features	Head match
	Head substring match
	Full NP match
Morphological features	Gender match
	Number match
	Animateness match
	First member is pronoun
	(if Yes, pronoun type)
	Second member is pronoun
	(if Yes, pronoun type)
	First member is proper
	Second member is proper
Relative location	Distance in sentences
	Distance in noun phrases

Based on this set of features a classifier was trained, using the 10-fold cross-validation method. Simple precision, recall and F-measure metrics were deemed enough for evaluation of the results, because the same markup was used as the gold standard and the base for classification.

Results of the first step of the work are shown in Table 2. They are on par with state-of-the-art results for Russian (see [2, 4, 5]), but might be partly attributed to over-matching due to limited corpus size. The main goal at this stage of research was to evaluate the amount of correctly classified pairs, and later compare it to the results of classification with added semantic features. Improving the system's performance in general is one of important steps on future stages of this work.

Table 2. Recall, precision, F-measure (first step)

Recall	Precision	F-measure
0.79	0.78	0.785

3.2 Adding Wikipedia Data

The next step of the experiment was extracting semantic features from Wikipedia data and updating the training set with them. Unfortunately, providing semantic data for all mentions in the corpus put too heavy load on the equipment used in the experiment, so the target for adding semantic features had to be narrowed.

To achieve that, firstly named entities were extracted from the texts of the corpus, using the Natasha Python library[2] that took part in the recent FactRuEval 2016 competition of Named Entity Recognition (NER)-systems for Russian [12]. It was one of the few participating systems to which open access was provided, and that was the deciding factor for choosing this system for the NER-task. It can extract People, Geography (toponyms and post addresses), Organizations and Events as entity types, as well as descriptors for People, including jobs and titles.

After extracting named entities from text, a procedure similar to the one described in [7] was adapted. Namely, the named entity, together with its descriptor if present, was queried to Wikipedia search API, and the first result obtained after resolving page redirects was used as the data source. Then, for each pair of mentions containing the extracted entity resulting page was analyzed: if the other member of the pair was contained in the first paragraph of the article or in its hyperlinks, the corresponding feature value was assigned to the pair. If none of the members of the pair was found in Wikipedia, the feature value was left as 'undefined'.

After updating the feature set with the results of Wikipedia analysis, the pairs were classified for the second time. The results are presented in Table 3. As can be seen from the table, adding these features has had a positive effect on quality of the performance, which falls in line with results described in other works.

Table 3. Recall, precision, F-measure (second step)

Recall	Precision	F-measure
0.79	0.80	0.795

4 Results

4.1 Discussion

While results, obtained in this preliminary research, are promising, they cannot yet be described as final. The achieved here increase in quality in 1% is comparable to gain from Wikipedia features, described in [7]. This gives reasons to assume that improving choice of features for the baseline set and increasing number of semantic features can provide even better results in future research.

The results above show that semantic features serve to increase precision, possibly acting as filters separating more closely connected pairs of mentions. Using them to increase recall as well is an important direction of improving the analysis. While the frame of this research allowed only working with proper nouns, adding semantic features to common nouns would certainly improve the quality of analysis even more.

Problems that arose at both steps of analysis were the large amount of mentions that did not corefer to any of the extracted entities, and presence of false positive semantic

---

[2] https://github.com/bureaucratic-labs/natasha.

feature values. While methods of solving the first problem are described e.g. in [10], the second one should be looked at more closely. Cases in which such false positives arise, are mostly pairs of a person and the institution they are occupied at, such as *António Guterres – the UN*. To avoid this, morphological filters by case or animateness can be applied to Wikipedia information before attributing this semantic feature.

4.2 Future Work

Several improvements and alternative approaches can be implemented on all steps of the described above process. Firstly, set of features should be improved. Semantic features should be defined for as many mentions as possible, and apposition and results of named entity recognition should be included in the set, as was done in [1, 7]. Another important task would be excluding singletons from list of coreferential candidates. This can be done based on methods suggested in [10]. Finally, approaches to coreference resolution, different from the pair-based one can be considered. One of such approaches, combining clustering and ranking methods, has been described in [8] as yielding better results than algorithms based on both of those methods.

Apart from that, extraction of semantic features can also be improved. Features not only for named entities, but also for common names should be added, and other knowledge bases, such as RuThes and RuWordNet [6], can be employed as sources of semantic data. Use of ontologies seems most preferable because, apart from providing semantic links between named entities, it could also give access to links between common names, absent in Wikipedia, e.g. *president – head of state*.

References

1. Bengtson, E., Roth, D.: Understanding the value of features for coreference resolution. In: Proceedings of the Conference on Empirical Methods in Natural Language Processing, pp. 294–303. Association for Computational Linguistics (2008)
2. Bogdanov, A.V., Dzhumaev, S.S., Skorinkin, D.A., Starostin, A.S.: Anaphora analysis based on ABBYY Compreno linguistic technologies. In: Computational Linguistics and Intellectual Technologies: Proceedings of the International Conference "Dialogue 2014", pp. 89–102. Moscow, Publishing Center RSUH (2014)
3. Bunescu, R., Marius, P.: Using encyclopedic knowledge for named entity disambiguation. In: Proceedings of the 11th Conference of the European Chapter of the Association for Computational Linguistics (EACL-06), pp. 9–16, Trento, Italy (2006)
4. Kamenskaya, M.A., Khramoin, I.V., Smirnov, I.V.: Data driven methods for anaphora resolution of Russian texts. In: Computational Linguistics and Intellectual Technologies: Proceedings of the International Conference "Dialogue 2014", pp. 241–250. Moscow, Publishing Center RSUH (2014)
5. Kutuzov, A.B., Ionov, M.: The impact of morphology processing quality on automated anaphora resolution for Russian. In: Computational Linguistics and Intellectual Technologies: Proceedings of the International Conference "Dialogue 2014", pp. 232–240. Moscow, Publishing Center RSUH (2014)

6. Loukachevitch, N., Dobrov, B., Chetviorkin, I.: RuThes-lite, a publicly available version of thesaurus of Russian language RuThes. In: Computational Linguistics and Intellectual Technologies: Papers from the Annual International Conference "Dialogue 2014" (2014)

7. Ponzetto, S.P., Strube, M.: Exploiting semantic role labeling, WordNet and Wikipedia for coreference resolution. In: Proceedings of the Main Conference on Human Language Technology Conference of the North American Chapter of the Association of Computational Linguistics, pp. 192–199. Association for Computational Linguistics (2006)

8. Rahman, A., Ng, V.: Supervised models for coreference resolution. In: Proceedings of the 2009 Conference on Empirical Methods in Natural Language Processing, vol. 2, pp. 968–977. Association for Computational Linguistics (2009)

9. Rahman, A., Ng, V.: Coreference resolution with world knowledge. In: Proceedings of the 49th Annual Meeting of the Association for Computational Linguistics: Human Language Technologies, vol. 1, pp. 814–824. Association for Computational Linguistics (2011)

10. Recasens, M., de Marneffe, M., Potts, C.: The life and death of discourse entities: identifying singleton mentions. In: Proceedings of NAACL-HLT, pp. 627–633 (2013)

11. Schmid, H.: Probabilistic part-of-speech tagging using decision trees. In: Proceedings of the International Conference on New Methods in Language Processing, vol. 12. pp. 44–49. Citeseer (1994)

12. Starostin, A.S., Bocharov, V.V., Alexeeva, S.V., Bodrova, A.A., Chuchunkov, A.S., Dzhumaev, S.S., Efimenko, I.V., Granovsky, D.V., Khoroshevsky, V.F., Krylova, I.V., Nikolaeva, M.A., Smurov, I.M., Toldova, S.Y.: FactRuEval 2016: evaluation of named entity recognition and fact extraction systems for Russian. In: Computational Linguistics and Intellectual Technologies: Proceedings of the International Conference "Dialogue 2016", pp. 702–720. Moscow, Publishing Center RSUH (2016)

13. Strube, M., Ponzetto, S.P.: WikiRelate! Computing semantic relatedness using Wikipedia. In: AAAI, vol. 6, pp. 1419–1424 (2006)

14. Sysoev, A.A., Andrianov, I.A., Khadzhiiskaia, A.Y.: Coreference resolution in Russian: state-of-the-art approaches application and evolvement. In: Computational Linguistics and Intellectual Technologies: Papers from the Annual International Conference "Dialogue 2017", vol. 1, pp. 327–338. Moscow, Publishing Center RSUH (2017)

15. Toldova, S.Ju., Roytberg, A., Ladygina, A.A, Vasilyeva, M.D., Azerkovich, I.L., Kurzukov, M., Sim, G., Gorshkov, D.V., Ivanova, A., Nedoluzhko, A., Grishina, Y.: Ru-Eval-2014: evaluating anaphora and coreference resolution for Russian. In: Computational Linguistics and Intellectual Technologies: Proceedings of the International Conference "Dialogue 2014", pp. 681–694. Moscow, Publishing Center RSUH (2014)

Building Wordnet for Russian Language from Ru.Wiktionary

Yuliya Chernobay[✉]

Saint Petersburg State University, Saint Petersburg, Russia
`jamsic@yandex.ru`

Abstract. This paper presents a method of fully-automatic transformation of the free-content Russian dictionary ru.wiktionary to WordNet-like thesaurus. The primary concern of this study is to describe a procedure of relating words to their meanings throughout Wiktionary pages and establish synonym and hyponym-hypernym relation between specific senses of words. The produced database contains 104696 synsets and is publicly available in alpha version as a python package wiki-ru-wordnet.

1 Introduction

WordNet is a lexical database that contains words and establishes lexical relations (synonyms, antonyms, hyponyms, hypernyms) between them according to word senses. Thus one word can be included in different synonym sets and have different hypernyms and hyponyms for each of its senses. WordNet is widely used in natural language processing tasks. Wiktionary is a multilingual dictionary that includes not only the definition of a word, but also etymologies, pronunciations, sample quotations, synonyms, antonyms and other lexical relations and translations to other languages. These two resources are pretty similar, however, Wiktionary does not provide disambiguation between different senses of the same word.

2 Related Work

Researchers used different approaches to create WordNet for Russian language. Some of them [8] used automated translation from Princeton WordNet [14]. RussNet [2] has been developed manually since 1999 and currently contains more than 5500 synsets. Braslavski [4] introduced a user interface for a crowd-sourced thesaurus based on Wiktionary approach that contains currently 73061 synsets, 104906 raw synonym pairs and 29764 raw hypernym-hyponym relations. The most recent study [13] introduced a semi-automatic process of transforming the Russian language thesaurus RuThes to WordNet-like thesaurus, called RuWordNet.

There are plenty of researches on automatic creation of WordNet from dictionaries and thesauri. In [9] authors proposed a fuzzy clustering algorithm called

© Springer International Publishing AG 2018
A. Filchenkov et al. (Eds.): AINL 2017, CCIS 789, pp. 113–120, 2018.
https://doi.org/10.1007/978-3-319-71746-3_10

ECO that was used to induce synsets for a Portuguese WordNet from several synonymy dictionaries. In [6,18] authors presented different approaches for disambiguation of word senses in synsets.

At the same time Wiktionary has gained a lot of attention as a sourse of semantic knowledge. [11] presents Wikokit, a convenient toolkit for the extraction of structured data from Wiktionary. The Russian Wiktionary parser in JWKTL [19] is based on it. Both toolkits make it possible to construct "ambiguous" WordNet from Wiktionary. [17,20] suggest that Wiktionary is a valuable alternative source of lexical knowledge.

3   Data

Wiktionary is a crowdsourced dictionary and thesaurus that exists for many languages. Wiktionary pages related to a specific word can contain a lot of useful information about word senses, including a list of lexical senses, definition and examples for a lexical sense, lexical relations (synonyms, antonyms, hyponyms, hypernyms), which are represented as links to Wiktionary pages. However, there are also some problems in word senses description, which can hamper creating a WordNet-like resource:

1. lexical links lead not to a specific sense, but to the whole word page,
2. synonyms can be described as partial synonyms, for example "gayser" and "fountain".
3. lexical relations are not symmetrical. For example, word w_1 might be indicated as a synonym to word w_2, but word w_2 might not be indicated as a synonym to word w_1 at all or might be indicated as a hypernym or a hyponym to word w_1.

A copy of Wiktionary database was downloaded and all category pages, redirects and articles written in languages other than Russian were removed, which gave 295599 articles. Then sets of (word, meaning, synonym words, hyponym words, hypernym words) were extracted from each article. These sets are further called "word senses". Word senses without any lexical links were removed and after it 115437 word senses with 94039 unique words remained in the database.

4   Algorithm Description

4.1   Synonym Relations Extraction

As it has already been mentioned earlier, lexical links from Wiktionary articles lead from specific word meaning to the whole synonym word's page. That means each word sense has a list of synonym words, but these words may be linked to more than one word sense. The aim of this step is to choose the right word senses among them.

Below is the pseudocode of the algorithm:

Input: word sense s

Output: synset S that contains all synonym word senses of s

intialize S with s;

intialize set of unresolved synonym words l_{unres} with all synonym words of s;

while *l_{unres} is not empty* **do**

> **foreach** *unresolved synonym word usw_i in l_{unres}* **do**
>> **if** *usw_i has only one word sense s_i* **then**
>>> add s_i to S and remove usw_i from l_{unres};
>>> add to l_{unres} all synonym words of s_i that were not previously resolved;
>>
>> **else**
>>> **foreach** *word sense s_{ij} of word usw_i* **do**
>>>> calculate $m(S, s_{ij})$;
>>>
>>> **end**
>>> **if** $\exists! s_{max} : m(S, s_{max}) = \max_{s_{ij}} m(S, s_{ij})$ **then**
>>>> add s_{max} to S and remove usw_i from l_{unres};
>>>> add to l_{unres} all synonym words of s_{max} that were not previously resolved;
>>>
>>> **else**
>>>> leave usw_i in l_{unres}
>>>
>>> **end**
>>
>> **end**
>
> **end**
> **if** *no word sense is added to S* **then**
>> exit;
>
> **end**

end

Algorithm 1. Synonym relations extraction

So for every unresolved synonym word we add to S either its only word sense or the one with maximum $m(S, s)$ measure. After it we add all synonym words of newly added word sense (that were not previously resolved) to the set of S's unresolved synonym words.

It is crucial to determine the measure $m(S, s)$ as it plays significant role in the algorithm. It takes into account only lexical links between word senses and does not require any additioal information:

$$m(S, s) = |Syn_s \cap Syn_S| + |Hyper_s \cap Hyper_S| + |Hypo_s \cap Hypo_S| , \qquad (1)$$

where S stands for the synset, s stands for the word sense. Syn_S denotes the set of all synonym words of all word senses in S, both resolved and unresolved, syn_s denotes the set of all synonym words of word sense s. $Hyper_*$ means hypernyms and $Hypo_*$ means hyponyms of S or s.

Thus every newly added to S word sense gives us information to resolve previously unresolved synonym words and suggests new synonym words to resolve and add to synset S. That means that if after going through all unresolved synonym words no new word sense is added to S, we do not get any additional information to resolve them, and repeating the procedure is meaningless.

4.2   Hierarchical Links Extraction

As hierarchical links are not symmetrical, this algorithm needs to be performed twice for hyponyms and hypernyms separately. And note that now links are determined between synsets, not word senses.

Below is the pseudocode of the algorithm:

Input: synset S
Output: set H of hyponym (hypernym) synsets of S
initialize set of unresolved hyponym (hypernym) words l_{unres} with all
  hyponym (hypernym) words of all word senses in S;
foreach *hyponym (hypernym) word w_i in set l_{unres}* **do**
  | **if** *w_i has only one synset S_i* **then**
  | | add S_i to H;
  | **else**
  | | **foreach** *synset S_i of word w_i* **do**
  | | | calculate $M(S, S_i)$;
  | | **end**
  | | **if** $\exists!$ *synset S_{max}: $M(S, S_{max}) = \max_{S_i} M(S, S_i)$* **then**
  | | | add S_{max} to H;
  | | **else**
  | | | leave w_i in l_{unres};
  | | **end**
  | **end**
end

Algorithm 2. Hierarchical links extraction

For every unresolved hyponym (hypernym) word of S we add to H either its only synset or the one with maximum $M(S, S_i)$ measure.

Hypernym measure $M(S_1, S_2)$ between synset S_1 and synset S_2 shows how likely S_2 is a hypernym of S_1. It is calculated as

$$M(S_1, S_2) = |Hypo_{S_1} \cap (Hypo_{S_2} \cup Syn_{S_1})| + |Hyper_{S_1} \cap (Syn_{S_2} \cup Hyper_{S_2})| \quad (2)$$

with all the notations the same as in Sect. 4.1. Obviously, hyponym measure between S_1 and S_2 equals hypernym measure between S_2 and S_1.

4.3   Links Cleaning

Newly created WordNet needed to be tested for cycles and excess links. However, only 3 cycles were found, so they were resolved manually and no special algorithm was needed. At the same time, there were plenty of excess links. Excess link is a link that shortens the hierarchical distance between words. For example, consider hypernym chains dog → mammal → animal and dog → animal. Here the first path is longer and contains all synsets from the second, and thus is preferable. To remove all excess links the transitive reduction algorithm was used.

5   Results

The produced database contains 104696 synsets and 53033 hypernym links between them. Although 50710 synsets do not have neither hypernyms nor hyponyms, they still remain in database to provide information about word meanings. The whole database is freely available as a python package and may be installed via pip install wiki-ru-wordnet command. However, among 104696 synsets 97555 contain only one word and only 1679 contain 3 or more synonyms. The synset with the greatest number of synonyms consists of 79 words, but it contains vulgarities so it will not be illustrated here.

To evaluate the quality of obtained lexical resource we used two publicly available datasets, RUSSE [15] and LRWC [5].

RUSSE dataset consists of 4 different datasets: HJ, RT, AE, AE2 [16].

HJ contains human judgements on 398 word pairs that were translated to Russian from the widely used benchmarks for English: MC [1], RG [10] and WordSim353 [7]. It quantifies how well a system predicts a similarity score of a word pair. It expects scores in the range [0;1] and uses Spearman's rank correlation coefficient (rho) between a vector of real human judgments and test similarity scores.

RT follows structure of the BLESS dataset [3]. Each target word has the same number of related and unrelated source words. The dataset contains 114,066 relations for 6,832 nouns. Half of these relations are synonyms and hypernyms from the RuThes Lite thesaurus [12] and half of them are unrelated words. It quantifies how well a system can distinguish related word pairs from unrelated ones. It expects similarity score between each pair in the range [0;1] and uses Average Precision to evaluate results.

In AE and AE2 datasets two words are considered similar if the second is an association of the first one. In AE relations were sampled from the Russian Associative Thesaurus, and in AE2 relations were sampled from the Sociation.org database. They contain overall 86772 word pairs.

Thus to evaluate dataset on RUSSE datasets binary answers were not enough, so we used measure of relatedness of two words:

$$ssm(w_1, w_2) = e^{-d(w_1, w_2)} \ , \tag{3}$$

where $d(w_1, w_2)$ is the distance to the common hypernym of w_1 and w_2. This way more close synsets would get higher rates of relatedness.

The creators of RUSSE dataset also shipped it with an open source tool for automatic evaluation of semantic similarity measures based on these datasets. The results of running these scripts are presented in Table 1.

LRWC represents both positive and negative human judgements for hyponymy and hypernymy relations for the Russian language and contains 10600 word pairs. Because the data is binary it was easy to calculate precision, results are presented in Table 1 too.

Table 1. Evaluation results on each dataset from RUSSE and LRWC

dataset	hj	rt-avep	ae-avep	ae2-avep	LRWC-precision
result	0.46100	0.76403	0.72419	0.78060	0.79969

6   Conclusion and Future Work

Table 1 shows that WordNet built from Wiktionary demonstrates pretty high values of average precision on each of test datasets. Although if we look at the other results we can see that the current results are not high. They would place us on 51th position among 106 participants of RUSSE contest. At the same time, the recall evaluated on LRWC dataset is pretty low, only 20%. Though the scripts provided with RUSSE datasets did not calculate recall, we assume it might be low as well. That suggests that the produced database provides some quality but needs further development.

Although the database achieves the main characteristic feature of WordNet database – lexical relations between words, the list of possible improvements includes:

1. further work on lexical links resolving;
2. providing additional relations like meronym, holonyms, antonyms etc.;
3. extraction parts of speech from Wiktionary article;
4. extraction of marks like "prof.", "biol." and so on;
5. future development of more precise synonym and hypernym measures;
6. application of the algorithm to other natural languages.

References

1. Miller, G.A., Charles, W.G.: Contextual correlates of semantic similarity. Lang. Cogn. Processes **6**(1), 1–28 (1991)
2. Azarova, I., Mitrofanova, O., Sinopalnikova, A., Yavorskaya, M., Oparin, I.: Russnet: building a lexical database for the Russian language. In: Proceedings Workshop on Wordnet Structures and Standardisation and How this Affect Wordnet Applications and Evaluation, Las Palmas, pp. 60–64 (2002)
3. Baroni, M., Lenci, A.: How we blessed distributional semantic evaluation. In: Proceedings of the GEMS 2011 Workshop on Geometrical Models of Natural Language Semantics GEMS 2011, pp. 1–10. Association for Computational Linguistics, Stroudsburg (2011)

4. Braslavski, P., Ustalov, D., Mukhin, M.: A spinning wheel for yarn: user inter-face for a crowdsourced thesaurus. In: Proceedings of the Demonstrations at the 14th Conference of the European Chapter of the Association for Computational Linguistics, pp. 101–104. Association for Computational Linguistics, Gothenburg, April 2014

5. Dmitry, U.: Expanding hierarchical contexts for constructing a semantic word net-work. In: Computational Linguistics and Intellectual Technologies: Papers from the Annual conference "Dialogue", vol. 1, pp. 369–381. RGGU (2017)

6. Faralli, S., Panchenko, A., Biemann, C., Ponzetto, S.P.: Linked disambiguated dis-tributional semantic networks. In: The Semantic Web - ISWC 2016–15th Interna-tional Semantic Web Conference Proceedings, Part II, Kobe, 17–21 October 2016, pp. 56–64 (2016)

7. Finkelstein, L., Gabrilovich, E., Matias, Y., Rivlin, E., Solan, Z., Wolfman, G., Ruppin, E.: Placing search in context: the concept revisited. In: Proceedings of the 10th International Conference on World Wide Web. WWW 2001, pp. 406–414. ACM, New York (2001)

8. Gelfenbeyn, I., Goncharuk, A., Lehelt, V., Lipatov, A., Shilo, V.: Automatic trans-lation of wordnet's semantic network into Russian. In: Proceedings of the Interna-tional Dialog Conference, pp. 193–198 (2003)

9. Gonçalo Oliveira, H., Gomes, P.: Eco and onto.pt: a flexible approach for creating a portuguese wordnet automatically. Lang. Res. Eval. **48**(2), 373–393 (2014)

10. Rubenstein, H., Goodenough, J.B.: Contextual correlates of synonymy. Commun. ACM **8**(10), 627–633 (1965)

11. Krizhanovsky, A.A., Smirnov, A.V.: An approach to automated construction of a general-purpose lexical ontology based on wiktionary. J. Comput. Syst. Sci. Int. **52**(2), 215–225 (2013)

12. Loukachevitch, N.V., Dobrov, B.V., Chetviorkin, I.I.: Ruthes-lite, a publicly avail-able version of thesaurus of Russian language ruthes. In: Computational Linguistics and Intellectual Technologies: Papers from the Annual conference "Dialogue", vol. 13, pp. 340–350. RGGU (2014)

13. Loukachevitch, N.V., Lashevich, G., Gerasimova, A.A., Ivanov, V.V., Dobrov, B.V.: Creating Russian wordnet by conversion. In: Komp'juternaja Lingvistika i Intellektual'nye Tehnologii, vol. 15, pp. 405–415. Rossiiskii Gosudarstvennyi Gumanitarnyi Universitet (2016)

14. Miller, G.A.: Wordnet: a lexical database for english. Commun. ACM **38**(11), 39–41 (1995)

15. Panchenko, A., Loukachevitch, N.V., Ustalov, D., Paperno, D., Meyer, C.M., Konstantinova, N.: RUSSE: the first workshop on Russian semantic similarity. In: Computational Linguistics and Intellectual Technologies: Papers from the Annual conference "Dialogue", vol. 2, pp. 89–105. RGGU, Moscow (2015)

16. Panchenko, A., Ustalov, D., Arefyev, N., Paperno, D., Konstantinova, N., Loukachevitch, N., Biemann, C.: Human and machine judgements for russian semantic relatedness. In: Ignatov, D.I., et al. (eds.) AIST 2016. CCIS, vol. 661, pp. 221–235. Springer, Cham (2017). https://doi.org/10.1007/978-3-319-52920-2_21

17. Perez, L.A., Gonçalo, H.O., Gomes, P.: Extracting lexical-semantic knowledge from the Portuguese wiktionary. In: 15th Portuguese Conference on Artificial Intelli-gence (EPIA 2011), Lisbon, Portugal (2011)

18. Ustalov, D., Panchenko, A., Biemann, C.: Automatic induction of synsets from a graph of synonyms. In: Proceedings of the 55th Annual Meeting of the Association for Computational Linguistics, ACL 2017, Vancouver, July 30 - August 4, vol. 1: Long Papers, pp. 1579–1590 (2017)

19. Zesch, T., Müller, C., Gurevych, I.: Extracting lexical semantic knowledge from wikipedia and wiktionary. In: Proceedings of the Conference on Language Resources and Evaluation (LREC), electronic proceedings. Ubiquitious Knowledge Processing, Universitt Darmstadt, Mai (2008)
20. Zesch, T., Müller, C., Gurevych, I.: Using wiktionary for computing semantic relatedness. In: Proceedings of AAAI, pp. 861–867 (2008)

Corpus of Syntactic Co-Occurrences: A Delayed Promise

Eduard S. Klyshinsky[1]([⌖]) and Natalia Y. Lukashevich[2]

[1] Keldysh IAM RAS, Miusskaya sq. 4, Moscow, Russia
klyshinsky@mail.ru
[2] Moscow State University, Moscow, Russia

Abstract. The paper gives a technical description of CoSyCo, a corpus of syntactic co-occurrences, which provides information on syntactically connected words in the Russian language. The paper includes an overview of the corpora collected for CoSyCo creation and the amount of collected combinations. In the paper, we also provide a short evaluation of the gathered information.

Keywords: Corpora creation · Shallow parsing · Grammatically ambiguous text
Words combinations · The Russian language

1 Online Resources on Word Combinations in Russian

Detailed information on a word's usage is often required in linguistic research. An electronic dictionary of word combinations for a language is also useful in other spheres, ranging from language learning, editing and translating texts, etc. to natural language processing.

At the moment, there is a whole range of online[1] resources which provide information about word combinations in Russian. They differ in such features as the size of the used corpus, its vocabulary, genre and style, what kind of data they offer, etc. Depending on the combination of such features, a resource is more suited for a particular sphere than the rest.

The size of the corpus and of its dictionary is of particular importance in many respects. As estimated in [3], to make a corpus sufficient for various types of linguistic and lexicographic research, it should contain not less than 10–100 bln words. Natural language processing tasks may require still more than that.

Current resources differ greatly in the type of information they offer. N-grams are relevant in various language processing tasks (e.g. term extraction, ambiguity resolution, etc.). Such resources as Google n-gramms, give statistics on frequencies of words sequences of a given length, n-gramms. They search for words which directly follow each other in the text and are unable to account for syntactic relations between words without additional processing. Other resources like Sketch Engine[2] [5], RNC Sketches[3],

---

[1] We are not discussing resources that are primarily published in paper format here despite the high quality of information they give, because they lack general accessibility.

[2] https://www.sketchengine.co.uk/.

[3] http://ling.go.mail.ru/synt/.

© Springer International Publishing AG 2018
A. Filchenkov et al. (Eds.): AINL 2017, CCIS 789, pp. 121–127, 2018.
https://doi.org/10.1007/978-3-319-71746-3_11

Collocations Colligations Corpora (CoCoCo)[4] [9], combine statistics on the search word's usage with lists of word combinations including it. However, in many other tasks (like language learning, translating, etc.) a user needs to see the whole sentence with the search word (and sometimes even more than that) to draw correct conclusions on the context where the word or word combination can be used. Resources like Russian National Corpus (RNC) [11], the General Internet Corpus of Russian (GICR) [2], Sketch Engine and its derivative RuSkELL[5] [1] offer lists of sentences with the search word or word combination, which is more relevant for such purposes.

As for syntactical relations between words, though few resources can actually give information on them, e.g. SynTagRus [4], or provide extensive lists of examples taking account of them, many resources use the intuition (but do not openly state) that some word combinations could be considered as syntactically connected (e.g. terms, collocations etc.).

Another important feature is the style and genre of texts included in the corpus. Not many resources offer much variety in this respect. As for the details of the style and genre of the source text, still fewer resources make such data available for each particular context, although the importance of taking it into account has been widely discussed ([3], [10]).

We are currently working on a database that would:

- be of a size big enough to carry out various kinds of research and to be used for NLP systems development,
- be collected over huge untagged Russian corpora with the help of simple and easily accessible methods,
- have a convenient interface,
- include texts of different styles and genres,
- give lists of word combinations with the search word (taking into account syntactic relations between words),
- give examples of whole sentences with the search word,
- provide enough information on the source text of each example to make correct inferences about how the word is normally used,
- provide general statistics on that,
- be freely accessible.

2   Method and Used Corpora

We have already partly described in [6, 8] the method and software that were used to gather CoSyCo database. The main idea in this project was that Russian texts contain sequences of PoS-unambiguous words that can be considered as syntactically unambiguous without any further processing. It is important that the percentage of such syntactically unambiguous phrases is high enough to collect information on a significant part of syntactically connected word combinations. In our previous experiments we

---

[4] http://cosyco.ru/cococo/.
[5] http://ruskell.sketchengine.co.uk/run.cgi/skell.

studied ambiguity in news texts for seven European languages [7]. We found out that the structure of homonymy/ambiguity differs significantly across languages: in Russian almost 50% or even 65% of words are PoS-unambiguous (as compared to less than 40% in English).

The structure of such unambiguous sequences can be represented in the form of shallow parsing templates, which will help to extract phrases with a clear and unambiguous syntactical structure.

For example, one of the templates was a sequence of adjectives and a noun, which agree in gender, number, and case. The sequence should be bordered by the beginning of the sentence, a verb or a preposition on the left. This template has a clear syntactical structure if such sequence contains only PoS-unambiguous words and helps to extract noun phrases of the given structure in the given context.

On the stage of template set selection we manually checked how each of the suggested templates worked: we studied roughly the first 200 and 100 random sentences extracted with its help from our corpus to check if it identified constructions as intended and assess the percentage of mistakes. The templates which gave unsatisfactory results were discarded and only those which did return the required constructions were left for further work.

As a result, we compiled a set of six shallow parsing templates for such combinations as verb + preposition + noun, noun + adjective, noun + participle, participle + preposition + noun, adjective + adverb[6]. The amount of templates was relatively small, but, as explained above, these were the ones which could guarantee high probability of error-free output.

The general assumption was that if we apply such set of templates to a very large corpus, it will allow us to find the most part of possible combinations for a representative amount of Russian words. It will also make it possible to collect representative and correct statistics of their use, with no mistakes or with an acceptable rate of them.

On the first stage when we applied the selected set of templates to our corpus, we received results which were rather disappointing: the amount of extracted phrases was high, the number of mistakes was low, but a big part of vocabulary was omitted. A closer look at the results showed that it was so (at least partly) because some words in Russian are grammatically ambiguous in all their forms (e.g. *больной* is ambiguous between a noun *'patient, a sick person'* and an adjective *'sick'* in every form), and their percentage turned out to be surprisingly high. Therefore, we decided to add several new templates which would include possibly ambiguous words into still clear and unambiguous syntactical constructions. For example, the last word in a prepositional phrase at the end of a sentence can be ambiguous between a noun and an adjective. This step slightly reduces the correctness of extracted phrases but, as it turned out, significantly increases the completeness of our vocabulary.

---

[6] We enumerate here the combinations in the way respective sections are named on the site. In each title the head constituent in the combination is placed first. The order of the elements in the title does not always follow the typical word order in sentences with such word combinations.

Therefore, at the moment we have about a dozen shallow parsing templates for the types of combinations named above.

We applied them to a collection of texts totalling about 17 bln words which we gathered from open sources. We deliberately included texts of different styles and genres, so that the obtained co-occurrence data gave a more precise picture of the chosen word's usage. Table 1 shows details of the six subcorpora collected for the project.

Table 1. Subcorpora in CoSyCo.

	CoSyCosubcorpora	mln words	%
1	News sites	1 352.3	8.0%
2	IT news sites	131.3	0.8%
3	Lib.rus.ec fiction collection	~15 000.0	88.3%
4	Science sites	102.2	0.6%
5	Wikipedia.ru texts (dump 01/05/2016)	400.9	2.4%
6	Governmental sources (mil.ru)	11.2	0.07%
Total		~17 000.0	100.0%

3   CoSyCo Database and Site

Using shallow parsing templates, we extracted a large database of syntactically connected words from the gathered corpora. The amount of adjective + noun pairs is as big as 13.4 mln unique combinations of lemmas and 28.5 mln of tokens with 450 mln of occurrences. This part of database contains 68000 nouns and 41000 adjectives. We gathered a database of verb + preposition + noun triplets containing 29.2 mln unique combinations of lemmas, 53.5 mln combinations of tokens, 349 mln occurrences, 28000 different verbs, and 73000 nouns. We also gathered 28.1 mln combinations of participles and nouns with 3.1 mln unique combinations of lemmas and 5.1 of tokens, 20000 participles, and 52000 nouns. The part for participle + preposition + noun triplets

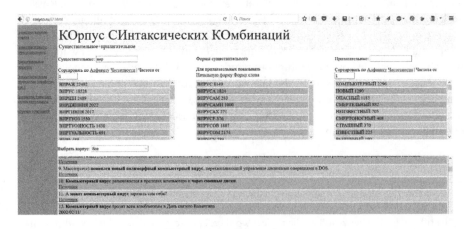

Fig. 1. CoSyCo search page.

contains 4.3 mln combinations, with 1.2 mln unique combinations for lemmas and 1.8 mln for tokens, 15000 participles, and 40000 nouns. All this information is freely available at the site. The interface of CoSyCo site is presented on Fig. 1.

4   Evaluation

Our first hypothesis was that using such simple templates for phrases containing only unambiguous word forms we will be able to collect a big database of syntactically connected words.

To check how complete the vocabulary extracted from CoSyCo database was we compared it with the vocabulary extracted from SynTagRus. We extracted all noun + adjective and verb + preposition + noun combinations from SynTagRus; then we tried to find these combinations in CoSyCo database. Results are shown in Table 2.

Table 2. Combinations from SynTagRus found in CoSyCo.

Combination	Found in SynTagRus	Among them in CoSyCo	% Found
Verb + Noun(+Prep)	100 125	81685	81.5
Noun + Adj	60485	58077	96.0

We also compared the extracted dictionary from CoSyCo with the dictionary of I-RU bigrams available at the site of CoCoCo corpus. We compared words that occur more than 1000 and more than 100 times in noun + adjective and noun + preposition + verb combinations (see Table 3). We found that some of the missing words were erroneously tagged in CoCoCo database; that is why they were not found in our list.

Table 3. Words from I-RU found in CoSyCo.

PoS	Frequency	Found in I-RU	Found in CoSyCo	% Found
noun + adjective				
noun	>1000	2528	2431	96.2
noun	>100	15388	12590	81.8
adj	>1000	402	392	97.5
adj	>100	3352	3046	90.9
verb + preposition +noun				
noun	>1000	2528	2405	95.1
noun	>100	15388	12486	81.1

Finally, we manually compared the order of adjectives connected to selected nouns according to their frequencies in RNC and CoSyCo databases. We have found that the order of adjectives differs between RNC and CoSyCo, but these changes can be explained by differences in used corpora.

Thus, we believe that the gathered database contains a representative part of the Russian lexis. Due to various technical problems which are still being resolved (e.g. duplicate sentences downloaded from news sites in the database) the frequencies of co-occurrence shown on the site at the moment may not reflect the real situation for some words, but these will hopefully be corrected in the nearest future. The examples given on the site have links to their source texts on the net (indicated where possible) and can be used for practical tasks.

5   Conclusion

In the paper, we give a brief description of CoSyCo, a corpus of syntactic co-occurrences that provides information on syntactically connected words for Russian. The article presents a numerical overview of the corpora collected for CoSyCo creation and the collected database of co-occurrences. We also evaluate the correctness of the collected data.

The database is freely available on web site http://cosyco.ru/ both for downloading and as a web-site. We are planning to improve the templates used and to increase their amount. We are also continuing work on the interface and contents of CoSyCo, and we hope to make it a resource which will be of use both for practical and theoretical tasks in Russian language research.

References

1. Apresjan, V., Baisa, V., Buivolova, O., Kultepina, O.: RuSkELL: online language learning tool for Russian language. In: Proceedings of the XVII EURALEX International congress, Tbilisi, Georgia, pp. 292–299 (2016)
2. Belikov, V., Selegey, V., Sharoff, S.: Preliminary considerations towards developing the General Internet Corpus of Russian, Computational Linguistics and Intelligent Technologies: Proceedings of the International Conference "Dialog" 2012", Bekasovo, vol. 1, pp. 37–49 (2012)
3. Belikov, V., Kopylov, N., Piperski, A., Selegey, V., Sharoff, S.: Corpus as language: from scalability to register variation [Korpus kak yazyk: ot masshtabiruemosti k differentsialnoi polnote] Computational Linguistics and Intellectual Technologies: Papers from the Annual International Conference "Dialog" (2013) [Komp'iuternaia Lingvistika i Intellektual'nye Tekhnologii: Po materialam ezhegodnoi Mezhdunarodnoii Konferentsii "Dialog" (2013)], Bekasovo, vol. 1, pp. 83–96 (2013)
4. Frolova, T.I., Podlesskaya, O.Y.: Tagging lexical functions in Russian texts of SynTagRus. In: Proceedings of Dialog 2011, pp. 207–218 (2011)
5. Kilgariff, A., Rychly, P., Smrz, P., Tugwell, D.: The sketch engine. In: Proceedings of the XI Euralex International Congress, Lorient, France, pp. 105–116 (2004)
6. Klyshinsky, E., Kochetkova, N., Litvinov, M., Maximov, V.: Method of POS-disambiguation using information about words co-occurrence (for Russian). In: Proceedings of the annual meeting of the GSCL, Hamburg, pp. 191–195 (2011)
7. Klyshinsky, E., et al.: Analysis of Words Ambiguity in European Languages, № 4, p. 31. Keldysh IAM Preprints 2015 (2015)
8. Klyshinsky, E., Ermakov, P., Lukashevich, N., Karpik, O.: The corpus of syntactic co-occurences: the first glance. In: Proceedings of the Fifth International Conference on Analysis of Images, Social Networks and Texts (AIST 2016), pp. 85–90 (2016)

9. Kormacheva, D., Pivovarova, L., Kopotev, M.: Automatic collocation extraction and classification of automatically obtained bigrams. In: Proceedings of Workshop on Computational, Cognitive, and Linguistic Approaches to the Analysis of Complex Words and Collocations (CCLCC 2014), pp. 27–33 (2014)
10. Lukashevich, N., Klyshinky, E., Kobozeva, I.: Lexical research in Russian: are modern corpora flexible enough?, Computational Linguistics and Intellectual Technologies: Proceedings of the Annual International Conference "Dialog" (2016) [Komp'iuternaia Lingvistika i Intellektual'nye Tekhnologii: Po materialam ezhegodnoi Mezhdunarodnoi Konferentsii "Dialog" (2016)], Moscow, pp. 385–397 (2016)
11. Lyashevskaja, O., Plungian, V.: Morphological annotation in Russian National Corpus: a theoretical feedback. In: Proceedings of the 5th International Conference on Formal Description of Slavic Languages (FDSL-5), Leipzig, pp. 26–28 (2003)

Computation Morphology and Word Embeddings

A Close Look at Russian Morphological Parsers: Which One Is the Best?

Evgeny Kotelnikov[(✉)], Elena Razova, and Irina Fishcheva

Vyatka State University, Kirov, Russia
kotelnikov.ev@gmail.com, razova.ev@gmail.com,
fishchevain@gmail.com

Abstract. This article presents a comparative study of four morphological parsers of Russian – mystem, pymorphy2, TreeTagger, and FreeLing – involving the two main tasks of morphological analysis: lemmatization and POS tagging. The experiments were conducted on three currently available Russian corpora which have qualitative morphological labeling – Russian National Corpus, Open-Corpora, and RU-EVAL (a small corpus created in 2010 to evaluate parsers). As evaluation measures, the authors use accuracy for lemmatization and F1-measure for POS tagging. The authors give error analysis, identify the most difficult parts of speech for the parsers, and analyze the work of parsers on dictionary words and predicted words.

Keywords: Morphological analysis · POS tagging · Comparison of parsers
Text corpora

1 Introduction

The task of morphological parsing in natural language processing systems is to define the lemma of a word (basic or canonical form) and its grammatical features [10, 15]. Morphological parsers accomplish this task, and they can also perform tokenization – separating text into tokens – words, numbers, punctuation marks, etc.

Parsers have the following characteristics: accuracy in recognizing lemmas and grammatical characteristics of words, operation speed, the possibility of tokenization, the possibility to take into account the context, the convenience to represent the analysis results, the possibility to rank hypotheses, the ability to connect the user dictionary.

As a rule, morphological analysis uses two main approaches [10, p. 65]: dictionary-based and dictionary-free. The first uses a dictionary of word stems or word forms, which for Russian is often based on the grammatical dictionary of Zaliznyak [22]. The second approach does not use a dictionary, and the morphological analysis is performed either with the help of a list of affixes (prefixes, suffixes, infixes), or on the basis of machine learning and a labeled text corpus. A hybrid approach is also widespread.

Morphological analysis raises two key problems: unfamiliar words and word ambiguity. If an unfamiliar word occurs, the parser can predict a possible lemma and its grammatical features. As a rule, predicting is carried out by analogy with known words

© Springer International Publishing AG 2018
A. Filchenkov et al. (Eds.): AINL 2017, CCIS 789, pp. 131–142, 2018.
https://doi.org/10.1007/978-3-319-71746-3_12

[10, p. 66]. The problem of word ambiguity is solved either on the basis of the context, or by choosing the most frequent variant [7, p. 133; 15, p. 18].

A morphological parser can perform lemmatization, i.e. defining the lemma of the word form (in this case it is called a lemmatizer), or stemming, i.e. selecting a stem of the word form (the analyzer will be called a stemmer) [7, p. 46]. It is more efficient in practice to use lemmatizers for Russian, since it is inflectional with rich morphology.

A morphological parser, as a rule, should also be able to implement part-of-speech tagging (POS tagging), although it is often a separate component of natural language processing systems, especially for English [14].

Currently, there are more than 20 morphological parsers of Russian[1]. With such a variety, selecting the parser which most adequately meets the needs of analyzing and processing a natural language poses great difficulties. In our work, we have conducted an experimental comparison of the four best-known morphological parsers – mystem, pymorphy2, TreeTagger, and FreeLing, and answer the question: which one is better?

This article has the following structure. The second section provides an overview of comparative studies of Russian-language morphological parsers conducted in recent years. The third section considers the four parsers compared in the work. The fourth section is devoted to the methodology of our research: corpora, POS tagsets, and evaluation measures. The fifth section presents and discusses the results of the experiments. The conclusion contains the major study findings and directions for further research.

2   Previous Work

There have been several comparative studies of Russian morphological parsers in recent years [5, 9, 11].

An independent comparative evaluation of Russian morphological parsers was conducted in 2010 [11]. Fifteen participants submitted an application for the competition[2]. The participants were offered tracks with and without disambiguation on lemmatization, POS tagging, grammatical tagging, and rare words. Due to the wide variety of formats of morphological analysis results for different parsers, it was decided to greatly reduce the estimated POS tagset: nouns, adjectives, verbs, including participles and adverbial participles, prepositions, conjunctions, and a combined category (adverbs, introductory words, particles, interjections). Pronouns (including adverbial and predicative pronouns), numerals, as well as compound prepositions and conjunctions were excluded from the labeling. A specially labeled corpus (RU–EVAL) was used for evaluation. The anonymized results of the participants were assessed according to the accuracy metric. The best results for lemmatization and POS tagging with disambiguation are 98.1% and 97.3%, respectively.

In [9] three analyzers were evaluated: FreeLing, pymorphy2, and TreeTagger. The author focused on potentially combining the parser outputs to improve the analysis results. The assessment was made on the accuracy metric based on the RU-EVAL corpus and the disambiguated subcorpus of the Russian National Corpus (RNC). The tagset of

---

[1]   https://nlpub.ru.
[2]   http://ru-eval.ru/go/morpho.html.

mystem and RNC was taken as the base POS tagset. As a result, the TreeTagger proved to be better on the RNC corpus (in fact, it was trained on this corpus), pymorphy2 – on the RU-EVAL corpus when lemmatizing, FreeLing – when POS tagging.

In [5] several probabilistic parsers (TreeTagger, TnT, HunPos, Lapos, Citar, Morfette, Stanford POS tagger, and SVMTool) were trained and tested on the RNC corpus for the POS tagging task. Mystem and pymorphy2 were evaluated as well. The best result was shown by TreeTagger (96.94% with 5-fold cross-validation).

The results of these works are rather difficult to use when choosing a parser for a project on natural language processing: in [11] the results are anonymized, and in addition, parsers of obsolete versions were used; mystem was not evaluated in [9], and the evaluation on RNC was biased for TreeTagger; [5] did not evaluate FreeLing, the evaluation of mystem and pymorphy2 was simplified (incomplete), and the rest of the parsers were compared only on the task of POS tagging. Our work carries out a more complete study of the four most common parsers (mystem, pymorphy2, TreeTagger, and FreeLing) on all of the three currently available morphologically labeled text corpora: RU-EVAL, RNC, and OpenCorpora. We solve the problems of lemmatization and POS tagging without taking into account grammatical tags such as gender, number, case, etc. Unlike [11], the extended tagset is used. We propose an automatic recognition method of different lemmas for perfective/imperfective verb forms. Unlike previous works, the study uses Information Retrieval measures – precision, recall, and F1-measure to assess the quality of POS tagging.

3   Russian Morphological Parsers

This section examines the morphological parsers under review: mystem, pymorphy2, TreeTagger, and FreeLing.

3.1   Mystem

The morphological parser mystem was developed by Ilya Segalovich and Vitaly Titov in the company Yandex [18]. Currently, mystem version 3.0 is a console application (for Windows, Linux, FreeBSD, and MacOS), has a closed source code, and is available for non-commercial use[3].

The parser mystem is based on the dictionary of Zaliznyak [22] and allows us to recognize unknown words by searching for the most similar vocabulary words. Also, mystem is able to use context to analyze homonyms (for word disambiguation), to derive the probability of hypotheses, and to take into account the user dictionary.

---

[3] https://tech.yandex.ru/mystem.

3.2 pymorhpy2

The parser pymorphy2 was created by Mikhail Korobov in the Python language [8]. At present, pymorphy2 version 0.8 has an open source code and is available for both non-commercial and commercial use[4].

The parser pymorphy2 uses the OpenCorpora project dictionary [2], which in turn is based on the AOT project dictionary [1]. Prediction of hypotheses for unknown words is carried out by using a rule set – prefixes and endings of words, hyphens, etc. Conditional probabilities are derived for all analysis results. Also, pymorphy2 allows you to generate word forms according to given grammatical features.

3.3 TreeTagger

The TreeTagger is a tool for POS tagging and lemmatization based on machine learning (decision trees) [17]. If there is a labeled text corpus for some language, one can train TreeTagger without resorting to any other linguistic information. For Russian, Tree-Tagger was trained by Serge Sharoff [20] on the disambiguated version of the Russian National Corpus [RNC]. Unknown words are recognized with the help of CST Lemmatiser [6], which generates lemmatization rules on the basis of matching common parts of known word forms and lemmas.

A specially created tagset[5], based on the morphosyntactic descriptions of the Multext East project [19], is used to encode parts of speech.

TreeTagger has a closed source code and is available for non-commercial use[6].

3.4 FreeLing

FreeLing is a software library written in C++ for text processing in multiple languages, including Russian [12]. Currently, FreeLing version 4.0 is open source and is available for non-commercial research[7].

Tokenization, lemmatization, POS tagging, and recognition of unknown words were implemented in FreeLing for Russian. Lemmatization works on the basis of the dictionary approach and the list of affixes. For POS tagging[8] two approaches are used – hidden Markov models [4] and Relaxation Labeling [13].

---

[4]   https://github.com/kmike/pymorphy2.

[5]   http://corpus.leeds.ac.uk/mocky/msd-ru.html.

[6]   http://www.cis.uni-muenchen.de/~schmid/tools/TreeTagger.

[7]   http://nlp.lsi.upc.edu/freeling.

[8]   https://talp-upc.gitbooks.io/freeling-user-manual/content/tagsets/tagset-ru.html.

4   Methodology

4.1   Corpora

To test the parsers, we used three corpora with morphological labeling – RU-EVAL, RNC, and OpenCorpora.

1. RU-EVAL is a corpus of 2,025 words, specially created and labeled for testing morphological parsers in 2010[9] [11].
2. Russian National Corpus (RNC) is a corpus of modern Russian containing over 600 million word forms. It was created in 2003 and has been built up since then by a large group of Russian linguists[10] [16]. As a rule, in studies on morphology, a disambiguated subcorpus is used. The organizers provided us with a fragment of this subcorpus, which contains 965,574 words.
3. OpenCorpora is a freely available, labeled corpus, which was created in 2009 [2]. We used a disambiguated subcorpus without unknown words, which included 38,537 words.

4.2   POS Tagsets and Verb Lemmas

One of the most serious problems, which was expected, is the problem of correlating tagsets of corpus and parsers. For example, OpenCorpora, mystem, and pymorphy2 single out predicate adverbs, and RNC, RU-EVAL, TreeTagger, and FreeLing do not distinguish this category. Due to the fact that mystem was used for RNC labeling, and OpenCorpora was used to develop pymorphy2, the tagsets inside these pairs almost completely correspond to each other. But it is a rather complicated task to find correspondences between these pairs, as well as other parsers. In total, it comes up to 29 different parts of speech for the three corpora and four parsers[11].

We decided to reach a compromise between the detail of the POS tagging analysis and the problem of POS tagsets correspondence. We combined 29 parts of speech into 14: nouns, adjectives, verbs, participles, adverbial participles, comparatives, numerals, adjective numerals, adverbs, pronouns, prepositions, conjunctions, particles, and interjections. As a result, it was possible to establish the correspondence of POS tagsets for all components of the study. The complete table of correspondence is given in the link[12]. It should be noted that an original limited POS tagset from [11] was used for the RU-EVAL corpus.

Another important, well-known problem is the issue of mismatch with the gold labeling of the resulting lemmas in different parsers for perfective/imperfective verbs. It is with this very problem that most of the "mistakes" in verbs, as well as in participles and adverbial participles, are associated in the process of lemmatization. We attempted to solve this problem by automatically searching for variants of verbs of perfective/

---

[9] http://ru-eval.ru/collections_index.html.
[10] http://www.ruscorpora.ru.
[11] There exist a tagset converter: https://github.com/kmike/russian-tagsets.
[12] https://goo.gl/tLrskC.

imperfective aspects in dictionaries, edited by Ushakov [21] and edited by Kuznetsov [3]. As a result, 73% of errors in lemmas of verbs, participles, and adverbial participles were corrected on average for all corpora and parsers. Corrections of the verbs took place mostly in mystem and FreeLing parsers: an average of 78% of errors were corrected there. Only 1.3% of the errors were corrected for pymorphy2 and TreeTagger parsers, since the resulting lemmas in them already basically coincided with the lemmas given in the corpora.

4.3   Evaluation Measures

To assess the quality of lemmatization, we used the standard accuracy metric as the ratio of correctly recognized lemmas to the total number of words. Three evaluations were computed: for dictionary lemmas, for predicted lemmas, and a general evaluation (all the parsers under study allow us to deduce for a given lemma whether it was taken from a dictionary or predicted).

To evaluate the results of POS tagging, we also calculated accuracy, but as the main ones we decided to use the typical measures of Information Retrieval and Machine Learning – *precision* (P), *recall* (R) and *F1-measure* (F1) [7, p. 455], each of the 14 parts of speech being considered as a separate class. A distribution of words over parts of speech is non-uniform, therefore accuracy is not the best choice for evaluation owing to mixing of true positive and true negative examples. The metrics for each separate class were calculated as follows:

$$P = \frac{tp}{tp + fp}, \ R = \frac{tp}{tp + fn}, \ F_1 = \frac{2 \cdot P \cdot R}{P + R}.$$

where tp – the number of words of a given part of speech identified by the parser correctly, fp – the number of words attributed by the parser to a given part of speech incorrectly, fn – the number of words of a given part of speech not specified by the parser.

We used a micro-averaging scheme to average the metrics through all parts of speech (in order not to give the same weight to parts of speech with different numbers of words):

$$P_{micro} = \frac{\sum_i tp_i}{\sum_i tp_i + \sum_i fp_i}, \ R_{micro} = \frac{\sum_i tp_i}{\sum_i tp_i + \sum_i fn_i}, \ F_{1micro} = \frac{2 \cdot P_{micro} \cdot R_{micro}}{P_{micro} + R_{micro}}.$$

5   Results and Discussion

5.1   Lemmatization

Table 1 summarizes the results of lemmatization for all parsers. Two evaluations are given for mystem: with and without context-based disambiguation (CBD).

Mystem demonstrated the best results for all corpora. The CBD option in mystem allows accuracy to be increased by an average of 1%. TreeTagger came in second place. It shows consistently high results for all collections, and is slightly behind mystem (an

Table 1. Accuracy of lemmatization (%)

Corpus	mystem		pymorphy2	TreeTagger	FreeLing
	w/o CBD	with CBD			
RU-EVAL	96.25	**96.94**	93.83	95.21	86.72
RNC	95.87	**97.54**	93.55	97.31	88.91
OpenCorpora	98.00	**98.47**	98.29	96.95	90.97
Average	96.71	**97.65**	95.22	96.49	88.87

average of 1.16%). Pymorphy2 came in third place: high accuracy in OpenCorpora was offset by relatively low values for RU-EVAL and RNC. The FreeLing parser proved to be much worse than the others on all corpora.

When analyzing Table 1, it should be taken into account that mystem was used for RNC labeling, TreeTagger was trained on RNC, and pymorphy2 uses the Open-Corpora dictionary. These circumstances lead, to some extent, to biased evaluations, but the average evaluation for the three corpora allows this effect to be diminished: consistently high results in all corpora both in mystem and TreeTagger speak of the high quality of these parsers.

Compared with study [9], accuracy of pymorphy2 and FreeLing for corpora RU-EVAL and RNC turned out to be higher on average by 6% for each of these parsers in our research. For FreeLing, this is most likely due to our method of processing lemmas of perfective/imperfective verbs. Accuracy of TreeTagger for RU-EVAL also increased (by 8.3%), while for RNC it remained at about the same level (97.31% vs. 97.00%). The differences in the results for pymorphy2 on both corpora and for TreeTagger on RU-EVAL may be due to the differences in the versions of the parsers and due to the procedures for calculating the evaluation measures (for example, processing of the letter "ё"). The maximum accuracy received in the competition from work [11] is 98.1%, exceeding by 1.2% the result of our leader, mystem (96.94%), and the results of other participants are comparable with the results of the parsers under investigation.

Table 2 lists the first three parts of speech with the highest percentage of errors in lemmatization.

Table 2. Contribution of parts of speech to the lemmatization error rate (% of total error count)

Corpus	mystem		pymorphy2	TreeTagger	FreeLing
	w/o CBD	with CBD			
RU-EVAL	verbs (44%)	nouns (40%)	nouns (39%)	nouns (45%)	nouns (63%)
	nouns (33%)	verbs (32%)	verbs (25%)	verbs (26%)	verbs (22%)
	comb. (15%)	comb. (19%)	comb. (16%)	comb. (16%)	comb. (8%)
RNC	pron. (35%)	nouns (26%)	nouns (27%)	pron. (48%)	nouns (45%)
	nouns (20%)	pron. (23%)	pron. (25%)	nouns (13%)	verbs (16%)
	adject. (15%)	adject. (19%)	adject. (15%)	adject. (13%)	pron. (12%)
Open-Corpora	nouns (25%)	nouns (29%)	nouns (44%)	nouns (42%)	nouns (59%)
	pron. (17%)	prep. (23%)	pron. (17%)	prep. (13%)	verbs (18%)
	prep. (17%)	adv. (12%)	verbs (12%)	verbs (12%)	pron. (5%)

[a]In Tables 2 and 6, the notation "comb." means "combined category" (adverbs, introductory words, particles, interjections) for RU-EVAL corpus (see Sect. 2).

Table 2 shows that in most cases, nouns, verbs, and pronouns are the most difficult lemmas to recognize. Noun errors are most often encountered[13]:

- in proper names (about 50% of all errors):
  - mystem: *Сена → сено* (*Sena → seno*)/RNC: *Сена* (*Sena*);
  - pymorphy2: *Анастасия → анастасий* (*Anastasiya → anastasij*)/OpenCorpora: *Анастасия* (*Anastasiya*);
- in compound words:
  - TreeTagger: *Интернет-форуме → интернет-форуме* (*Internet-forume → internet-forume*)/RNC: *интернет-форум* (*internet-forum*);
- due to incorrect word disambiguation:
  - FreeLing: *дне → день* (*dne → den'*)/OpenCorpora: *дно* (*dno*);
- due to a mismatch of normal forms in the parser and in the corpus:
  - mystem: *гуляний → гуляние* (*gulyanij → gulyanie*)/RNC: *гулянье* (*gulyan'e*).

In verbs, errors are associated with invalid word disambiguation: FreeLing: *вели → велеть* (*veli → velet'*)/RNC: *вести* (*vesti*), as well as with mismatch of normal forms in the parser and in the corpus: mystem: *существующей → существующий* (*sushhestvuyushhej → sushhestvuyushhij*)/RU–EVAL: *существовать* (*sushhestvovat'*).

In pronouns, errors occur due to the mismatch of normal forms: mystem: *все → все* (*vse → vse*)/RNC: *весь* (*ves'*), and in compound pronouns: TreeTagger: *какие–то → какие–то* (*kakie-to → kakie-to*)/RNC: *какой–то* (*kakoj-to*).

Table 3 lists the percentages of lexicon lemmas among the total number of all lemmas.

Table 3. Parts of lexicon lemmas (%)

Corpus	mystem	pymorphy2	TreeTagger	FreeLing
RU-EVAL	97.0	98.9	93.1	85.7
RNC	98.9	99.5	99.5	90.4
OpenCorpora	99.1	99.6	94.0	88.1
Average	99.0	99.6	95.5	88.1

As the table shows, mystem and pymorphy2, for all corpora, have a very high proportion of lexical lemmas, exceeding 97%, which indicates that internal dictionaries of these parsers are quite complete. TreeTagger has a large percentage of dictionary words for the RNC corpus, on which it has been trained, and is significantly lower for the other two corpora. FreeLing has a limited dictionary compared to other parsers.

Table 4 shows accuracy for dictionary and predicted lemmas.

---

[13] The format of the examples: the name of the parser: *the initial word form → the parser response* (*transliteration*)/the name of the corpus: *the gold answer in the corpus* (*transliteration*).

Table 4. Accuracy of lemmatization: lexicon lemmas/predicted lemmas (%)

Corpus	mystem		pymorphy2	TreeTagger	FreeLing
	w/o CBD	with CBD			
RU-EVAL	97.4/78.3	98.1/78.3	94.4/78.6	97.4/78.0	96.6/28.6
RNC	96.2/80.1	97.9/81.9	93.7/66.6	97.4/80.0	95.5/27.8
OpenCorpora	98.2/87.0	98.6/89.0	98.6/72.1	98.4/79.8	97.6/43.5
Average	97.3/81.8	98.2/83.1	95.6/72.4	97.7/79.3	96.6/33.3

As the table indicates, for all parsers, accuracy on the predicted lemmas is significantly lower than on the dictionary lemmas. Accuracy ranking of parsers on the predicted lemmas coincides with accuracy ranking as a whole (see Table 1), but does not coincide for dictionary lemmas: pymorphy2 comes in last place, behind FreeLing. However, the very low quality of the predictive component of FreeLing, given the relatively small number of lexical lemmas, leads to its overall low accuracy.

5.2 POS Tagging

Table 5 shows the F1-measure values for the POS tagging task.

Table 5. F1-measure of POS tagging (%)

Corpus	mystem		pymorphy2	TreeTagger	FreeLing
	w/o CBD	with CBD			
RU-EVAL	97.06	**97.65**	95.66	96.82	96.87
RNC	95.12	**96.19**	91.14	95.66	92.08
OpenCorpora	97.14	97.49	**98.22**	96.51	97.01
Average	96.44	**97.11**	95.01	96.33	95.32

As in the case of lemmatization, the results for RU-EVAL and RNC are on average higher for mystem. Pymorphy2 was the winner only for OpenCorpora (here again, it must be taken into account that this parser uses OpenCorpora tagset), which at the same time turned out to be the worst for RU-EVAL and RNC. TreeTagger again demonstrates stable results for all corpora, thus coming in second place (the difference with mystem is only 0.8%). FreeLing also shows good results, falling behind the TreeTagger by only 1%.

Table 6 lists the parts of speech that make the biggest contribution to POS tagging errors ($fp + fn$ – the sum of the number of words incorrectly assigned by the parser to the given part of speech and the number of words of that part of speech not defined by the parser).

Table 6. Contribution of parts of speech to the POS tagging error rate (% of total error count)

Corpus	mystem		pymorphy2	TreeTagger	FreeLing
	w/o CBD	with CBD			
RU-EVAL	nouns (29%)	nouns (29%)	adj. (32%)	nouns (30%)	adj. (26%)
	adj. (24%)	adj. (21%)	nouns (28%)	adj. (27%)	comb. (25%)
	verbs (24%)	comb. (20%)	verbs (19%)	comb. (20%)	conj. (17%)
RNC	pron. (18%)	nouns (20%)	pron. (23%)	numer. (26%)	nouns (17%)
	adv. (16%)	pron. (18%)	adv. (19%)	pron. (15%)	pron. (16%)
	nouns (16%)	adv. (15%)	nouns (13%)	conj. (11%)	numer. (14%)
Open-Corpora	adv. (32%)	adv. (31%)	nouns (30%)	adv. (23%)	nouns (24%)
	pron. (20%)	pron. (21%)	particl. (13%)	pron. (21%)	adv. (21%)
	nouns (12%)	nouns (14%)	conj. (12%)	nouns (20%)	pron. (14%)

Unlike the lemmatization task, in which nouns caused the main problems, there are difficulties with different parts of speech in POS tagging for different corpora. For RNC and OpenCorpora, the most difficult were:

- adverbs:
  - mystem, noun: *хором* → *хор* (*khorom* → *khor*)/OpenCorpora, adverb: *хором* (*khorom*);
  - pymorphy2, adjective: *важно* → *важный* (*vazhno* → *vazhnyj*)/RNC, adverb: *важно* (*vazhno*);
- pronouns:
  - TreeTagger, adverb: *нечего* → *нечего* (*nechego* → *nechego*)/OpenCorpora, pronoun: *нечего* (*nechego*);
  - FreeLing, numeral: *немногих* → *немного* (*nemnogikh* → *nemnogo*)/RNC, pronoun: *немногие* (*nemnogie*);
- nouns:
  - pymorphy2, numeral: *семью* → *семь* (*sem'yu* → *sem'*)/OpenCorpora, noun: *семья* (*sem'ya*);
  - mystem, adjective: *горячее* → *горячий* (*goryachee* → *goryachiy*)/RNC, noun: *горячее* (*goryachee*).

TreeTagger in the RNC analysis makes mistakes most often with numerals: Tree-Tagger, pronoun: *один* → *один* (*odin* → *odin*)/RNC, numeral: *один* (*odin*). Nouns are much of the problem for RU-EVAL: FreeLing, verb: *обвиняемому* → *обвинять* (*obvinyaemomu* → *obvinyat'*)/RU–EVAL, noun: *обвиняемый* (*obvinyaemyj*), and adjectives: pymorphy2, noun: *серой* → *сера* (*seroy* → *sera*)/RU–EVAL, adjective: *серый* (*seryj*).

Unlike lemmatization, the best results in POS tagging for non-dictionary words were shown by FreeLing – the average accuracy for all corpora is 89.9%. TreeTagger (86%) and mystem (79.5%) rank second and third. Pymorphy2 (58.3%) ranks last, which determined its relatively low overall result (see Table 5).

It should be noted that in the POS tagging task for the RNC using the accuracy metric, we obtained almost the same results as in [5]: the difference by modulo does not exceed 1%. Only the results for pymorphy2 on RNC completely coincided with work [9]. The average difference for RNC is 1.7%, and for RU-EVAL – 6%. This is probably related to different tables of POS tagsets correspondence. In work [11] the maximum result on the RU-EVAL corpus according to accuracy is 97.3%, in our work – 97.43% (mystem).

6   Conclusion

In our study, the mystem parser demonstrated the best results for both lemmatization and POS tagging, which makes it possible to recommend its use in those natural language processing tasks that correspond to its license. Selecting the context-based disambiguation option allows the results to be slightly increased (within 1%).

The TreeTagger parser shows stable performance on all corpora and only slightly falls behind mystem (by 1.5% in lemmatization and 0.5% in POS tagging).

The pymorphy2 parser also has high accuracy for lemmatization (falling behind TreeTagger only by 0.5%) and F1-measure for POS tagging (behind TreeTagger by 1.7%), but unlike mystem and TreeTagger, it is less stable when working on different corpora, especially in the task of POS tagging due to poor part-of-speech recognition of non-dictionary words.

Finally, the FreeLing parser does not yet accurately recognize the lemmas of Russian words (7% behind pymorphy2), but it ensures high quality when determining parts of speech.

Subsequent studies are intended to expand the spectrum of analyzed Russian parsers and include more characteristics in the study.

Acknowledgments.   The reported study was funded by RFBR according to research project No. 16-07-00342a.

References

1. AOT – Avtomaticheskaja obrabotka teksta (Automatic processing of text). http://www.aot.ru. Accessed 20 June 2017
2. Bocharov, V., Granovsky, D., Bichineva, S., Ostapuk, N., Stepanova, M.: Quality assurance tools in the OpenCorpora project. In: Proceedings of Dialogue-2011, pp. 109–114 (2011)
3. Kuznetsov, S.A. (ed.): Bol'shoj tolkovyj slovar' russkogo jazyka (The Large Explanatory Dictionary of the Russian Language). Norint, Saint-Petersburg (1998)
4. Brants, T.: TnT: a statistical part-of-speech tagger. In: Proceedings of the 6th Conference on Applied Natural Language Processing (ANLP), pp. 224–231 (2000)
5. Dereza, O.V., Kayutenko, D.A., Fenogenova, A.S.: Automatic morphological analysis for Russian: a comparative study. In: Proceedings of Student Session of Dialogue-2016 (2016)
6. Jongejan, B., Dalianis, H.: Automatic training of lemmatization rules that handle morphological changes in pre-, in- and suffixes alike. In: Proceedings of the Joint Conference of the 47th Annual Meeting of the ACL and the 4th International Joint Conference on Natural Language Processing of the AFNL, vol. 1, pp. 145–153 (2009)
7. Jurafsky, D., Martin, J.H.: Speech and Language Processing, 2nd edn. Prentice-Hall Inc., Upper Saddle River (2009)

8. Korobov, M.: Morphological analyzer and generator for Russian and Ukrainian languages. In: Proceedings of 4th International Conference on Analysis of Images, Social Networks and Texts (AIST-2015), pp. 320–332 (2015)

9. Kuzmenko, E.: Morphological analysis for Russian: integration and comparison of taggers. In: Proceedings of 5th International Conference on Analysis of Images, Social Networks and Texts (AIST-2016), pp. 162–171 (2016)

10. Leont'eva, N.N.: Avtomaticheskoe ponimanie tekstov: sistemy, modeli, resursy (Automatic text comprehension: systems, models, resources). Akademia Publ., Moscow (2006)

11. Lyashevskaya, O., Astaf'eva, I., Bonch-Osmolovskaya, A., Garejshina, A., Grishina, J., D'jachkov, V., Ionov, M., Koroleva, A., Kudrinsky, M., Lityagina, A., Luchina, E., Sidorova, E., Toldova, S., Savchuk, S., Koval', S.: NLP evaluation: Russian morphological parsers. In: Proceedings of Dialogue-2010, pp. 318–326 (2010)

12. Padró, L., Stanilovsky, E.: FreeLing 3.0: towards wider multilinguality. In: Proceedings of the Language Resources and Evaluation Conference (LREC 2012), pp. 2473–2479. ELRA. Istanbul, Turkey (2012)

13. Padró, L.: A hybrid environment for syntax-semantic tagging. PhD thesis, Dept. Llenguatges i Sistemes Informàtics. Universitat Politècnica de Catalunya (1998)

14. POS Tagging (State of the art). https://www.aclweb.org/aclwiki/index.php?title=POS_Tagging_(State_of_the_art). Accessed 20 June 2017

15. Nikolaev, I.S., Mitrenina, O.V., Lando, T.M. (eds.): Prikladnaja i komp'juternaja lingvistika (Applied and computational linguistics). Lenand, Moscow (2016)

16. Russian National Corpus. http://www.ruscorpora.ru. Accessed 20 June 2017

17. Schmid, H.: Probabilistic part-of-speech tagging using decision trees. In: Proceedings of International Conference on New Methods in Language Processing, pp. 44–49, Manchester, UK (1994)

18. Segalovich, I.: A fast morphological algorithm with unknown word guessing induced by a dictionary for a web search engine. In: Proceedings of MLMTA-2003, pp. 273–280 (2003)

19. Sharoff, S., Kopotev, M., Erjavec, T., Feldman, A., Divjak, D.: Designing and evaluating Russian tagsets. In: Proceedings of LREC-2008, pp. 279–285, Marrakech (2008)

20. Sharoff, S., Nivre, J.: The proper place of men and machines in language technology: processing Russian without any linguistic knowledge. In: Proceedings of Dialogue-2011, pp. 657–670 (2011)

21. Ushakov, D.N. (ed.): Tolkovyj slovar' russkogo jazyka (Dictionary of the Russian Language). Four volumes. State Publishing House of Foreign and National Dictionaries, Moscow (1940)

22. Zaliznyak, A.A.: Grammaticheskij slovar' russkogo jazyka (Russian Grammar Dictionary), Moscow (1977)

Morpheme Level Word Embedding

Ruslan Galinsky[1](✉), Tatiana Kovalenko[2], Julia Yakovleva[2],
and Andrey Filchenkov[1]

[1] ITMO University, Kronverksky Pr. 49, St. Petersburg, Russia
galinskyifmo@gmail.com, afilchenkov@corp.ifmo.ru
[2] Peter the Great St. Petersburg Polytechnic University,
Polytechnicheskaya, 29, St. Petersburg, Russia
tanyakovalenko1994@gmail.com, juli-jakovleva@yandex.ru

Abstract. Modern NLP tasks such as sentiment analysis, semantic analysis, text entity extraction and others depend on the language model quality. Language structure influences quality: a model that fits well the analytic languages for some NLP tasks, doesn't fit well enough the synthetic languages for the same tasks. For example, a well known Word2Vec [27] model shows good results for the English language which is rather an analytic language than a synthetic one, but Word2Vec has some problems with synthetic languages due to their high inflection for some NLP tasks. Since every morpheme in synthetic languages provides some information, we propose to discuss morpheme level-model to solve different NLP tasks. We consider the Russian language in our experiments. Firstly, we describe how to build morpheme extractor from prepared vocabularies. Our extractor reached 91% accuracy on the vocabularies of known morpheme segmentation. Secondly we show the way how it can be applied for NLP tasks, and then we discuss our results, pros and cons, and our future work.

Keywords: NLP · Word2Vec · Word embedding · Morphemes
Synthetic language · Semantic analysis · Sentiment analysis
Text entity extraction · Naive bayesian classifier

1 Introduction

Language modelling is one of the main techniques for the natural language processing. Language modelling is building a language model from samples of the "real world" texts - corpora.

The common type of language modeling is word embedding, where words or phrases from the vocabulary are mapped to vectors of real numbers. The input of a word embedding model is a corpus, the output is a set of word-vector pairs. The basic principle of constructing vectors is based on the assumption that words appearing in similar contexts with a similar frequency are semantically close. Thus, these words are located close to each other in the vector space.

© Springer International Publishing AG 2018
A. Filchenkov et al. (Eds.): AINL 2017, CCIS 789, pp. 143–155, 2018.
https://doi.org/10.1007/978-3-319-71746-3_13

The semantic similarity is calculated as a cosine distance between the vectors of two words and can take values in the interval $[-1 ... 1]$ (in practice only values above 0 are often used). Zero value approximately means that the words have no similar contexts and their values are not related to each other. On the contrary, value equalling to 1 indicates the complete identity of their contexts and a close meaning, respectively [7]. For example, these are words *"coffee"* and *"tea"* which are semantically close. Their cosine similarity is close to 0,7 and they are not far from each other in the vector space.

There are many of different word embedding techniques exist: Word2Vec [27], GloVe [17] etc. However, the word embedding technique is not the only one model, and some models are based on morphemes [25] and even characters [30]. There are several reasons to do that, and some authors describe it in details. One will consider the reasons. Firstly, it is quite challenging to produce high-quality word representations for rare or unknown words due to the insufficient context information in the training data. Secondly, it is difficult to obtain word embedding for merged words as they are not included in the vocabulary of the training data. The second issue is very topical for the synthetic languages (i.e. inflection-rich languages), because due to the high number of word forms, many perfectly valid word forms could not be observed at all in the corpus. For example, if a corpus had the word *"intellectual"*, but did not have a word *"intellectualism"*, the result model would not contain a vector for the word *"intellectualism"*. Here is another example: there are words *"establish"* and *"establishment"*. The same situation occurs here, too. Moreover, some specific domain texts often use words gluing, chemistry here is an example. Last but not least, the analyzed text can contain neologisms - words invented by the author, and are made from existing morphemes.

The problem with not observed word forms can be solved by returning words to their base forms, however neologisms either require to rebuild a model from a scratch in order to estimate the corresponding context or how it uses an existing model.

The word form problem is more relevant for inflection-rich languages, and returning word to its base form doesn't solve the problem well, because it ignores meaningful information of a word in it. According to a linguistic theory, morphemes are considered to be the smallest meaning-bearing element of a language. Sometimes an inflection influences the word meaning a lot. For example, a word *"unprofessional"* has a prefix *"un"*, the meaning of which is *"absence"*. Here is another example: a word *"pianist"* has suffix *"ist"*, the meaning of which is *"profession"*.

We propose to work with a model based on morphemes along with the corresponding words to solve the described problems. In Sect. 4 we present an algorithm to segment a word into morphemes built on known segmentation from vocabularies, because we have not found any existing algorithm for the Russian language. The developed algorithm reached accuracy 91% and found a wrong segmentation in original vocabularies. We think that a practical purpose of segmentation is also important, because it provides a vocabulary of language units that is smaller than a vocabulary consisting of words as they appear in text, especially for synthetic languages.

In order to get morpheme representations, build the morpheme model and construct the algorithm, we present the vocabularies. The way of getting morpheme representations is described in Sect. 5. We chose some the existing vocabularies as sources, converted them into one parsable format and merged them together by the type. Also there were fixed a lot of different types of mistakes because a quality of dictionaries influences a quality of the model. More details about the composed dictionaries could be found in Sect. 3. Then we propose the approach to construct and adjust a word vector representation using underlying morphemes, which is described in Sect. 6. In our research we consider that model is already built for a word level, and we build a morpheme vector representation from the vocabularies and existing word representation model. We used the Word2Vec as the base model for our experiments because it is one of the most common, popular and easy-to-use set of algorithms.

Finally, we tested our assumptions. We used already classified words and looked on their density before and after our adjustment with assumption that words in clusters are distributed normally. The results, that could be found in Sect. 7, turned out to be good.

In Sect. 8, we outline our further work, ideas and thoughts for the discussion.

We also made an overview of the existing works on the topic and described them in Sect. 2 of this work.

2   Related Work

A word level language model representation is widely used in different NLP tasks. However, some authors show [30] that a character level model on some NLP tasks outperform word level models like Word2Vec [27] or GloVe [17]. Ling et al. [28] report state of the art results in language modeling and part-of-speech tagging, especially for synthetic languages. Good results for sentiment analysis using character level model were shown for the Russian language as well [22]. It shows that word level models are not able to capture a whole structure of a morphologically rich language, however, from our point of view, a character level model is not natural and at least requires to be compared with morpheme level models.

The approximated automated morphological analysis seems to be beneficial for many natural language applications [21,29]. For example, in text retrieval it is customary to preprocess texts by return words to their base forms, especially for morphologically rich languages. The big progress here was made by Morfessor [19,26]. Its creates proposed the general purpose algorithm to find segmentations. It's unsupervised or semi-supervised algorithm, based on the EM [14] iterative method and the Minimum Description Length principle [20,24], that tries to find the common and different parts of words that look similar. They have a very good idea and impressive results, however, due to a general purpose, they lose a precision. Also their algorithm produces non existing morphemes, and the largest problem that the algorithm doesn't mark segments with a type. It means that resulting segments should be clustered to determine the type. These problems don't allow to apply it's results to adjust words somehow using the knowledge of the underlying morphemes, and also the resulting morphemes lexicon should be reviewed manually to verify morphemes on existence. Anyway, the Morfessor algorithm is good for building morpheme embedding from scratch, because this task doesn't need to have existing morphemes and only requires some technique to do the segmentation. A lot of research groups use the Morfessor for their works for automatically builder morphological segmentations, for example [25].

3   Vocabularies

We considered five types of morphemes for Russian language: prefix, root, connecting vowel, suffix and ending (we suppose that postfix and ending are the same here). In order to describe them and use further, we composed four vocabularies. These vocabularies are presented in Table 1.

Table 1. Vocabularies' description

#	Name	Size
1	Vocabulary of prefixes' meanings	88
2	Vocabulary of suffixes' meanings	293
3	Vocabulary of roots' meanings	1195
4	Vocabulary of the words segmentation into their morphemes	137827

Column *name* contains types of the vocabularies, column *size* contains the number of items (morphemes/words) in the vocabularies. Three first vocabularies are the vocabularies of the morpheme (prefixes' (1), suffixes' (2) and roots' (3)) meanings. Each line in them contains a morpheme itself, a list of morpheme meanings and corresponding examples for every meaning. The model of every line in these vocabularies is

morpheme: *meaning1* [*example1, example2, ...*]; *meaning2* [*example1, ...*]; *....*

For example,

un: *not* [*unable, unhappy, unnecessary, unemployment, unrest*]; *reverse* [*undo, unlock, unpack, unplug, unwrap*].

The fourth vocabulary is the vocabulary of the words segmentation into their morphemes. The model of each line in this vocabulary is

word: *morpheme type - morpheme; morpheme type - morpheme; ....*

For example,

goodness: *root - good; suffix - ness.*

We made vocabularies only for prefixes, roots and suffix. In our opinion, only these type of morphemes significantly influence the meaning of the word. However, the vocabulary of segmentations includes ending to make our algorithm correctly distinguish root's and suffixes' boundaries.

We manually compiled all these vocabularies from several sources, the main sources are vocabularies of Alexander Tikhonov [13] and Tatiana Efremova [8] and other information sources are [4–6,10–12,15]. We found out that the digital versions of the vocabularies have a large number of semantic mistakes and formatting errors, and we had to fix most of them to make them correctly parsable. Then converted them to the same format and merged sources together by the type. Also we augmented presented vocabularies with examples. We implemented parsers for compiled vocabularies and then we published our parsers and the vocabularies at the public Github[1] repository to share it with everyone who is interested in.

4   Algorithm for Segmentation a Word to Morphemes

Application of the morpheme-level model requires a technique to segment a word to the corresponding morphemes. We were not satisfied with the general purpose algorithm [19,26], because it does not allow to use built morpheme lexicon to adjust word embedding with it's corresponding underlying morphemes. As the result, we implemented our own algorithm based on the prepared vocabularies, since vocabularies have a finite and not large (roughly 137 k words) number of words. Let us describe our algorithm here.

Our algorithm is based on the described vocabularies of known words segmentation. Firstly, the algorithm finds a distribution for every morpheme from the dictionary of segmentations. Then, for a given word, the algorithm looks through the all possible segmentations for the segmentation with maximum probability.

4.1   Learning

We make an assumption that morphemes in segmentation are distributed independently, which is a naive Bayes assumption. It means that a prefix does not influence an ending or a suffix and vice versa. Upon first glance, it looks naturally on different examples, at least for Russian. Truly, this assumption does not reflect the real nature, because some prefixes can appear in a word only with

---

[1] https://github.com/TanyaKovalenko/Morpheme.

certain suffixes and endings. But our experiments show that our assumption is a good enough to reach 91% accuracy for the word segmentation task.

The independence assumption leads to some issues. Here are three major problems that we solved for our algorithm. The first problem is that a prefix can be placed only before a root, when a suffix can be only after a root. The second problem is that some morphemes could be applied only with a certain adjacent letter. For example, some prefixes with a consonant at the end are applicable only for certain consonants at the beginning of the root. The third problem is that some morphemes could influence (i.e. change) letters in adjacent morphemes. For example, some prefixes with a consonant at the end could change the first letter in the root.

These problems are very frequent to ignore them, and we propose to solve them in a certain way: we take previous and next letters into account when learn probability of a morpheme. If the adjacent letter doesn't exist (in case of the beginning and the end of a word), then we use a specific letter $ in order to distinguish it from other letters. As a result, we propose to learn conditional probability $p(m|b, a)$ of the morpheme m for the given letter a after the morpheme and letter b before the morpheme. It's easy to see that our augmentation solves all mentioned problems before. Our model would allow to occur a prefix in the middle of a word without the augmentation, however an updated model has almost a zero probability $p(m|b, a)$ where m is a prefix, and b and a are non $ letters. We use "almost" in the last two sentences because for synthetic languages, especially in the Russian language, an ending (and therefore suffixes and prefixes as well) could appear, for example, in the middle of a word, in case of a multi-roots word.

Eventually, we propose to learn a distribution for every morpheme from the vocabularies of known words segmentation taking everything above into account. Note that we can learn $p(m|b, a)$ directly when the size of the alphabet is not large.

4.2   Segmentation

Let us define the likelihood of a segmentation as a product of probabilities of underlying morphemes in the segmentation. Since a word has a limited number of different segmentations, we find the segmentation that have the maximum likelihood among all the possible segmentations. We suppose that the segmentation with the maximum likelihood is the right segmentation.

One final issue remains here. For instance, if we don't have a right root in the dictionaries, then a probability of every segmentation will be zero, and therefore the algorithm won't know which segmentation is right, even other known morphemes have non zero probabilities. We assume that the minimum probability is $\varepsilon > 0$ instead of zero in order to resolve this problem

The final algorithm is presented in listing 2.

Input: The corpus and the vocabulary of segmentation words to
      morphemes
Result: Words segmentation
foreach *word in the corpus* **do**
    foreach *word segmentation as a tuple of substrings* **do**
        foreach *substring m* **do**
           | calculate and save $p(m|b, a)$;
        end
        calculate and save the product of all $p(m|b, a)$ and the
          corresponding segmentation;
    end
    choose the largest product among all of them and save the
      corresponding segmentation;
    return the saved segmentation;
end

Algorithm 1. Words segmentation algorithm

The described algorithm to segment words to morphemes was tested on known segmentations from the vocabulary (4): it reaches 91% segmentation matching. Roughly 1–2% of missings were caused not by the algorithm mistakes, but by incorrectness of the original vocabularies, most of them were multi-roots words.

5 Morpheme Embedding

Here we discuss the way to make a morpheme embedding using the presented compiled vocabularies. One morpheme could have several meanings expressed as a certain word or a simple phrase. We map every meaning to the corresponding embedding using the word embedding for a certain word or average words embedding for a simple phrase. As a result, we map every morpheme to the corresponding list of word embeddings.

Then in Sect. 6 we use the morpheme embedding to correct the existing Word2Vec word embedding in our tests. To do that, we firstly segment a given word to the morphemes, and then for every morpheme we choose the most suitable meaning. Suitability here is the cosine similarity between the given vectors. So we got morpheme embedding. The pseudocode below shows the main steps of the algorithm for getting morpheme embedding.

Input: The word, the morpheme, morphemes vocabularies and a word
      embedding model
Result: Morpheme embedding
identify the type of the morpheme (prefix, root or suffix);
find the line with description of the morpheme in the corresponding
 vocabulary;
foreach *meaning in all possible meanings of the morpheme* **do**
    | get the vector of the meaning;
    | calculate the cosine similarity between the vector of the meaning and
    | the vector of the word;
    | save the calculated cosine similarity in the temporary array;
end
find the maximum value in the array of the cosine similarities;
get the vector of the meaning corresponding to the maximum cosine
 similarity;
return the result vector as morpheme embedding;

Algorithm 2. Morpheme embedding

6   Word Embedding Correction

We applied our morpheme segementation algorithm to build the word embedding model based on the Word2Vec and vectors of the morpheme meanings. At first, we downloaded some different models of the Russian language from RusVectōrēs [7] built by Word2Vec. Then we developed the algorithm to form a new model based on the morpheme meanings using the following formula:

$$v'_{\tau_i} = \frac{1}{2}(v_{\tau_i} + \sum_{w \in M_i} sim(v_{\tau_i}, v_w) \cdot v_w), \tag{1}$$

where τ_i is a word from corpus with index i, v_{τ_i} is a classic word embedding for a word τ_i, v'_{τ_i} is a new word embedding for a word τ_i, M_i is the set of all morphemes of word τ_i, w is a morpheme from word τ_i, v_w is a word embedding for meaning of a morpheme w, $sim(v_{\tau_i}, v_w)$ is the function used to denote the similarity between τ_i and its morphemes' meanings. We define a function $cos(v_i, v_j)$ to denote the cosine distance between v_i and v_j. $sim(v_{\tau_i}, v_w)$ is normalized as follows.

$$sim(v_{\tau_i}, v_w) = \frac{cos(v_{\tau_i}, v_w)}{\sum_{w \in M_i} cos(v_{\tau_i}, v_w)}, \tag{2}$$

As it can be seen from the first formula, each vector of a word from the corpus is adjusted with the vectors of the morpheme meanings. The coefficient $1/2$ is used, the result word embedding contains two components - the Word2Vec word embedding and the morpheme meanings. But since each morpheme of the word makes a different contribution to the meaning of the whole word, we added a coefficient $sim(v_{\tau_i}, v_w)$ to the vector v_w, which is a measure of the similarity of the meaning of the morpheme and of the word itself. The described formulae were taken from [29], where the authors resolve the similar task, but for the English language.

Our algorithm allows to set the limit on the minimum acceptable value of the parameter $cos(v_{\tau_i}, v_w)$. If the $cos(v_{\tau_i}, v_w)$ is less than the minimum value, we use 0 instead of $cos(v_{\tau_i}, v_w)$ and do not take into account meanings of the suitable morphemes.

The main steps of the algorithm are presented in the listing 3.

Input: Word2Vec model of the Russian corpus
Result: New model of the Russian language
load Word2Vec model;
foreach *word in the corpus* **do**
    split the word into morphemes;
    foreach *morpheme in the word* **do**
        execute Algorithm 2 to get the morpheme embedding and the cosine similarity between word and vector of the morpheme meaning;
    end
    recalculate the result vector of the word, using the old vector from Word2Vec model and morpheme vectors;
    save the word and its new word embedding into new morpheme model;
end

Algorithm 3. New model of the Russian language

7 Experiments

The resulting vector model of the Russian language was tested in the word classification task. For our experiments we used Word2Vec model *"ruwikiruscorpora"* from [7], that contains RNC and Wikipedia from November, 2016. The corpus size is 600 million words and vocabulary size is 392 339 words. Visualization of the vectors from three different classes of words is presented in Figs. 1 and 2. The first group of words refers to the medical topic and have blue color. The dataset was taken from [3]. The second green group refers to the financial topic, that was taken from [1]. The last, red group refers to the automotive topic from [2]. In order to demonstrate an advantage of our model over a Word2Vec model, we needed in datasets with complex words, composed in many different morphemes. So, we did not use standard conventional datasets such as semantic relatedness from [9] or [16]. Figure 2 illustrates vectors computed with the proposed model and Fig. 1 illustrates vectors computed with the Word2Vec model.

We used the t-SNE algorithm [18] to visualize clusters. The full name of the algorithm is *t-distributed stochastic neighbor embedding*. This is a machine learning algorithm for dimensionality reduction of the vectors. It embeds high-dimensional data into a space of two or three dimensions, which can then be visualized in a scatter plot.

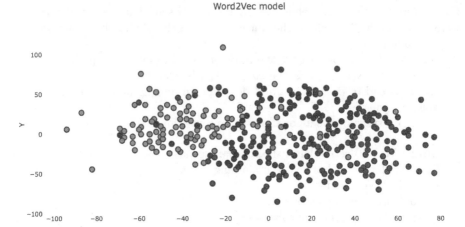

Fig. 1. Visualization of vectors from Word2Vec model, divided into three topics (green dots corresponds to the financial topic, blue dots corresponds to the medical topic and red dots corresponds to the automotive topic) (Color figure online)

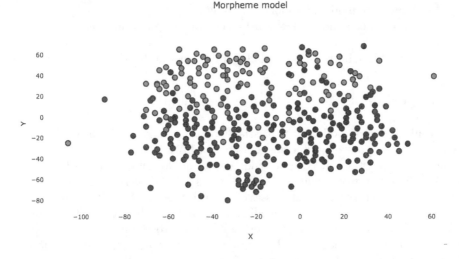

Fig. 2. Visualization of vectors from morpheme model, divided into three topic (green dots corresponds to the financial topic, blue dots corresponds to the medical topic and red dots corresponds to the automotive topic) (Color figure online)

We used this technique to map each high-dimensional vector to the corresponding two-dimensional point. We apply to word vectors.

The results of the cluster visualization may be compared using a visual method, and also by comparing the values of covariance matrix determinants (generalized determinants) for the Gaussian mixture model constructed for each of the word topic. The determinant of the covariance matrix reflects the degree of random disperse of the elements of the system [23]. The determinants are presented in Table 2. For the new model, the determinants are 386320.81 (green), 860356.17 (red) and 414574.62 (blue) for each of the topics, respectively. For the Word2Vec model the determinants are 387458.26 (green), 664700.42 (red), 615093.79 (blue). For two of the three word topics, the determinants of the covariance matrices became smaller, and for one — almost the same as the original. This fact shows that the new model could be considered as a base to solve the word classification problems.

Table 2. Generalized determinants

Topic type	Word2Vec model	Morpheme model
Medical	615093.79	414574.62
Financial	387458.26	386320.81
Automotive	664700.42	860356.17

In the course of our experiments we found some words which are not in the Word2Vec model, but they are in our. It means that our algorithm made the vectors for these words without the Word2Vec vectors. The vectors were made on the basis of morpheme meanings. Among the new words there are: "*моднявый*" that means "*fashionable*" in the Russian slang; "*безнравственная*", its meaning is close to "*girl who is immoral*"; "*подвыпивший*" that translated as "*tipsy*", etc. We checked the vectors locating close to ours in order to understand that the vectors in our model are correct. We got the following results: for "*fashionable*" - "*fashion*", "*fashionista*", "*newfangled*"; for "*girl who is immoral*" - "*immorality*", "*immoral*", "*debauchery*"; for "*tipsy*" - "*beer*", "*alehouse*", "*beery*", "*vodka*". These results show these vectors are good.

8   Conclusions and Future Work

As a result, we present:

- the digital version of updated and fixed vocabularies of the Russian morphemes,
- the vocabulary of the words segmentation,
- the parsers for the vocabularies,
- the algorithm for words segmentation to morphemes based on the vocabularies,
- the method for adjusting and clarifying word embedding using underlying morphemes.

In the further work, we are going to expand the vocabularies and fix errors in them that our algorithm found. We are planning to test our method for adjusting on other different practical tasks to figure out strong and weak sides. Also we plan to try the approach using the Markov chains as an algorithm to segment a word to morphemes. Further we plan to learn morpheme level model directly (something like "Morpheme2Vec") for the Russian language and test it in the common tasks since character level models showed good results for morphological rich languages.

References

1. Dictionary of financial terms and economic concepts. http://www.fingramota.org/servisy/slovar. Accessed 28 Aug 2017
2. Great automotive dictionary. http://www.perfekt.ru/dictionaries/avto/s_rus.html. Accessed 28 Aug 2017
3. Medical dictionary. http://www.medslv.ru. Accessed 28 Aug 2017
4. Words segmentation to morphemes. http://www.sostavslova.ru. Accessed 28 Aug 2017
5. Words segmentation to morphemes. http://www.morphemeonline.ru. Accessed 28 Aug 2017
6. Kamchatov, A., et al.: Russian "drevoslov". http://www.drevoslov.ru/word creation/morphem. Accessed 28 Aug 2017
7. Kutuzov, A., Kuzmenko, E.: Rusvectores: distributional semantic models for the Russian. http://www.rusvectores.org. Accessed 28 Aug 2017
8. Kuznetsova, A.: Efremova publisher=Springer T. Dictionary of the Morphemes of Russian
9. Panchenko, A., Loukachevitch, N.V., Ustalov, D., Paperno, D., Meyer, C.M., Konstantinova, N.: Russe: the first workshop on russian semantic similarity. In: Proceedings of the International Conference on Computational Linguistics DIA-LOGUE, pp. 89–105 (2015)
10. Safyanova, A.: The meaning of Latin morphemes. http://www.grammatika-rus.ru/znachenie-latinskih-morfem. Accessed 28 Aug 2017
11. Safyanova, A.: The meanings of prefixes in Russian. http://www.spelling.siteedit.ru/page51. Accessed 28 Aug 2017
12. Safyanova, A.: The meanings of suffixes in the Russian. http://www.spelling.siteedit.ru/page50. Accessed 28 Aug 2017
13. Tikhonov, A.: Morphology and spelling dictionary of the Russian language. Astrel, AST (2002)
14. Do, C.B., Batzoglou, S.: What is the expectation maximization algorithm? Nature Biotechnol. **26**, 897–899 (2008)
15. Garshin, I.: Slavic roots of the Russian. http://www.slovorod.ru/slavic-roots. Accessed 28 Aug 2017
16. Leviant, I., Reichart, R.: Separated by an un-common language: towards judgment language informed vector space modeling. https://arxiv.org/abs/1508.00106. Accessed 28 Aug 2017
17. Pennington, J., Socher, R., Manning, C.D.: Glove: global vectors for word representations (2014)

18. Van Der Maaten, L., Hinton, G.: Visualizing data using t-sne (2008). http://jmlr. org/papers/volume9/vandermaaten08a/vandermaaten08a.pdf. Accessed 28 Aug 2017

19. Creutz, M., Lagus, K.: Unsupervised morpheme segmentation and morphology induction from text corpora using morfessor 1.0. Publications in Computer and Information Science and Report A81 and Helsinki University of Technology, March 2005

20. Grunwald, P.: The Minimum Description Length Principle. MIT Press, Cambridge (2007)

21. Cotterell, R., Schutze, H.: Morphological word-embeddings. https://ryancotterell. github.io/papers/cotterell+schuetze.naacl15.pdf. Accessed 28 Aug 2017

22. Galinsky, R., Alekseev, A., Nikolenko, S.: Improving neural networks models for natural language processing in Russian with synonyms (2016)

23. Aivazyan, S., Enukov, I., Meshalkin, L.: Applied Statistics: Basics of Modeling and Primary Data Processing. Finance and Statistics, Moscow (1983)

24. Bordag, S.: Unsupervised knowledge-free morpheme boundary detection. http://nlp.cs.swarthmore.edu/~richardw/papers/bordag2005-unsupervised.pdf. Accessed 28 Aug 2017

25. Qiu, S., Cui, Q., Bian, J., Gao, B., Liu, T.-Y.: Co-learning of word representations and morpheme representations. https://www.aclweb.org/anthology/ C14-1015. Accessed 28 Aug 2017

26. Virpioja, S., Smit, P., Gronroos, S.-A., Kurimo, M.: Morfessor 2.0: Python implementation and extensions for morfessor baseline. Aalto University publication series SCIENCE + TECHNOLOGY (2013)

27. Mikolov, T., Chen, K., Corrado, G., Dean, J.: Efficient estimation of word representations in vector space. https://arxiv.org/pdf/1301.3781.pdf. Accessed 28 Aug 2017

28. Ling, W., Tiago, L., Marujo, L., Fernandez Astudillo, R., Amir, S., Dyer, C., Black, A.W., Trancoso, I.: Finding function in form: compositional character models for open vocabulary word representation. https://arxiv.org/abs/1508.02096. Accessed 28 Aug 2017

29. Xu, Y., Liu, J.: Implicitly incorporating morphological information into word embedding. https://arxiv.org/abs/1701.02481. Accessed 28 Aug 2017

30. Xiang, Z., Junbo, Z., Yann, L.: Character-level convolutional networks for text classification. http://papers.nips.cc/paper/5782-character-level-convolutional-networks-for-text-classification.pdf. Accessed 28 Aug 2017

Comparison of Vector Space Representations of Documents for the Task of Information Retrieval of Massive Open Online Courses

Julius Klenin[✉], Dmitry Botov, and Yuri Dmitrin

Information Technologies Institute, Chelyabinsk State University,
Chelyabinskaya oblast', Chelyabinsk 454001, Russian Federation
jklen@yandex.ru

Abstract. One of the important issues, arising in development of educational courses is maintaining relevance for the intended receivers of the course. In general, it requires developers of such courses to use and borrow some elements presented in similar content developed by others. This form of collaboration allows for the integration of experience and points of view of multiple authors, which tends to result in better, more relevant content. This article addresses the question of searching for relevant massive open online courses (MOOC) using a course programme document as a query. As a novel solution to this task we propose the application of language modelling. Presented results of the experiment, comparing several most popular models of vector space representation of text documents, such as the classical weighting scheme TF-IDF, Latent Semantic Indexing, topic modeling in the form of Latent Dirichlet Allocation, popular modern neural net language models word2vec and paragraph vectors. The experiment is carried out on the corpus of courses in Russian, collected from several popular MOOC-platforms. The effectiveness of the proposed model is evaluated taking into account opinions of university professors.

Keywords: Vector space model · Educational course programme
Document modelling · Information retrieval · Word embedding
Mooc-platform · Educational data mining

1 Introduction

Today ensuring the relevance and quality of educational content is one of the more time-consuming tasks for developers of educational courses. It is performed in the conditions of highly competitive market. The volume of open educational content on the Internet is growing rapidly, with many leading universities around the world providing free access to their programmes and courses, sharing them on the massive open online course (MOOC) platforms. These and other documents act as sources for programme developers, to be used in order to keep the other courses relevant. And to do so, developers have to regularly search for

© Springer International Publishing AG 2018
A. Filchenkov et al. (Eds.): AINL 2017, CCIS 789, pp. 156–164, 2018.
https://doi.org/10.1007/978-3-319-71746-3_14

and analyze a large quantities of such sources. Typically, the process of finding these courses is reduced to using various web-search engines or to search systems, provided by platforms themselves. This usually is done by using the name of the course, keywords and broad topics or categories. In the end such approach tends to waste a lot of time and not always yields a sufficiently relevant result.

In this paper we propose a simpler process for retrieving relevant Russian courses from most popular MOOC platforms [1], which uses the contents of the target programme as a search query. The main objective of the study is to compare different models of vector representation of texts. This research is part of the effort to create an intelligent decision support system (IDSS) for the processes of creating and maintenance of documentation in universities.

2   Related Work

Most major approaches to the tasks of comparison and search of educational content use ontological models, knowledge bases, systems based on rules or reasoning on precedents. The use of these methods for search and comparison of educational courses and programmes is widely described in literature. However, despite having potential for high quality results, these approaches all share the main drawback - high complexity of building and updating ontologies of different subject areas in various knowledge domains and developing a complete system of rules.

For example [2] proposes the use of ontological models for comparison of educational courses and determines the measures of semantic similarity of educational courses based on keyword analysis of discipline contents and learning outcomes, using the taxonomy of educational objectives.

Another approach is to use vector space models of texts which represent each document from the collection in the form of a vector from the common multidimensional feature space.

One of the most historically popular approaches, Tf-idf is a simplistic weighting scheme, that combines the frequency of word's occurrences in the document with the inverse frequency of its occurrences in collection overall. If every word in the document collection is weighted using this technique, the result can be used to form a set of vectors for each document, consisting of the number of dimensions, equal to the total number of unique words, encountered in the collection. However, that does mean, that the vectors are extremely sparse.

Latent Semantic Indexing or Latent Semantic Analysis [6] applies Singular Value Decomposition to the term-document matrix to create a low-rank approximation of it, providing reduced vectors in comparison to the TF-IDF method, which means these vectors are denser and easier to use in calculations. Also, since this approach tries to replace multiple similar dimensions with one, while preserving the existing trends between documents in the collection, it results in merging of distributionally similar words.

Latent Dirichlet Allocation is a topic model, which considers the collection of texts and words to be a mixture of a certain number of to topics. LDA applies

Dirichlet's prior to initialize such distribution and then iteratively improves upon it. Document vectors for LDA consist of the topic probabilities for document.

Neural net language model Word2Vec [3] has recently became overwhelmingly popular. For instance, Word2Vec has proven its quality in the task of semantic similarity in Russian [7]. Word2Vec is a shallow neural network with a single hidden layer, tasked with learning the correlation between words and their contexts.

The applicability of the word2vec model for information retrieval purposes is considered in [9]. On the standard TREC corpus, suggested word2vec approach shows a better result than the LM and LDA-smoothed LM baselines.

In general, in order to represent entire documents, word2vec vectors need to be combined for all the words in the document.

Another application of word embedding in the IR [10] suggests construction of Dual Embedding Space Model, which includes a combination of separate word embedding models for search queries and search results.

The logical successor to word2vec, paragraph2vec [4] is a similar approach, that works with entire documents.

Overall, our research, as far as we know, presents the first attempt at applying various machine learning techniques to the task of modelling and retrieval of educational course programmes based on their semantic similarity.

A popular next step in improving the quality of information retrieval is application of various neural network models on top of learned vectors. [11] presents an extended look at different deep neural network (DNN) architectures utilized in information retrieval.

3   Approach to Comparison of Vector Representations

In this paper we apply the a simplistic approach of processing documents. Initially, we collect course programmes to serve as a search result. All of them are then preprocessed and then used to calculate vector space models. Each query-document is then passed through the same preprocessing steps and used by each model to generate a vector from the corresponding vector space.

3.1   Corpus

Since there are no available data corpora suitable for our goals, we had to create our own dataset of online courses. To create such corpus we applied crawling and scraping methods to MOOC-platform web-pages. For each course the following attributes were extracted:

- Course name;
- Course language;
- Source university;
- Specialization;
- Course category (knowledge domain);

- Annotation;
- Goals and objectives of the course;
- Entrance requirements;
- Learning outcomes;
- Formed competences;
- Course structure (Lecture topics);

The following MOOC-platforms were selected for this experiment: openedu.ru, coursera.org, intuit.ru, stepik.org, universarium.org. We have downloaded available courses for each platform and ended up with the corpus of 1300+ courses.

Lion's share of the corpora comes from INTUIT courses - more than 900. Other platforms contribute approximately the same number of courses. Subject-wise, about half of the courses are devoted to information technologies. The second place belongs to economics and management roughly one-fifth.

Prior to pre-processing, each document contained an average of 62 sentences or 953 words, totalling at 103K+ sentences and 1.39M+ words.

3.2   Preprocessing

Preprocessing step of our approach involves removal of stopword sequences from the documents, followed by tokenization and lemmatization of each token.

Considering variability of word forms in Russian, lemmatization is much more important than in corpora consisting of texts in English. Otherwise, most models would not be able to draw parallels between different forms of the word.

Corpus statistics (average estimates are given within $+/-$ 3 standard deviations) after preprocessing:

- corpus consists of 1.02M+ tokens;
- 80% of programs contain less than 1000 tokens after preprocessing;
- 20% contain less than 200;
- The average number of tokens is 693;
- The average number of unique tokens - 240;

3.3   Vector Space Models

In our research, we have focused on various statistical vector space representations for documents. This decision was made due to the fact, that the knowledge domain of educational courses is far too wide to be properly explained by the ontology of a reasonable size. Thus, in order to resolve this issue, we needed to use models that can work on any topics without requiring manual labeling or other supervision.

TF-IDF. Forming a term-document matrix and filling it with TF-IDF weights results in following document vectors [5]

$$d_i = [w_1, w_2, \ldots, w_N] \tag{1}$$

where

$$tfidf_{i,j} = frequency(j,i) * \log_2 \frac{K}{document frequency(j)} \qquad (2)$$

Here, K is the total number of documents and N is the amount of unique words. $frequency(j,i)$ and $document frequency(j)$ stand for the frequency of the word w_j in the document d_i, and in the collection in general, respectively.

In our experiment we ended up with document vectors, made up of 22K+ unique word tokens as dimensions. As is with all models used in our experiment, tokens here are lemmas left after preprocessing step, and term-document matrix is assembled as described in Sect. 2.

LSI. The overall process of single-value decomposition leads to following vectors

$$X = U\Sigma V^T \qquad (3)$$

the resulting document vectors are contained in rows of V^T

$$d_i = \Sigma_{i,i} * V_i^T \qquad (4)$$

For our purposes, we have tested LSI models with various numbers of dimensions and have found the target of 25 dimensions to provide the best results for our dataset.

LDA. As was the case with the LSI we have experimentally found that using 25 topics yields the best results.

Word Vectors. In order to calculate document vectors we apply tf-idf weights to word vectors and calculate weighted averaged word2vec vectors, as was suggested in [8]

$$d_i = \frac{\sum\limits_{j \in N} TFIDF(w_{i,j}) * w_{i,j}}{|d_i|} \qquad (5)$$

For this experiment, we have experimented with dimensionality of the model and in the end settled for skip-gram model, with negative sampling for optimization and 200 dimensions.

We have also attempted to utilize the set of word2vec vectors, trained on Russian Wikipedia, with similar dimensionality.

Paragraph Vectors. Here, we train distributed memory version of paragraph2vec, since previous experiments showed barely any difference in results of two options with PV-DM pulling slightly ahead. Similarly to previous model we have tested out different dimensionalities and the best results were achieved with vectors containing 50 dimensions.

3.4   Processing Query

In our approach we use the target programme as a query. To use it for the search, we first need to perform the same preprocessing steps, as with the courses in our corpus. This includes removal of stopword subsequences, tokenization and lemmatization. Now that the query is represented in the same way as possible results, we use each model to generate a vector for the query document in the same vector space as the possible results.

The actual set of results is then acquired by calculating the cosine similarity between query vector and every possible result vector, arranging results in order of decreasing similarity and selecting the top 5 results.

3.5   Human Judgment

To assess the quality of the tested models, we assembled the team of 13 experts, who were tasked with evaluating the search results, produced by our system for each of 27 programmes, serving as queries, with the total of 917 evaluations. Grading of courses used a 5-point scale

- 1 - No similarities at all;
- 2 - Far from the topic of the query;
- 3 - Partially covers the topic of the query;
- 4 - Matches topics of the query-programme;
- 5 - Mostly matches topics of the query-programme;

Each of the experts is a specialist, teaching the disciplines they were asked to evaluate the results for. Overall, an average of 2–3 experts were tasked with evaluation of each query.

In order to ensure the existence of coherent agreement between experts for each course, we have calculated Fleiss' kappa for each case. This value averaged at 0,861 across the entire experiment dataset, which corresponds to almost perfect agreement among raters. This means, that human judgment, used to evaluate the performance of our approach is statistically solid.

3.6   Evaluation Metrics

In order to evaluate the quality of selected models, we use two popular metrics, used for evaluation of search systems: mean average precision (MAP) and normalized discounted cumulative gain (nDCG).

MAP is calculated for N result sets at rank k by calculating the mean of average precision (AP) metrics

$$MAP@k = \frac{\sum_{i \in N} AP@k_i}{N} = \frac{\sum_{i \in N} \frac{\sum_{j \in k} rel_j * P@j}{k}}{N} \qquad (6)$$

Here, rel_j is a relevance of the result at position j and $P@j$ is precision at rank j, which equals the number of relevant results above j.

Relevance here is considered binary - document is either relevant or not. To convert our 5-point scale to this binary format, we have chosen the level of 3 to be the border between relevant and not relevant documents.

nDCG specifies relation between the actual DCG of the result set and the ideal DCG for the same set - calculated as a regular DCG for the set reordered according to descending grades

$$nDCG@k = \frac{DCG@k}{IDCG@k} \tag{7}$$

Here discounted cumulative gain is

$$DCG@k = \sum_{i \in k} \frac{2^{rel_i} - 1}{log_2(i+1)} \tag{8}$$

4   Results and Discussion

MAP and nDCG metrics, calculated as described above are shown in Table 1.

Table 1. MAP@5 and nDCG@5 scores

Model	TF-IDF	Paragraph vectors	LDA	LSI	Weighted avr. W2V	WikiW2V
MAP@5	**0.85**	0.561	0.28	0.439	0.663	0.425
nDCG@5	**0.837**	0.665	0.64	0.6	**0.773**	0.705

Since nDCG normalizes actual DCG with "ideal" DCG, calculated over the same set in another order, it would vary severely, depending on the order of the relevant items. As we can see, both TF-IDF and weighted averaged word2vec have achieved similarly high results, meaning that both models mostly return sets of results ordered by decreasing relevance. Other models showed worse quality.

Of course, it is worth noting, that if one of tested models shows extremely poor performance and scores low with all returned results (all results are scored 2, for example), the "ideal" DCG is going to be quite similar (if not the same) to the actual value and thus, normalization will produce high nDCG, despite actual results being not very high.

MAP, on the other hand, reflects specifically the average ability of the model to return relevant results above all else. Here we can see that highest result is achieved by TF-IDF model, followed by weighted averaged word2vec vectors and others.

We attribute such results to the specifics of the corpora. Despite having completely different knowledge domains covered in each document, the base for all of them is educational course programme. This means, that these texts consist not only of topic-specific terms, but also terms from the general course programme lexicon.

While most models, including paragraph vectors, have issues with mitigating this, TF-IDF is capable of filtering them out. It is also worth noting, that TF-IDF model has over 20 thousand dimensions, making it easier to discern differences (albeit harder in terms of computation time) between documents in comparison to weighted word2vec, for instance, which, while being capable of filtering common words, limits the overall dimensionality to that of a original word2vec model.

There is also a possibility of corpora being not big enough for training of word2vec and paragraph vector models to be sufficient.

While word2vec trained on russian Wikipedia had enough tokens to be trained better, the difference in styles of language used for different documents resulted in worse overall results for this model.

In general, that means, that to properly train our models, we need to assemble a bigger corpora of various educational documents, not necessarily course programmes.

5   Conclusion

We have presented results of our evaluation of a variety of statistical vector space language models in the task of search and comparison of educational courses. The results have shown that TF-IDF model has taken the first place with significant gap from other models. Word2Vec, weighted using TF-IDF achieved the second place, while the rest lag behind.

We propose that the possible issues of our approach are both the need for a model to be able to consider importance of each word, and the requirement to have a bigger training dataset for more advanced models, such as word2vec and paragraph2vec.

In future research we are considering:

- growing the corpora further, up to at least 10M tokens, or about 14 K course programmes;
- improving filtering techniques;
- using structure of the document in analysis;
- introducing more complex models on top of tested ones, such as various neural network models (CNN, for instance);
- using learning to rank to improve the quality of the returned results.

References

1. Class Central. By The Numbers: MOOCS in 2016. https://www.class-central.com/report/mooc-stats-2016/
2. Chernikova, E.: A Novel Process Model-driven Approach to Comparing Educational Courses using Ontology Alignment (2014)
3. Mikolov, T., Sutskever, I., Chen, K., Corrado, G.S., Dean, J.: Distributed representations of words and phrases and their compositionality. In: Advances in Neural Information Processing Systems, pp. 3111–3119 (2013)

4. Mikolov, T., Sutskever, I., Chen, K., Corrado, G.S., Dean, J.: Distributed represe-
   nations of sentences and documents. In: Proceedings of ICML 2014, pp. 1188–1196
   (2014)
5. Manning, C.D., Raghavan, P., Schütze, H.: Introduction to Information Retrieval.
   Cambridge University Press, Cambridge (2008)
6. Deerwester, S., et al.: Indexing by latent semantic analysis. J. Am. Soc. Inf. Sci.
   41(6), 391–407 (1990)
7. Panchenko, A., et al.: RUSSE: the first workshop on Russian semantic similarity. In:
   Computational Linguistics and Intellectual Technologies Papers from the Annual
   International Conference Dialogue, RGGU 2015, Moscow, vol. 2, pp. 89–105 (2015)
8. Lilleberg, J., Zhu, Y., Zhang, Y.: Support vector machines and word2vec for text
   classification with semantic features. In: IEEE 14th International Conference on
   Cognitive Informatics & Cognitive Computing (ICCI*CC) (2015)
9. Ganguly, D.: Word embedding based generalized language model for information
   retrieval. In: Proceedings of the 38th International ACM SIGIR Conference on
   Research and Development in Information Retrieval, pp. 795–798 (2015)
10. Nalisnick, E., et al.: Improving document ranking with dual word embeddings.
    In: Proceedings of WWW. International World Wide Web Conferences Steering
    Committee (2016)
11. Mitra, B., Craswell, N.: Neural text embeddings for information retrieval. In: Pro-
    ceedings of WSDM. ACM, pp. 813–814 (2017)

Machine Learning

Interpretable Probabilistic Embeddings: Bridging the Gap Between Topic Models and Neural Networks

Anna Potapenko[1]([⊠]), Artem Popov[2], and Konstantin Vorontsov[3]

[1] National Research University Higher School of Economics, Moscow, Russia
anna.a.potapenko@gmail.com
[2] Lomonosov Moscow State University, Moscow, Russia
popov.artem.s@yandex.ru
[3] Moscow Institute of Physics and Technology, Moscow, Russia
vokov@forecsys.ru

Abstract. We consider probabilistic topic models and more recent word embedding techniques from a perspective of learning hidden semantic representations. Inspired by a striking similarity of the two approaches, we merge them and learn probabilistic embeddings with online EM-algorithm on word co-occurrence data. The resulting embeddings perform on par with Skip-Gram Negative Sampling (SGNS) on word similarity tasks and benefit in the interpretability of the components. Next, we learn probabilistic document embeddings that outperform paragraph2vec on a document similarity task and require less memory and time for training. Finally, we employ multimodal Additive Regularization of Topic Models (ARTM) to obtain a high sparsity and learn embeddings for other modalities, such as timestamps and categories. We observe further improvement of word similarity performance and meaningful inter-modality similarities.

1 Introduction

Recent progress in deep natural language understanding prompted a variety of word embedding techniques that work remarkably well for capturing semantics. These techniques are usually considered as general neural networks that predict context words given an input word [3,17,27]. Although this perspective is convenient to generalize to more complex neural network architectures, e.g. skip-thought vectors [16], we believe that it is also important to establish connections between neural embeddings and more traditional models of distributional semantics. It gives theoretical insights about certain models and enables to use previous work as a grounding for further advances.

One of the first findings in this line of research is interpreting Skip-Gram Negative Sampling (SGNS [27]) as an implicit matrix factorization of the shifted Pointwise Mutual Information (PMI) matrix [20]. It brings SGNS to the context of various vector space models (VSMs) developed during the last decades. Pantel and Turney [40] provide a thorough survey of VSMs dividing them into word-word, word-context and word-document categories based on the type of the co-occurrence matrix. According to the distributional hypothesis [12], similar words tend to occur in similar contexts; thus the rows of any of these matrices can be used for estimating word similarities [9]. Gentner [11] defines attributional similarity (e.g. *dog* and *wolf*) and relational similarity (e.g. *dog:bark* and *cat:meow*), which are referred to as similarity and analogy tasks in more recent papers. While Baroni et al. [26] argue that word embeddings inspired by neural networks significantly outperform more traditional count-based approaches for both tasks, Levy et al. [21] tune a shared set of hyperparameters and show that two paradigms give a comparable quality.

We follow this line of research and demonstrate how principle ideas of the modern word embedding techniques and probabilistic topic models can be mutually exchanged to take the best of the two worlds. So far, topic modeling has been widely applied to factorize word-document matrices and reveal hidden topics of document collections [4,15]. In this paper we apply topic modeling to a word-word matrix to represent words by probabilistic topic distributions. Firstly, we discover a number of practical learning tricks to make the proposed model perform on par with SGNS on word similarity tasks. Secondly, we show that the obtained probabilistic word embeddings (PWE) inherit a number of benefits from topic modeling.

One such benefit is interpretability. Interpretability of each component as a coherent topic is vital for many downstream NLP tasks. To give an example, exploratory search aims not only to serve similar documents by short or long queries, but also to navigate a user through the results. If a model can explain why certain items are relevant to the query in terms of distinct topics, then these topics can be used to arrange the results by categories. Murphy et al. [30] motivated the importance of interpretability and sparsity from the cognitive plausibility perspective and introduced Non-Negative Sparse Embeddings (NNSE), which is a variation of Non-Negative Sparse Coding matrix factorization. State-of-the-art techniques, such as SGNS or GloVe [35] lack both sparsity and interpretability. To address this problem, more recent models [23,39] extend SGNS and CBOW [27] respectively. However, they do that with explicit modifications of optimization procedure, such as project gradient for SGD. A benefit of topic modeling framework is that interpretability comes naturally with a probabilistic interpretation of parameters.

Furthermore, probabilistic word embeddings can be easily extended with Additive Regularization of Topic Models, ARTM [43]. This is a general framework to combine multiple requirements in one topic model. In this work we use ARTM to obtain sparsity and to learn embeddings for additional *modalities*, such as timestamps, authors, categories, etc. It enables us to investigate inter-modality similarities, because all the embeddings are in the same space. Interestingly, additional modalities also improve performance on word similarity task. Finally, we build probabilistic document embeddings and show that they outperform DBOW architecture of paragraph2vec [17] on a document similarity task. Thus, we get a powerful framework for learning probabilistic embeddings for various items and with various requirements. We train these models with online EM-algorithm similar to [14] in BigARTM open-source library [41].

Related work includes Word Network Topic Model (WNTM [45]) and Biterm Topic Model (BTM [44]) that use word co-occurrence data for analyzing short and imbalances texts. However, they do not consider their models as a way to learn word representations. There are also a number of papers on building hybrids of topic models and word embeddings. Gaussian LDA [8] imposes Gaussian priors for topics in a semantic vector space produced by word embeddings. The learning procedure is obtained via Bayesian inference, however a similar idea is implemented more straightforwardly in [38]. They use pre-built word vectors to perform clustering via Gaussian Mixture Model and apply the model to Twitter analysis. Pre-built word embeddings are also used in [33] to improve quality of topic models on small or inconsistent datasets. Another model, called Topical Word Embeddings (TWE [22]) combines LDA and SGNS. It infers a topic for each word occurrence and learns different embeddings for the same word occurred under different topics. Unlike all these models, we do not combine the models as separate mechanisms, but highlight a striking similarity of optimization objectives and *merge* the models.

The rest of the paper is organized as follows. In Sect. 2 we remind the basics of word embeddings and topic models. In Sects. 3 and 4 we discuss theoretic insights and introduce our generalized approach. In the experiments section we use 3 text datasets (Wikipedia, ArXiv, and Lenta.ru news corpus) to demonstrate high quality on word similarity and document similarity tasks, drastic improvement of interpretability and sparsity, and meaningful inter-modality similarities.

2   Related Work

Definitions and Notation. Here we introduce the notation that highlights a common nature of all methods and will be used throughout the paper. Consider a set of documents D with a vocabulary W. Let n_{wd} denote a number of times the word w occurs in the document d. The document can be treated as a *global context*. We will be also interested in a *local context* of each word occurrence, which is a bag of words in a window of a fixed size. Let n_{uv} denote a number of co-occurrences of words u and v in a sliding window, $n_u = \sum_v n_{uv}$, $n_v = \sum_u n_{uv}$, and $n = \sum_u n_u$.

All the models will be parametrized with the matrices Φ and Θ, containing $|T|$-dimensional embeddings.

Skip-Gram Model. Skip-gram model learns word embeddings by predicting a local context for each word in a corpus. The probability of word u from a local context of word v is modeled as follows:

$$p(u|v) = \frac{\exp \sum_t \phi_{ut}\theta_{tv}}{\sum_{w \in W} \exp \sum_t \phi_{wt}\theta_{tv}}, \tag{1}$$

where $\Phi^{|W| \times |T|} = (\phi_{ut})$ and $\Theta^{|T| \times |W|} = (\theta_{tv})$ are two real-valued matrices of parameters. According to the bag-of-words assumption, each word in the local context is modeled independently, thus one can derive the log-likelihood as follows:

$$\mathcal{L} = \sum_{v \in W} \sum_{u \in W} n_{uv} \ln p(u|v) \rightarrow \max_{\Phi, \Theta}. \tag{2}$$

where n_{uv} denotes the number of times the two terms co-occurred in a sliding window. However, normalization over the whole vocabulary in formula (1) prevents from learning the model effectively on large corpora. Skip-Gram Negative Sampling (SGNS) is one of possible ways to tackle this problem. Instead of modeling a conditional probability $p(u|v)$, SGNS models the probability of a co-occurrence for a pair of words (u, v). The model is trained on word pairs from the corpus (positive examples) as well as randomly sampled pairs (negative examples):

$$\sum_{v \in W} \sum_{u \in W} n_{uv} \log \sigma \left(\sum_t \phi_{ut} \theta_{tv} \right) + k \, \mathbb{E}_{\bar{v}} \log \sigma \left(-\sum_t \phi_{ut} \theta_{tv} \right) \rightarrow \max_{\Phi, \Theta}, \tag{3}$$

where σ is a sigmoid function, $\bar{v}$ are sampled from unigram distribution and k is a parameter to balance positive and negative examples. SGNS model can be effectively learned via Stochastic Gradient Descent.

SGNS model can be extended to learn document representations if the probabilities in (1) are conditioned on a document instead of a word. This architecture is called DBOW [7] and it is one of the modifications of the popular paragraph2vec approach.

Topic Model. Probabilistic Latent Semantic Analysis, PLSA [15] is a topic model that describes words in documents by a mixture of hidden topics:

$$p(w|d) = \sum_{t \in T} p(w|t) p(t|d) = \sum_{t \in T} \phi_{wt} \theta_{td}, \tag{4}$$

where $\Phi^{|W| \times |T|}$ contains probabilities ϕ_{wt} of words in topics and $\Theta^{|T| \times |D|}$ contains probabilities θ_{td} of topics in documents. The distributions are learned via maximization of the likelihood given normalization and non-negativity constraints:

$$\mathcal{L} = \sum_{d \in D} \sum_{w \in W} n_{wd} \log p(w|d) \rightarrow \max_{\Phi, \Theta} \tag{5}$$

$$\phi_{wt} \geq 0, \quad \sum_w \phi_{wt} = 1 \tag{6}$$

$$\theta_{td} \geq 0, \quad \sum_t \theta_{td} = 1. \tag{7}$$

This task can be effectively solved via EM-algorithm [9] or its online modification [14]. The most popular Latent Dirichlet Allocation [4] topic model extends PLSA by using Dirichlet priors for Φ and Θ distributions.

Additive Regularization of Topic Models, ARTM [43] is a non-Bayesian framework for learning multiobjective topic models. The optimization task (5) is extended with n additive regularizers $R_i(\varPhi, \varTheta)$ that are balanced with τ_i coefficients:

$$\mathcal{L} + R \to \max_{\varPhi, \varTheta}; \quad R = \sum_{i=1}^{n} \tau_i R_i(\varPhi, \varTheta) \tag{8}$$

This approach addresses the problem of the non-uniqueness of the likelihood maximization (5) solution and imposes additional criteria to choose $\varPhi$ and $\varTheta$. The optimization is still done with online EM-algorithm, where M-step is modified to use the derivatives of the regularization terms [43].

3   Probabilistic Word Embeddings

Consider a modification of PLSA to predict the word u in a local context of the word v:

$$p(u|v) = \sum_{t \in T} p(u|t)p(t|v) = \sum_{t \in T} \phi_{ut}\theta_{tv} \tag{9}$$

In this formulation the topic model approximates a word co-occurrence matrix instead of a word-document matrix. Unlike in PLSA, $\varTheta^{|T| \times |W|}$ contains probabilities θ_{tv} of topics for *words*. However, from the topic modeling perspective, those words can be treated as *pseudo-documents*. One may think of a pseudo-document *derived by a word* v as a concatenation of all local contexts for all occurrences of the word v in the corpus. A local context is still defined as a fixed-size window, but this definition can be easily extended to use syntactic patterns, sentences, or any other structure.

Interestingly, this approach appears to be extremely similar to Skip-Gram model (1). Both models predict the same probabilities $p(u|v)$ and make use of the observed data by optimizing exactly the same likelihood (2). Both models are parametrized with matrices of hidden representations of words. The only difference is the space of the parameters: while Skip-Gram has no constraints, the topic model learns non-negative and normalized vectors that have a probabilistic interpretation. As a benefit, word probabilities can be predicted with a mixture model of the parameters with no need in explicit *softmax* normalization.

Learning probabilistic word embeddings (PWE) can be treated as a stochastic matrix factorization of probabilities $p(u|v)$ estimated from a corpus. This makes a perfect analogy with matrix factorization formulations of SGNS [19], GloVe, NNSE, and other similar techniques. GloVe uses a squared loss with a weighting function $f(n_{uv})$ that penalizes too frequent co-occurrences. Apart from two real-valued matrices of parameters, it introduces bias terms b_u and $\tilde{b}_v$. NNSE also uses a squared loss, but imposes additional constraints to obtain sparse non-negative embeddings ϕ_u and guarantees the limited $l2$-norm for $\varTheta$ rows, which are called *dictionary* entries.

We summarize the connections between all mentioned models in Table 1. Each method is decomposed into several components: the type of raw co-occurrence data $F = (f_{uv})^{W \times W}$, the matrix factorization loss, the constraints for a parameter space, and the optimization technique. From this point of view, there is no big difference between so called *count-based* and *predictive* approaches. On the one hand, each method counts f_{uv} values (probably implicitly) and performs dimensionality reduction by a matrix factorization. On the other hand, each matrix factorization objective can be treated as a loss, which is used to train the model from data. More importantly, the unified view provides a powerful tool to analyze a diverse set of existing models and exchange components across them.

Table 1. Learning word embeddings with a low-rank matrix factorization.

PWE	Data type	$F_{uv} = \frac{n_{uv}}{n_v} = \hat{p}(u|v)$
	Objective	$\sum_{v \in W} n_v \, \mathrm{KL} \left(\hat{p}(u|v) \,\middle\|\, \langle \phi_u \theta_v \rangle \right) \to \min_{\Phi, \Theta}$
	Constrains	$\phi_{ut} > 0, \quad \sum_u \phi_{ut} = 1; \quad \theta_{tv} > 0, \quad \sum_t \theta_{tv} = 1$
	Technique	EM-algorithm (online by F columns)
SGNS	Data type	$F_{uv} = \log \frac{n_{uv} n}{n_u n_v} - \log k$
	Objective	$\sum_{u \in W} \sum_{v \in W} n_{uv} \log \sigma \left(\langle \phi_u \theta_v \rangle \right) + k \, \mathbb{E}_{\bar{v}} \log \sigma \left(- \langle \phi_u \theta_v \rangle \right) \to \max_{\Phi, \Theta}$
	Constrains	No constraints
	Technique	SGD (online by corpus)
GloVe	Data type	$F_{uv} = \log n_{uv}$
	Objective	$\sum_{v \in W} \sum_{u \in W} f(n_{uv}) \left(\langle \phi_u \theta_v \rangle + b_u + \tilde{b}_v - \log n_{uv} \right)^2 \to \min_{\Phi, \Theta, b, \tilde{b}}$
	Constrains	No constraints
	Technique	AdaGrad (online by F elements)
NNSE	Data type	$F_{uv} = max(0, \log \frac{n_{uv} n}{n_u n_v})$ or SVD low-rank approximation
	Objective	$\sum_{u \in W} \left(\| f_u - \phi_u \Theta \|^2 + \| \phi_u \|_1 \right) \to \min_{\Phi, \Theta}$
	Constrains	$\phi_{ut} \geq 0, \forall u \in W, t \in T \quad \theta_t \theta_t^T \leq 1, \forall t \in T$
	Technique	Online algorithm from [25]

4 Additive Regularization and Embeddings for Multiple Modalities

The proposed probabilistic embeddings can be easily extended as a topic model. First, there is a natural way to learn document embeddings. Second, additive regularization of topic models [43] can be used to meet further requirements. In this paper we employ it to obtain a high sparsity with no reduction in the accuracy of matrix factorization. The regularization criteria is a sum of cross-entropy terms between the target and fixed distributions:

$$R = -\tau \sum_{t \in T} \sum_{u \in W} \beta_u \ln \phi_{ut} \tag{10}$$

where β_u can be set to the uniform distribution.

Furthermore, we extend the topic model to incorporate meta-data or *modalites*, such as timestamps, categories, authors, etc. Real data often has such type of information associated with each document and it is desirable to build representations for these additional tokens as well as for the usual words.

Recall that each *pseudo-document* v in our training data is formed by collecting words u that co-occur with word v within a sliding window. Now we enrich it by the tokens u of some additional modality m that co-occur with the word v within a document. The only difference here is in using *global* document-based co-occurrences for additional modalities as opposed to *local* window-based co-occurrences for the modality of words. Once the *pseudo-documents* are prepared, we employ Multi-ARTM approach [42] to learn topic vectors for tokens of each modality:

$$\sum_{m\in M} \lambda_m \underbrace{\sum_{v\in W^0} \sum_{u\in W^m} n_{uv} \ln p(u|v)}_{\text{modality log-likelihood } \mathcal{L}_m(\Phi,\Theta)} \to \max_{\Phi,\Theta}, \tag{11}$$

$$\phi_{ut} \geq 0, \quad \sum_{u\in W^m} \phi_{ut} = 1, \forall m \in M; \tag{12}$$

$$\theta_{tv} \geq 0, \quad \sum_{t\in T} \theta_{tv} = 1. \tag{13}$$

where $\lambda_m > 0$ are *modality weights*, W^m are modality vocabularies, and $m = 0$ for the basic text modality. Optionally, the tokens of other modalities can also form pseudo-documents and this would restore the symmetric property of the factorized matrix. Regularizers can be still added to the multimodal optimization criteria.

Online EM-algorithm. Regularized multimodal likelihood maximization is performed with online EM-algorithm implemented in BigARTM library [41]. First, we compute all necessary co-occurrences and build the *pseudo-documents* as described before. We store this corpus on disk and process it by batches of $B = 100$ pseudo-documents. The algorithm starts with random initialization of Φ and Θ matrices. The E-step estimates posterior topic distributions $p(t|u,v)$ for words u in a pseudo-document v. These updates are alternating with θ_{tv} updates for the given pseudo-document. After a fixed number of iterations through the pseudo-document, θ_{tv} are thrown away, while $p(t|u,v)$ are used to compute incremental unnormalized updates for ϕ_{ut}. These updates are applied altogether when the whole batch of pseudo-documents is processed. Importantly, these procedure does not overwrite the previous value of Φ, but slowly forgets it with an exponential moving average. The detailed formulas for the case of usual documents can be found in [41]. Note that the only matrix which has to be always stored in RAM is Φ. The number of epochs (runs through the whole corpus) in our experiments ranges from 1 to 6.

Table 2. Spearman correlation for word similarities on Wikipedia.

Model	Data	Optimization	Metric	WordSim Sim.	WordSim Rel.	WordSim	Bruni MEN	SimLex-999
LDA	n_{wd}	Online EM	hel	0.530	0.455	0.474	0.583	0.220
PWE	n_{uv}	Offline EM	dot	0.709	0.635	0.654	0.658	0.240
PWE	pPMI	Offline EM	dot	0.701	0.615	0.647	0.707	0.276
PWE	n_{uv}	Online EM	dot	0.718	**0.673**	**0.685**	0.669	0.263
SGNS	sPMI	SGD	cos	**0.752**	0.632	0.666	**0.745**	**0.384**

5   Experiments

We conduct experiments on three different datasets. Firstly, we compare the proposed Probabilistic Word Embeddings (PWE) to SGNS on Wikipedia dump by word similarities and interpretability of the components. Secondly, we learn probabilistic document embeddings on ArXiv papers and compare them to DBOW on the document similarity task [7]. Finally, we learn embeddings for multiple modalities on a corpus of Russian news Lenta.ru and investigate inter-modality similarities. All topic models are learnt in BigARTM[1] open source library [41] using Python interface[2]. SGNS is taken from Hyperwords[3] package and DBOW is taken from Gensim[4] library.

Word Similarity Tasks. We use Wikipedia 2016-01-13 dump and preprocess it with Levy's scripts[2] to guarantee equal conditions for SGNS and topic modeling [21]. We delete top 25 stop-words from the vocabulary, keep the next 100000 words, and delete the word pairs that co-occur less than 5 times. We performed experiments for *windows of size 2, 5, and* 10, but report here only window-5 results, as the others are analogous. We use *subsampling* with the constant 10^{-5} for all models. While common for SGNS, subsampling has never been used for topic modeling. However, our experiments show that it slightly improves topic interpretability by filtering out too general terms and therefore might be a good preprocessing recommendation. Also, we tried using *dynamic* window, which is a weighting technique based on the distance of the co-occurred words, but we didn't find it beneficial.

Following a traditional benchmark for word similarity tasks, we rank word pairs according to our models and measure Spearman correlation with the human ratings from WordSim353 dataset [10] partitioned into WordSim Similarity and WordSim Relatedness [1], MEN dataset [5], and SimLex-999 [13]. We consider SGNS model as a baseline and investigate if probabilistic word embeddings (PWE) are capable of providing the comparable quality. We start with LDA and Hellinger distance for word vectors as this is the default choice from many

---

[1] bigartm.org.
[2] github.com/bigartm/bigartm-book/blob/master/applications/word_embeddings. ipynb.
[3] bitbucket.org/omerlevy/hyperwords.
[4] radimrehurek.com/gensim/.

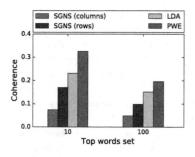

Fig. 1. Coherence scores.

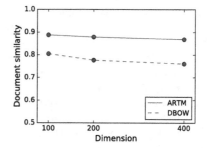

Fig. 2. Document similarities.

papers, e.g. [30]. Table 2 shows that SGNS dramatically outperforms LDA. Our further experiments demonstrate how to make topic models work.

First, we get an improvement by modeling the word-word matrix instead of the word-document matrix. Second, we investigate how to compute word similarity in the obtained space of probabilistic embeddings. We find the topic distributions should be normalized using Bayes' rule $p(t|u) = \frac{\phi_{ut} p(t)}{\sum_t \phi_{ut} p(t)}$ and that dot-product performs better than Hellinger distance or cosine similarity. Third, we find that online EM-algorithm with incremental Φ updates performs better than its offline analogue, where Φ is overwritten once per epoch. We also find that it is beneficial to initialize Θ randomly each time rather than store the values from the previous epoch. This combination of tricks gives the accuracy comparable to SGNS.

To obtain *sparsity*, we add the regularizer at the last iterations of EM-algorithm and observe **93%** of zeros in word embeddings *with the same* performance on word similarity tasks. We also try different co-occurrence scores instead of raw counts such as $\log n_{uv}$ to penalize frequent co-occurrences or normalized $\frac{n_{uv}}{\sum_u n_{uv}}$ values to obtain a sum of *non-weighted* KL-divergences in the optimization criteria. While most of these weighting schemes give worse results, positive PMI values appear to be beneficial for some testsets.

Interpretability of Embedding Components. We characterize each component by a set of words with the highest values in the embedding matrix and check if those sets correspond to some aspects that can be named by a human. *Word intrusion* [6] technique is based on the idea that for well formed sets, a human expert can easily detect an intruder, randomly sampled from the vocabulary. This technique has been widely used in topic modeling and also for Non-Negative Sparse Embeddings [30] and Online Interpretable Word Embeddings [24]. Word intrusion requires experts, but it can be automated by the *coherence score*, which is shown to have high correlations with human judgements [32]. It averages pairwise similarities across the set of words. For similarities one can use PMI scores from an external corpus [31], log-conditional probabilities from the same corpus [29], distributional similarities [2], or other variants [36].

Table 3. Interpretability of topics.

PWE		SGNS	
art	arbitration	transports	rana
painting	ban	recon	walnut
museum	requests	grumman	rashid
painters	arbitrators	convoys	malek
gallery	noticeboard	piloted	aziz
sculpture	block	stealth	khalid
painter	administrators	flotilla	yemeni
exhibition	arbcom	convoy	andalusian
portraits	sanctions	supersonic	bien
drawings	mediation	bomber	gcc

Table 4. Event timestamps.

2015-12-18 SW release	2016-02-29 The Oscars	2015-05-09 Victory Day
jedi	statuette	great
sith	award	anniversary
fett	nomination	normandy
anakin	linklater	parade
chewbacca	oscar	demonstration
film series	birdman	vladimir
hamill	win	celebration
prequel	criticism	concentration
awaken	director	auschwitz
boyega	lubezki	photograph

In our experiments we use the PMI-based coherence for top-10 and top-100 words for each component. The score is averaged over the components and reported in Fig. 1. For SGNS we consider two different schemes of ranking words within each component. First, using the raw values; second, applying softmax *by rows* and using Bayes' rule to convert $p(t|w)$ into $p(w|t)$ probabilities. We show that the coherence for probabilistic word embeddings is consistently higher than that of LDA or SGNS for a range of embedding sizes. Also, this result is confirmed by visual analysis of the obtained components (see Table 3 for the examples).

Table 5. Spearman correlation for word similarities on Lenta.ru.

Model	WordSim Sim	WordSim Rel	MC	RG	HJ	SimLex
SGNS	0.630	0.530	0.377	0.415	0.567	**0.243**
CBOW	0.625	0.513	0.403	0.370	0.551	0.170
PWE	0.649	0.565	0.605	**0.594**	0.604	0.123
Multi-PWE	**0.682**	**0.58**	**0.607**	0.584	**0.611**	0.144

Document similarity task. In this experiment we learn probabilistic document embeddings on ArXiv corpus and test them on a document similarity task. The testset released by Dai et al. [7] contains automatically generated triplets of a query paper, a similar paper that shares key words, and a dis-similar paper that does not share any key words. The quality is evaluated by the accuracy of identifying the similar one within each triplet. We preprocess[5] plain texts of 963564 ArXiv papers with a total of 1416554733 tokens and reduce the vocabulary size to 122596 words with a frequency-based filtering. The restored mapping between the plain texts and the URLs from the testset[6] covers 15853 triplets out of 20000.

---

[5] https://github.com/romovpa/arxiv-dataset.
[6] http://cs.stanford.edu/~quocle/triplets-data.tar.gz.

We train embeddings with 1 epoch of online EM-algorithm. Note that the matrix Θ is not stored, so memory consumption does not grow linearly with the number of documents. Afterwards, we infer test embeddings with 10 passes on each document. As a baseline, we train DBOW [7] with 15 epochs and use linear decay of learning rate from 0.025 to 0.001; afterwards we infer test embeddings with 5 epochs. Unlike online EM-algorithm, DBOW needs in-memory storage of document vectors and also takes much longer to train (several hours instead of 30 min on the same machine). We do not facilitate training word vectors in DBOW, because it slows down the process dramatically.

Figure 2 shows that our ARTM model consistently outperforms DBOW for a range of embedding sizes. The absolute numbers are also better than for all other methods reported in [7], thus giving a new state-of-the-art on this dataset.

Multimodal Embedding Similarities. The experiments are held on Russian *lenta.ru* corpus, that contains 100033 news with a total of 10050714 tokens. The corpus has additional modalities of timestamps (825 unique tokens), categories (22 unique tokens) and sub-categories (97 unique tokens). The basic text modality has 54963 unique words.

We produce a collection of pseudo-documents using the window of size 5 and subsampling. For evaluation we use HJ testset [34] with human judgments on 398 word pairs translated to Russian from the widely used English testsets: MC [28], RG [37], and WordSim353 [10]. We also use SimLex-999 testset translation [18].

Table 5 shows that probabilistic word embeddings (PWE) outperform SGNS for most of the testsets even without using additional modalities. One can note that this corpus is relatively small and it might be a reason for poor SGNS performance. We have also tried CBOW [27] following a common recommendation to use it for small data, but it performed even worse. Generally, we observe that topic modeling requires less data for a good performance, thus the proposed PWE approach might be beneficial for applications with limited data.

Next, we use additional modalities and optimize the modality weights in the objective (11). With this approach we observe a further boost in the performance for the word similarity task (see Multi-PWE in Table 5). Finally, we experiment with two different modes: using modalities only as tokens (a non-symmetric case) and both as tokens and pseudo-documents (a symmetric case). While word similarities are better for the non-symmetric case, we observe better inter-modality similarities for the symmetric case. Table 4 provides several examples of remarkable timestamps and their closest words. The words are manually translated from Russian to English for reporting purposes only. Each column is easily interpretable as a coherent event, namely the release of Star Wars, the Oscars 2016, and Victory Day in Russia.

6   Conclusions

In this work we revisited topic modelling techniques in the context of learning hidden representations for words and documents. Topic models are known to provide interpretable components but perform poorly on word similarity tasks. However, we have shown that topic models and neural word embeddings can be made to predict the same probabilities with the only difference in the probabilistic nature of parameters. This theoretical insight enabled us to merge the models and get practical results. First, we obtained probabilistic word embeddings (PWE) that work on par with SGNS on word similarity tasks, but have high sparsity and interpretability of the components. Second, we learned document embeddings that outperform DBOW on a document similarity task and require less memory and time for training. Furthermore, considering the task as a topic modeling, enabled us to adapt Multi-ARTM approach and learn embeddings for multiple modalities, such as timestamps and categories. We observed meaningful inter-modality similarities and a boost of the quality on the basic word similarity task. In future we plan to apply the proposed probabilistic embeddings to a suite of NLP tasks and take even more advantage of the additive regularization to incorporate task-specific requirements into the models.

Acknowledgements. The work was supported by Government of the Russian Federation (agreement 05.Y09.21.0018) and the Russian Foundation for Basic Research grants 17-07-01536, 16-37-00498.

References

1. Agirre, E., Alfonseca, E., Hall, K., Kravalova, J., Paşca, M., Soroa, A.: A study on similarity and relatedness using distributional and wordnet-based approaches. In: Proceedings of Human Language Technologies: The 2009 Annual Conference of the North American Chapter of the Association for Computational Linguistics, NAACL 2009, pp. 19–27. Association for Computational Linguistics, Stroudsburg (2009)
2. Aletras, N., Stevenson, M.: Evaluating topic coherence using distributional semantics. In: IWCS (2013)
3. Bengio, Y., Ducharme, R., Vincent, P., Janvin, C.: A neural probabilistic language model. J. Mach. Learn. Res. **3**, 1137–1155 (2003)
4. Blei, D.M., Ng, A.Y., Jordan, M.I.: Latent dirichlet allocation. J. Mach. Learn. Res. **3**, 993–1022 (2003)
5. Bruni, E., Boleda, G., Baroni, M., Tran, N.K.: Distributional semantics in technicolor. In: Proceedings of the 50th Annual Meeting of the Association for Computational Linguistics: Long Papers, ACL 2012, vol. 1, pp. 136–145. Association for Computational Linguistics, Stroudsburg (2012)
6. Chang, J., Boyd-Graber, J., Wang, C., Gerrish, S., Blei, D.M.: Reading tea leaves: how humans interpret topic models. In: Neural Information Processing Systems (2009)
7. Dai, A.M., Olah, C., Le, Q.V.: Document embedding with paragraph vectors. CoRR abs/1507.07998 (2015)
8. Das, R., Zaheer, M., Dyer, C.: Gaussian LDA for topic models with word embeddings. In: ACL (1), pp. 795–804. The Association for Computer Linguistics (2015)
9. Deerwester, S., Dumais, S.T., Furnas, G.W., Landauer, T.K., Harshman, R.: Indexing by latent semantic analysis. J. Am. Soc. Inf. Sci. **41**, 391–407 (1990)

10. Finkelstein, L., Gabrilovich, E., Matias, Y., Rivlin, E., Solan, Z., Wolfman, G., Ruppin, E.: Placing search in context: the concept revisited. ACM Trans. Inf. Syst. **20**(1), 116–131 (2002)
11. Gentner, D.: Structure-mapping: a theoretical framework for analogy. Cogn. Sci. **7**(2), 155–170 (1983)
12. Harris, Z.: Distributional structure. Word **10**(23), 146–162 (1954)
13. Hill, F., Reichart, R., Korhonen, A.: Simlex-999: evaluating semantic models with genuine similarity estimation. Comput. Linguist. **41**(4), 665–695 (2015)
14. Hoffman, M.D., Blei, D.M., Bach, F.R.: Online learning for latent dirichlet allocation. In: Lafferty, J.D., Williams, C.K.I., Shawe-Taylor, J., Zemel, R.S., Culotta, A. (eds.) NIPS, pp. 856–864. Curran Associates, Inc. (2010)
15. Hofmann, T.: Probabilistic latent semantic analysis. In: Proceedings of the Fifteenth Conference on Uncertainty in Artificial Intelligence, UAI 1999, pp. 289–296. Morgan Kaufmann Publishers Inc., San Francisco (1999)
16. Kiros, R., Zhu, Y., Salakhutdinov, R., Zemel, R.S., Torralba, A., Urtasun, R., Fidler, S.: Skip-thought vectors. In: Proceedings of the 28th International Conference on Neural Information Processing Systems, NIPS 2015, pp. 3294–3302. MIT Press, Cambridge (2015)
17. Le, Q.V., Mikolov, T.: Distributed representations of sentences and documents. CoRR abs/1405.4053 (2014)
18. Leviant, I., Reichart, R.: Judgment language matters: towards judgment language informed vector space modeling. arXiv (arXiv:1508.00106) (2015)
19. Levy, O., Goldberg, Y.: Linguistic regularities in sparse and explicit word representations. In: Morante, R., Yih, W. (eds.) CoNLL, pp. 171–180. ACL (2014)
20. Levy, O., Goldberg, Y.: Neural word embedding as implicit matrix factorization. In: Ghahramani, Z., Welling, M., Cortes, C., Lawrence, N.D., Weinberger, K.Q. (eds.) Advances in Neural Information Processing Systems, vol. 27, pp. 2177–2185. Curran Associates, Inc. (2014)
21. Levy, O., Goldberg, Y., Dagan, I.: Improving distributional similarity with lessons learned from word embeddings. TACL **3**, 211–225 (2015)
22. Liu, Y., Liu, Z., Chua, T.S., Sun, M.: Topical word embeddings. In: AAAI, pp. 2418–2424 (2015)
23. Luo, H., Liu, Z., Luan, H.B., Sun, M.: Online learning of interpretable word embeddings. In: Màrquez, L., Callison-Burch, C., Su, J., Pighin, D., Marton, Y. (eds.) EMNLP, pp. 1687–1692. The Association for Computational Linguistics (2015)
24. Luo, H., Liu, Z., Luan, H.B., Sun, M.: Online learning of interpretable word embeddings. In: EMNLP (2015)
25. Mairal, J., Bach, F., Ponce, J., Sapiro, G.: Online learning for matrix factorization and sparse coding. J. Mach. Learn. Res. **11**, 19–60 (2010)
26. Baroni, M., Dinu, G., Kruszewski, G.: Don't count, predict! A systematic comparison of context-counting vs. context-predicting semantic vectors. In: Proceedings of the Conference 52nd Annual Meeting of the Association for Computational Linguistics, ACL 2014, vol. 1, pp. 238–247 (2014)
27. Mikolov, T., Sutskever, I., Chen, K., Corrado, G.S., Dean, J.: Distributed representations of words and phrases and their compositionality. In: Burges, C.J.C., Bottou, L., Ghahramani, Z., Weinberger, K.Q. (eds.) NIPS, pp. 3111–3119 (2013)
28. Miller, G.A., Charles, W.G.: Contextual correlates of semantic similarity. Lang. Cogn. Process. **6**(1), 1–28 (1991)
29. Mimno, D., Wallach, H.M., Talley, E., Leenders, M., McCallum, A.: Optimizing semantic coherence in topic models. In: Proceedings of the Conference on Empirical Methods in Natural Language Processing, EMNLP 2011, pp. 262–272. Association for Computational Linguistics, Stroudsburg (2011)

30. Murphy, B., Talukdar, P.P., Mitchell, T.M.: Learning effective and interpretable semantic models using non-negative sparse embedding. In: Kay, M., Boitet, C. (eds.) COLING, pp. 1933–1950. Indian Institute of Technology Bombay (2012)

31. Newman, D., Bonilla, E.V., Buntine, W.L.: Improving topic coherence with regularized topic models. In: NIPS (2011)

32. Newman, D., Lau, J.H., Grieser, K., Baldwin, T.: Automatic evaluation of topic coherence. In: Human Language Technologies: The 2010 Annual Conference of the North American Chapter of the Association for Computational Linguistics, HLT 2010, pp. 100–108. Association for Computational Linguistics, Stroudsburg (2010)

33. Nguyen, Q.D., Billingsley, R., Du, L., Johnson, M.: Improving topic models with latent feature word representations. Trans. Assoc. Comput. Linguist. **3**, 299–313 (2015)

34. Panchenko, A., Ustalov, D., Arefyev, N., Paperno, D., Konstantinova, N., Loukachevitch, N., Biemann, C.: Human and machine judgements for Russian semantic relatedness. In: Ignatov, D.I., Khachay, M.Y., Labunets, V.G., Loukachevitch, N., Nikolenko, S.I., Panchenko, A., Savchenko, A.V., Vorontsov, K. (eds.) AIST 2016. CCIS, vol. 661, pp. 221–235. Springer, Cham (2017)

35. Pennington, J., Socher, R., Manning, C.D.: Glove: global vectors for word representation. In: EMNLP, vol. 14, pp. 1532–1543 (2014)

36. Röder, M., Both, A., Hinneburg, A.: Exploring the space of topic coherence measures. In: Proceedings of the Eighth ACM International Conference on Web Search and Data Mining, WSDM 2015, pp. 399–408. ACM, New York (2015)

37. Rubenstein, H., Goodenough, J.B.: Contextual correlates of synonymy. Commun. ACM **8**(10), 627–633 (1965)

38. Sridhar, V.K.R.: Unsupervised topic modeling for short texts using distributed representations of words. In: Blunsom, P., Cohen, S.B., Dhillon, P.S., Liang, P. (eds.) VS@HLT-NAACL, pp. 192–200. The Association for Computational Linguistics (2015)

39. Sun, F., Guo, J., Lan, Y., Xu, J., Cheng, X.: Sparse word embeddings using 1 regularized online learning. In: Kambhampati, S. (ed.) IJCAI, pp. 2915–2921. IJCAI/AAAI Press (2016)

40. Turney, P.D., Pantel, P.: From frequency to meaning: vector space models of semantics. J. Artif. Intell. Res. **2010**(37), 141–188 (2010)

41. Vorontsov, K., Frei, O., Apishev, M., Romov, P., Dudarenko, M.: BigARTM: open source library for regularized multimodal topic modeling of large collections. In: Khachay, M.Y., Konstantinova, N., Panchenko, A., Ignatov, D.I., Labunets, V.G. (eds.) AIST 2015. CCIS, vol. 542, pp. 370–381. Springer, Cham (2015)

42. Vorontsov, K., Frei, O., Apishev, M., Romov, P., Suvorova, M., Yanina, A.: Non-Bayesian additive regularization for multimodal topic modeling of large collections. In: Aletras, N., Lau, J.H., Baldwin, T., Stevenson, M. (eds.) TM@CIKM, pp. 29–37. ACM (2015)

43. Vorontsov, K., Potapenko, A.: Additive regularization of topic models. Mach. Learn. **101**(1), 303–323 (2015)

44. Yan, X., Guo, J., Lan, Y., Cheng, X.: A biterm topic model for short texts. In: Schwabe, D., Almeida, V.A.F., Glaser, H., Baeza-Yates, R.A., Moon, S.B. (eds.) WWW, pp. 1445–1456. International World Wide Web Conferences Steering Committee/ACM (2013)

45. Zuo, Y., Zhao, J., Xu, K.: Word network topic model: a simple but general solution for short and imbalanced texts. Knowl. Inf. Syst. **48**(2), 379–398 (2016)

Multi-objective Topic Modeling for Exploratory Search in Tech News

Anastasia Ianina[1(✉)], Lev Golitsyn[2], and Konstantin Vorontsov[1]

[1] Moscow Institute of Physics and Technology, Moscow, Russia
yanina@phystech.edu, vokov@forecsys.ru
[2] Integrated Systems, Moscow, Russia
lvgolitsyn@gmail.ru

Abstract. Exploratory search is a paradigm of information retrieval, in which the user's intention is to learn the subject domain better. To do this the user repeats "query–browse–refine" interactions with the search engine many times. We consider typical exploratory search tasks formulated by long text queries. People usually solve such a task in about half an hour and find dozens of documents using conventional search facilities iteratively. The goal of this paper is to reduce the time-consuming multi-step process to one step without impairing the quality of the search. Probabilistic topic modeling is a suitable text mining technique to retrieve documents, which are semantically relevant to a long text query. We use the additive regularization of topic models (ARTM) to build a model that meets multiple objectives. The model should have sparse, diverse and interpretable topics. Also, it should incorporate metadata and multimodal data such as n-grams, authors, tags and categories. Balancing the regularization criteria is an important issue for ARTM. We tackle this problem with coordinate-wise optimization technique, which chooses the regularization trajectory automatically. We use the parallel online implementation of ARTM from the open source library BigARTM. Our evaluation technique is based on crowdsourcing and includes two tasks for assessors: the manual exploratory search and the explicit relevance feedback. Experiments on two popular tech news media show that our topic-based exploratory search outperforms assessors as well as simple baselines, achieving precision and recall of about 85–92%.

Keywords: Information retrieval · Exploratory search
Relevance feedback · Topic modeling
Additive regularization for topic modeling · ARTM · BigARTM

1 Introduction

Exploratory search is a relatively new paradigm in information retrieval. It aims to satisfy advanced information needs of people for education, self-education, knowledge acquisition and discovery [11,21]. Potential users of exploratory search are students, teachers, researchers and professionals. In knowledge society, the

© Springer International Publishing AG 2018
A. Filchenkov et al. (Eds.): AINL 2017, CCIS 789, pp. 181–193, 2018.
https://doi.org/10.1007/978-3-319-71746-3_16

information needs of users increase constantly and become more and more complicated. This leads to the emergence of new search paradigms and tools.

In exploratory search, the user may not be familiar with the terminology and may assume that there are many correct answers. The user's search intent may be just learning the basics of the subject domain and defining the most important topics within it. In such cases it is difficult or even impossible to formulate an exact short query. The user of a conventional search system has to enter many queries iteratively, gradually learning the terminology and refining his or her knowledge and intentions. The iterative "query–browse–refine" process [21] may require a lot of time and experience. The alternative way is to indicate a broad search direction by a long text query, such as a whole document, a set of copy-pasted text fragments, or a document folder, and give the user a set of semantically similar documents. There are two obstacles along this way. The first one is in elaborating a semantic similarity measure appropriate for the purposes of exploratory search. The second one is in evaluating both precision and recall, which is a difficult task for human assessors. In order to address these challenges, we propose a topic-based approach to exploratory search and a three-stage model evaluation and selection technique based on crowdsourcing.

Topic modeling is often used for searching semantically similar documents [1,20,22] and has become more popular in exploratory search community in recent years [8,12,13,15]. The *probabilistic topic model* reveals the latent thematic structure of a text collection. It determines each topic as a discrete probability distribution over words and then represents each document by a discrete probability distribution over topics [5,6,9]. The conventional full text search is usually based on the inverted index and looks for documents, which contain all the words from the query [10]. So, if the query is long, it's most likely that nothing will be found. Topic-based search overcomes this problem by using compact topic vector representations for the query and documents instead of their bag-of-words representations. This way, one can use the same mechanisms of indexing and ranking for searching topically similar documents, it's just that now topics take the place of words.

To be used in the exploratory search system, the topic model has to meet multiple requirements. Topics should be significantly different and well interpretable to capture semantics appropriately. Vector representations of documents should be highly sparse to make the inverted index as compressed as possible. The model should take into account the modalities of authors, time stamps, categories, tags, named entities etc. to get the most out of the available meta-information. We use a multi-objective approach called *additive regularization of topic models* (ARTM) [17] to satisfy all these requirements. ARTM learns models with desired properties by maximizing a weighted sum of the log-likelihood and additional regularization criteria. We use an effective parallel implementation of the expectation-maximization (EM) algorithm from open source project BigARTM.org [7]. Our experiments show that the combination of the above requirements in a form of regularization criteria significantly improves not only the model itself, but also precision and recall of the exploratory search.

Two popular tech news media are used for the evaluation: techcrunch.com in English and habrahabr.ru in Russian. Our evaluation technique consists of three stages. At the first stage we ask assessors to find the documents relevant to the long-text queries using any search utilities of their choice. At the second stage we ask assessors to give *explicit relevance feedback* [4] for the topic-based search results on the same queries. At the third stage we join for each query all sets of relevant documents found at the previous stages. These enriched assessor data enables us to estimate precision and recall for new models. In addition, we get the opportunity to compare and select models without asking assessors.

Assessors spend about 30 min on average per a query. For this reason we afford to collect a limited amount of assessor data sufficient for model validation and selection. Learning the supervised topic model would require much more assessor data. However, this is not necessary, since the multi-objective unsupervised topic model already provides a high quality exploratory search.

The paper is organized as follows. In Sect. 2 we introduce the ARTM framework and describe the strategy of choosing regularization coefficients. In Sect. 3 we describe the evaluation technique for the topic-based exploratory search. In Sect. 4 we reports the experimental results of comparing topic-based search with baselines. In Sect. 5 we use assessor data for model selection. In Sect. 6 we conclude that topic-based exploratory search is much faster than assessors' iterative search, having better recall and comparable precision.

2   Probabilistic Topic Modeling and Additive Regularization

Let us denote a finite set (collection) of texts by D, a finite set of topics by T, and a finite set of modalities by M. Here are some examples of modalities: words, bigrams, tags, categories, authors, etc. Each modality $m \in M$ has a finite set (dictionary) of tokens W_m. Each document $d \in D$ is a sequence of n_d tokens from $W = \bigcup_w W_m$. We accept the bag-of-words hypothesis and take into account how many times n_{dw} the token w appears in the document d.

Given the $(n_{dw})_{D \times W_m}$ matrix, a probabilistic topic model finds its approximate matrix factorization by $\Phi_m = (\phi^m_{wt})_{W_m \times T}$ matrix of token probabilities for the topics and $\Theta = (\theta_{td})_{T \times D}$ matrix of topic probabilities for the documents:

$$\frac{n_{dw}}{n_d} \approx p(w \mid d) = \sum_{t \in T} p(w \mid t)\, p(t \mid d) = \sum_{t \in T} \phi_{wt} \theta_{td},$$

where $|T|$ is a user-defined number of topics in the model.

Usually, the problem of matrix factorization has infinitely many solutions. Additive regularization [17,19] narrows the set of solutions by maximizing the weighted sum of modality log-likelihoods and regularizers $R_i(\Phi, \Theta)$:

$$\sum_m \tau_m \sum_{d \in D} \sum_{w \in W_m} n_{dw} \ln \sum_{t \in T} \phi_{wt} \theta_{td} + \sum_{i=1}^{r} \tau_i R_i(\Phi, \Theta) \to \max_{\Phi, \Theta}$$

under non-negativity and normalization constraints for all columns of Φ_m and Θ matrixes. This optimization problem can be solved using the EM-algorithm [17]. Many topic models can be considered as special cases of additive regularization (ARTM) with appropriate choice of regularizers [16,17]. Regularization coefficients τ_m and τ_i are usually chosen empirically.

Probabilistic Latent Semantic Analysis (PLSA) [9] corresponds to the absence of regularization, $R(\Phi, \Theta) = 0$.

Latent Dirichlet Allocation (LDA) [6] corresponds to the *smoothing* regularizer, which minimizes the cross-entropy between columns ϕ_t and a fixed distribution $\beta = (\beta_w : w \in W)$ as well as the cross-entropy between columns θ_d and a fixed distribution $\alpha = (\alpha_t : t \in T)$:

$$R(\Phi, \Theta) = \beta_0 \sum_{t \in T} \sum_{w \in W} \beta_w \ln \phi_{wt} + \alpha_0 \sum_{d \in D} \sum_{t \in T} \alpha_t \ln \theta_{td}, \tag{1}$$

where positive vectors $\beta_0 \beta$ and $\alpha_0 \alpha$ are interpreted as hyperparameters of Dirichlet prior distributions in the Bayesian topic modeling framework. Scalars β_0 and α_0 are interpreted as regularization coefficients in the ARTM framework. Choosing uniform distributions for β and α corresponds to symmetric Dirichlet priors, which are often used in experiments with the LDA model.

The *sparsing regularizer* has the same form as in (1), but differs in that the coefficients β_0 and α_0 are negative [17]. Sparsing maximizes the cross-entropy enforcing columns ϕ_t and θ_d to be as far as possible from distributions β and α respectively. This regularizer can not be interpreted in terms of Dirichlet priors.

The *decorrelation regularizer* makes topics as different as possible by minimizing the sum of covariances between topic vectors ϕ_t:

$$R(\Phi) = -\sum_{t,s \in T} \sum_{w \in W} \phi_{wt} \phi_{ws}.$$

Diversifying the term distributions of topics is known to make the resulting topics more interpretable [14]. Also, this regularizer stimulates sparsity and tends to group stop-words and common words into separate topics.

The combination of three regularizers above improves the interpretability of topics [2,3,17,18]. In our experiments we also use the combination of three regularizers: decorrelation of term distributions in topics with the coefficient τ, sparsing topic distributions in documents with the coefficient α, smoothing term distributions in topics with the coefficient β.

We subsequently add regularizers to the model following empirical recommendations from [17]: decorrelation goes first, then smoothing and sparsing. Generally, the sequential strategy enables a regularizer to prepare data for the following ones or to compensate side-effects of the previous ones. In our case, decorrelation rotates topic vectors ϕ_t to make them more distinct, Φ-smoothing compensates for the excessive sparsing after decorrelation, and Θ-sparsing nullifies insignificant probabilities when the process is close to convergence.

For each regularizer we choose its regularization coefficient from a grid of values using multiple criteria. In our experiments we use the following criteria: perplexity, Φ-sparsity, and Θ-sparsity. We perform 8 iterations of the EM-algorithm for each value of each coefficient. Thus, every model is trained along its regularization trajectory, which consists of $3 \cdot 8 = 24$ iterations. From all regularization trajectories we choose the one that yields an improvement in at least one of the criteria without a significant impairment in the others. So, our technique for tuning the regularization coefficients is a particular case of coordinate-wise optimization with grid search along each coordinate. An example of a regularization trajectory is shown in Fig. 1 for the Habrahabr collection.

The optimization of the regularization trajectory is fully automated for further model selection. In Sect. 5 it will be used for the selection of the number of topics, the set of modalities, and the semantic similarity measure.

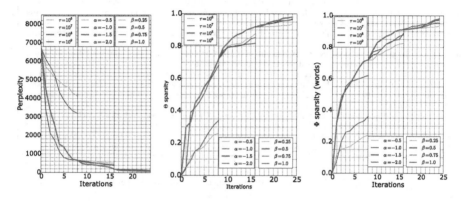

Fig. 1. Choosing regularization coefficients on Habrahabr collection. Perplexity, Θ and Φ sparsity depending on iteration count.

3   Topic-Based Exploratory Search

An exploratory search query q is a long text, so we learn its topic vector θ_q in the same way as it was done for the documents in the collection. Next, among topic vectors of documents θ_d, we find k documents closest to the query and return them as a search result.

Similarity between queries and documents can be measured using cosine similarity, Euclidean distance, Manhattan distance, Hellinger distance, Kullback–Leibler divergence, or others. In Sect. 5, we will empirically compare the search quality they yield.

For evaluating the results of topic-based exploratory search we simulate situations that analysts might encounter in practice when preparing reviews or digests of technical news. We form a set of long thematically focused text queries relevant to the collection (Fig. 2). On average, a query consists of roughly a single A4 page of text (Fig. 3). Each query is composed of fragments copy-pasted from texts both inside and outside the collection. The query should be sufficiently

3D-printers	Internet of things
AB-testing in huge IT corporations	Hadoop MapReduce
Algorithms for searching a minimal spanning tree	Healthcare devices
New Amazon Kindle products	How to write a good CV
Apple product presentations	LogService (Facebook system for storing logs)
Best-known Y Combinator projects	Main educational sources for data scientists
CERN-cluster	MIT MediaLab research
Communication within employees in large companies	Online education
Cryptosystems with public keys	Self-driving cars
Daily planners (mobile applications)	Seq2seq neural networks

Fig. 2. Examples of titles for 20 exploratory search queries

Title: SpaceX Falcon Launch

SpaceX has successfully launched a Falcon 9 to orbit during its BulgariaSat-1 mission Friday. The launch reused a first stage booster first employed during an Iridium Communications mission in January of this year, after that Falcon 9 first stage was recovered and refurbished.

Elon Musk has shared a new animation created by SpaceX to demonstrate the planned launch process for its Falcon Heavy rocket, which it hopes to test fly for the first time this coming November. The animation depicts launch of the three-booster heavy rocket, separation of the first and second stages, and the return flight and landing of the three booster cores used to get the rocket to space.

SpaceX has completed the other key ingredient of its historic flight, recovering its Falcon 9 rocket via its floating drone barge. This is a huge accomplishment because it already did this once before – with the same rocket, on the same barge, when it landed last year following a successful launch during a resupply mission to the International Space Station.

The recovery of the Falcon 9 means that not only did SpaceX reuse its rocket with this launch – it can also potentially use it again, after more stress testing and evaluation.

Its hard to underscore the significance of this milestone, but theres still ample work to do: SpaceXs goal is to eventually be able to relaunch rockets within the same day, which is obviously a feat on a different scale.

Fig. 3. An example of an exploratory search query

---

complete, so as to minimize discrepancies in its interpretation by different assessors. On the other hand, the query should be short enough for an assessor to understand its essence quickly.

For each query we ask an assessor to perform two sequential tasks.

In the first task, an assessor is asked to find within the collection as many documents relevant to the query as possible. The assessor may use any search tools available: a built-in search line, hyperlinks, tags or categories, a conventional search system such as Google, Bing, Yandex etc. This task is rather creative, usually taking a person about half an hour to complete. The time taken to process a query is recorded.

In the second task, the assessor is asked to look through the list of documents retrieved by the topic-based search for the same query and mark each document as relevant or irrelevant. Thus, we get the explicit relevance feedback for the topic-based search.

Each query is processed by 3 assessors to reduce the variance of the result and to find more relevant documents.

For each query we measure the quality by two metrics: Precision@k and Recall@k. Precision@k is the fraction of relevant documents among the first k documents found. Recall@k is the fraction of found relevant documents among all the relevant documents. We take the average Precision@k and Recall@k over all queries and over all assessors to evaluate the topic search quality.

The calculation of Recall requires knowing the set of all relevant documents for each query. We approximate this set by joining all the documents that were found by all assessors during the first task and all the documents that were found by topic-based search and confirmed by the majority of assessors as relevant during the second task. We also expanded the sets of relevant documents with the search results returned by baseline algorithms. However, this expansion has given very few relevant documents. From here we conclude that the obtained sets of relevant documents are close to being complete, and that they are suitable for comparing the search algorithms.

4 Experiments with Topic-Based Search

Datasets. The experiments were conducted on two tech news collections — TechCrunch.com in English and Habrahabr.ru in Russian. Text pre-processing included deleting punctuation, bringing the upper case letters down to the lower case and lemmatizing using the morphological analyzer pymorphy2.

The TechCrunch collection consists of 759324 articles. Articles contain tokens of four modalities: 11523 word unigrams, 1.2 mln. bigrams (the tail of rare bigrams was deleted), 605 authors and 184 categories.

The Habrahabr collection consists of 175143 articles. Articles contain tokens of six modalities: 10552 word unigrams, 742000 word bigrams, 524 authors, 10000 commentators (authors of comments to the articles), 2546 tags, 123 hubs (categories). We exclude 5 percent of the most frequent words in the collection.

Topic-based search vs. assessors. We applied the evaluation method described above to the Habrahabr and Techcrunch collections. For Habrahabr we constructed 100 queries by copying and merging fragments of text taken from sources outside Habrahabr such as other IT-oriented blogs, posts from stackoverflow.com, articles from ixbt.com, etc. The length of a query ranges from 93 to 455 words with the average of 262 words.

The experiment results for the Habrahabr collection are presented in Fig. 4. The points on the plot correspond to queries. We compare precision and recall of the search performed by the assessors with the topic-based search for the best of our models. On average, precision is a bit higher for assessors' search, whilst recall is higher for the topic-based search. The highest recall we got for the topic-based search is 1.0 for 26 queries out of 100. From the right chart in Fig. 4 it can be seen that there is no obvious dependence between the time spent by an assessor and the quality of the search. On average, it took assessors about 30 min to process a single query. The number of relevant articles ranges from 5 to 55, the average being 25.

The experiment for the TechCrunch collection is presented in Fig. 5. There were 100 queries and each of them was processed by 3 assessors. The length of the query ranges from 75 to 392 words, the average being 195 words. The average number of articles found by assessors per query is 32.

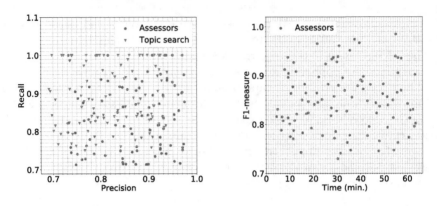

Fig. 4. The quality of assessors' and topic-based exploratory search (Habrahabr)

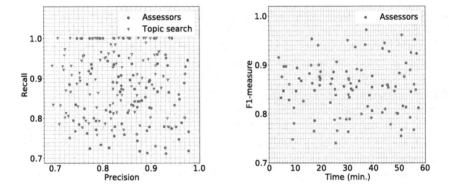

Fig. 5. The quality of assessors' and topic-based exploratory search (TechCrunch)

Thus, topic-based exploratory search obtains higher recall and produces the results significantly faster than human assessors. In some cases, topic-based search finds relevant documents that all three assessors have missed during the first task.

The significance of the difference in precision/recall between assessors' search and topic-based search was tested using the Wilcoxon signed-rank test. For all tests the p-value was less than 0.01. From here we conclude that the dataset of 100 queries is sufficient to compare the search quality.

Topic-based search vs. baselines. We use a simple but strong full-text TF-IDF search as a first baseline. We apply lemmatization to Russian texts and stemming to English texts. Then we get TF-IDF vectors from documents and queries using a simple vectorizer from the sklearn library. As a search result, we return those k documents that have TF-IDF vectors closest to the query. The TF-IDF search is a strong competitor for the topic-based search because it uses full information from word-document frequency matrix, whilst the topic-based search uses the low-rank approximation of this matrix. To make the baseline even stronger we take into consideration not only words, but also tags and categories. According to Figs. 6 and 7, topic-based search gives better results in terms of precision and recall than the TF-IDF search. This fact confirms that the topic model gives a rich semantic representation of documents and queries.

Another advantage of the topic-based search in comparison to TF-IDF search is that the low-dimensional sparse topical representation of documents can be converted into a highly compressed inverted index. Hence, an effective topic-based search engine can be implemented at low cost.

Also we introduce two additional baselines based on PLSA and LDA topic models respectively. Experiments show that they both perform worse than the ARTM-based search, see Figs. 6 and 7.

The Wilcoxon signed-rank test test has confirmed that the differences between our search and the baselines are significant: p-values were less than 0.0004 in 48 tests for Precision@k, Recall@k, $k \in \{5, 10, 15, 20\}$, all three baselines, and both collections.

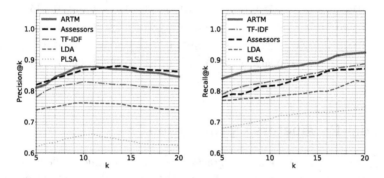

Fig. 6. Comparison of assessors' and topic-based search with regularization (ARTM) and baselines (TF-IDF, PLSA, LDA) for Habrahabr

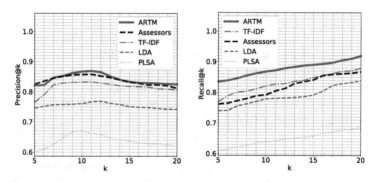

Fig. 7. Comparison of assessors' and topic-based search with regularization (ARTM) and baselines (TF-IDF, PLSA, LDA) for TechCrunch

The importance of regularizers. To show that each regularizer is important and significantly improves the search quality we carry out one more experiment. Table 1 shows that the decorrelation regularizer contributes the most to the search quality, but the other regularizers are also necessary. The model with no regularization gives the worst result.

Table 1. Topic-based search with different sets of regularizers: $\underline{D}$ecorrelation, Θ-sparsing, Φ-smoothing

	Habrahabr				TechCrunch			
	no reg	D	DΘ	D$\Theta\Phi$	no reg	D	DΘ	D$\Theta\Phi$
Pr@5	0.628	0.748	0.771	**0.810**	0.652	0.775	0.779	**0.819**
Pr@10	0.653	0.776	0.812	**0.879**	0.679	0.787	0.819	**0.867**
Pr@15	0.642	0.765	0.792	**0.868**	0.669	0.773	0.798	**0.833**
Pr@20	0.643	0.759	0.783	**0.847**	0.673	0.777	0.792	**0.825**
R@5	0.692	0.784	0.805	**0.840**	0.673	0.812	0.812	**0.835**
R@10	0.714	0.814	0.834	**0.870**	0.685	0.821	0.845	**0.868**
R@15	0.725	0.835	0.867	**0.891**	0.712	0.859	0.869	**0.890**
R@20	0.735	0.862	0.891	**0.925**	0.723	0.882	0.895	**0.919**

5   Model Parameters Optimization

Sets of relevant documents found by assessors for every query allow us to evaluate new topic models or new search algorithms without any additional assessment. Below we describe three experiments in which three hyperparameters were selected alternately (the similarity measure, the set of modalities, and the number of topics), while the other two were fixed to be optimal.

Table 2 shows that cosine similarity is the best similarity measure between query and document topic vectors. The topic model used in this experiment has the optimal number of topics and the full set of modalities.

Table 3 shows that the use of all modalities together improves both recall and precision of the search. Terms and tags contribute the most. Models with only one modality show the worst results. All the models used in this experiment have the optimal number of topics.

Table 4 shows that an optimal number of topics $|T|$ for the model having the full set of modalities equals 200 for Habrahabr, 475 for TechCrunch.

The whole set of experiments shows that the optimal number of topics stays the same for all similarity measures, and the optimal set of modalities stays the same for all similarity measures and all values of $|T|$.

Table 2. Topic-based search with different similarity measures: Euclidean, Cosine, Manhattan, Hellinger, Kullback-Leibler

	Habrahabr					TechCrunch				
	Eu	**cos**	Ma	He	KL	Eu	**cos**	Ma	He	KL
Pr@5	0.612	**0.810**	0.682	0.709	0.721	0.635	**0.819**	0.673	0.732	0.715
Pr@10	0.657	**0.879**	0.697	0.735	0.749	0.665	**0.867**	0.683	0.752	0.732
Pr@15	0.627	**0.868**	0.635	0.727	0.711	0.643	**0.833**	0.642	0.742	0.724
Pr@20	0.619	**0.847**	0.627	0.728	0.707	0.638	**0.825**	0.638	0.729	0.708
R@5	0.672	**0.840**	0.692	0.721	0.803	0.658	**0.835**	0.669	0.733	0.775
R@10	0.682	**0.870**	0.707	0.775	0.856	0.671	**0.868**	0.682	0.753	0.787
R@15	0.705	**0.891**	0.725	0.791	0.878	0.715	**0.890**	0.708	0.785	0.809
R@20	0.703	**0.925**	0.732	0.812	0.888	0.712	**0.919**	0.715	0.808	0.812

Table 3. Topic-based search using different modalities **Habrahabr**: Assessors, Words, Bigrams, Comments, Tags, Hubs, Authors **TechCrunch**: Assessors, Words, Bigrams, Authors, Categories

	Habrahabr						TechCrunch					
	As	W	C	WB	WBTH	**All**	As	W	C	WB	WBC	**All**
Pr@5	0.821	0.612	0.549	0.654	0.737	**0.810**	0.822	0.711	0.557	0.767	0.808	**0.819**
Pr@10	0.869	0.635	0.568	0.701	0.752	**0.879**	0.851	0.721	0.581	0.783	0.818	**0.867**
Pr@15	0.875	0.625	0.532	0.685	0.682	**0.868**	0.835	0.733	0.594	0.793	0.833	**0.833**
Pr@20	0.863	0.616	0.533	0.682	0.687	**0.847**	0.813	0.727	0.566	0.772	0.822	**0.825**
R@5	0.780	0.722	0.636	0.797	0.827	**0.840**	0.762	0.752	0.657	0.775	0.825	**0.835**
R@10	0.817	0.744	0.648	0.812	0.875	**0.870**	0.792	0.776	0.669	0.808	0.855	**0.868**
R@15	0.850	0.778	0.677	0.842	0.893	**0.891**	0.835	0.782	0.684	0.825	0.877	**0.890**
R@20	0.873	0.803	0.685	0.852	0.898	**0.925**	0.867	0.825	0.702	0.837	0.901	**0.919**

Table 4. Topic-based search using a different number of topics

	Habrahabr						TechCrunch					
	As	100	150	**200**	250	400	As	350	400	450	**475**	500
Pr@5	0.821	0.662	0.721	**0.810**	0.761	0.693	0.822	0.653	0.725	0.752	**0.819**	0.777
Pr@10	0.869	0.761	0.812	**0.879**	0.825	0.673	0.851	0.663	0.732	0.762	**0.867**	0.811
Pr@15	0.875	0.733	0.795	**0.868**	0.791	0.651	0.835	0.682	0.743	0.787	**0.833**	0.793
Pr@20	0.863	0.724	0.795	**0.847**	0.792	0.642	0.813	0.650	0.743	0.773	**0.825**	0.793
R@5	0.780	0.732	0.807	**0.840**	0.821	0.721	0.762	0.731	0.762	0.793	**0.835**	0.817
R@10	0.817	0.771	0.843	**0.870**	0.851	0.751	0.792	0.763	0.793	0.812	**0.868**	0.855
R@15	0.850	0.824	0.895	**0.891**	0.871	0.773	0.835	0.782	0.807	0.855	**0.890**	0.882
R@20	0.873	0.857	0.905	**0.925**	0.892	0.771	0.867	0.792	0.823	0.862	**0.919**	0.903

6   Conclusions

In this paper, we propose an additively regularized topic model for exploratory search of relevant documents by long text queries. We show that the combination of decorrelation, sparsing and smoothing regularizers originally designed to improve the model interpretability also improves the search quality. We also confirm that the model should incorporate all available meta-data and modalities, such as bigrams, authors, tags, and categories.

For evaluating both precision and recall of the search we use an empirical technique based on human assessments. We achieve high quality results on realistic tasks of exploratory search in tech news. It seems that this level of quality would be enough for applications, such as automation of writing reviews and information consolidation. The topic-based search instantly performs the work that people typically complete in about 30 min. Another advantage of topic-based search over conventional full-text search is in reduction of the size of the inverted index, which enables an effective and low-cost implementation.

Acknowledgements. The work was supported by the Government of the Russian Federation (agreement 05.Y09.21.0018), the Ministry of Education and Science of the Russian Federation (project RFMEFI57915X0117), and the Russian Foundation for Basic Research (grants 17-07-01536, 16-37-00498). We are thankful to Maria Veretennikova for her help.

References

1. Andrzejewski, D., Buttler, D.: Latent topic feedback for information retrieval. In: Proceedings of the 17th ACM SIGKDD International Conference on Knowledge Discovery and Data Mining. KDD 2011, pp. 600–608 (2011)
2. Apishev, M., Koltcov, S., Koltsova, O., Nikolenko, S., Vorontsov, K.: Additive regularization for topic modeling in sociological studies of user-generated texts. In: Sidorov, G., Herrera-Alcántara, O. (eds.) MICAI 2016. LNCS (LNAI), vol. 10061, pp. 169–184. Springer, Cham (2017). https://doi.org/10.1007/978-3-319-62434-1_14
3. Apishev, M., Koltcov, S., Koltsova, O., Nikolenko, S., Vorontsov, K.: Mining ethnic content online with additively regularized topic models. Computacion y Sistemas **20**(3), 387–403 (2016)
4. Baeza-Yates, R., Ribeiro-Neto, B.: Modern Information Retrieval: The Concepts and Technology Behind Search (ACM Press Books), vol. 2. Addison-Wesley Professional, Harlow (2011)
5. Blei, D.M.: Probabilistic topic models. Commun. ACM **55**(4), 77–84 (2012)
6. Blei, D.M., Ng, A.Y., Jordan, M.I.: Latent dirichlet allocation. J. Mach. Learn. Res. **3**, 993–1022 (2003)
7. Frei, O., Apishev, M.: Parallel non-blocking deterministic algorithm for online topic modeling. In: Ignatov, D.I., Khachay, M.Y., Labunets, V.G., Loukachevitch, N., Nikolenko, S.I., Panchenko, A., Savchenko, A.V., Vorontsov, K. (eds.) AIST 2016. CCIS, vol. 661, pp. 132–144. Springer, Cham (2017). https://doi.org/10.1007/978-3-319-52920-2_13
8. Grant, C.E., George, C.P., Kanjilal, V., Nirkhiwale, S., Wilson, J.N., Wang, D.Z.: A topic-based search, visualization, and exploration system. In: FLAIRS Conference, pp. 43–48. AAAI Press, Massachusetts (2015)

9. Hofmann, T.: Probabilistic latent semantic indexing. In: Proceedings of the 22nd Annual International ACM SIGIR Conference on Research and Development in Information Retrieval, pp. 50–57. ACM, New York (1999)

10. Manning, C.D., Raghavan, P., Schütze, H.: Introduction to Information Retrieval. Cambridge University Press, New York (2008)

11. Marchionini, G.: Exploratory search: from finding to understanding. Commun. ACM **49**(4), 41–46 (2006)

12. Rönnqvist, S.: Exploratory topic modeling with distributional semantics. In: Fromont, E., De Bie, T., van Leeuwen, M. (eds.) IDA 2015. LNCS, vol. 9385, pp. 241–252. Springer, Cham (2015). https://doi.org/10.1007/978-3-319-24465-5_21

13. Scherer, M., von Landesberger, T., Schreck, T.: Topic modeling for search and exploration in multivariate research data repositories. In: Aalberg, T., Papatheodorou, C., Dobreva, M., Tsakonas, G., Farrugia, C.J. (eds.) TPDL 2013. LNCS, vol. 8092, pp. 370–373. Springer, Heidelberg (2013). https://doi.org/10.1007/978-3-642-40501-3_39

14. Tan, Y., Ou, Z.: Topic-weak-correlated latent dirichlet allocation. In: 7th International Symposium Chinese Spoken Language Processing (ISCSLP), pp. 224–228 (2010)

15. Veas, E.E., di Sciascio, C.: Interactive topic analysis with visual analytics and recommender systems. In: 2nd Workshop on Cognitive Computing and Applications for Augmented Human Intelligence, CCAAHI 2015, International Joint Conference on Artificial Intelligence, IJCAI, Buenos Aires, Argentina, July 2015. CEUR-WS.org, Aachen (2015)

16. Vorontsov, K., Potapenko, A.: Tutorial on probabilistic topic modeling: additive regularization for stochastic matrix factorization. In: Ignatov, D.I., Khachay, M.Y., Panchenko, A., Konstantinova, N., Yavorskiy, R.E. (eds.) AIST 2014. CCIS, vol. 436, pp. 29–46. Springer, Cham (2014). https://doi.org/10.1007/978-3-319-12580-0_3

17. Vorontsov, K.V., Potapenko, A.A.: Additive regularization of topic models. Mach. Learn. **101**(1), 303–323 (2015). Special issue on data analysis and intelligent optimization with applications

18. Vorontsov, K., Potapenko, A., Plavin, A.: Additive regularization of topic models for topic selection and sparse factorization. In: Gammerman, A., Vovk, V., Papadopoulos, H. (eds.) SLDS 2015. LNCS (LNAI), vol. 9047, pp. 193–202. Springer, Cham (2015). https://doi.org/10.1007/978-3-319-17091-6_14

19. Vorontsov, K., Frei, O., Apishev, M., Romov, P., Suvorova, M., Yanina, A.: Nonbayesian additive regularization for multimodal topic modeling of large collections. In: Proceedings of the 2015 Workshop on Topic Models: Post-Processing and Applications, pp. 29–37. ACM, New York (2015)

20. Wei, X., Croft, W.B.: Lda-based document models for ad-hoc retrieval. In: Proceedings of the 29th Annual International ACM SIGIR Conference on Research and Development in Information Retrieval. SIGIR 2006, pp. 178–185. ACM, New York (2006)

21. White, R.W., Roth, R.A.: Exploratory Search: Beyond the Query-Response Paradigm. Synthesis Lectures on Information Concepts Retrieval, and Services. Morgan and Claypool Publishers, San Rafael (2009)

22. Yi, X., Allan, J.: A comparative study of utilizing topic models for information retrieval. In: Boughanem, M., Berrut, C., Mothe, J., Soule-Dupuy, C. (eds.) ECIR 2009. LNCS, vol. 5478, pp. 29–41. Springer, Heidelberg (2009). https://doi.org/10.1007/978-3-642-00958-7_6

A Deep Forest for Transductive Transfer Learning by Using a Consensus Measure

Lev V. Utkin$^{(\boxtimes)}$ and Mikhail A. Ryabinin

Peter the Great St.Petersburg Polytechnic University, St.Petersburg, Russia
lev.utkin@gmail.com, mihail-ryabinin@yandex.ru
http://pml.spbstu.ru

Abstract. A Transfer Learning Deep Forest (TLDF) is proposed in the paper. It is based on the Deep Forest or gcForest proposed by Zhou and Feng and can be viewed as a gcForest modification whose aim is to implement the transductive transfer learning. The transfer learning is based on introducing weights of trees in forests which impact on the forest class probability distributions. The weights can be regarded as training parameters of the deep forest and are determined in order to maximize the agreement on target and source domains. The convex quadratic optimization problem with linear constraints is obtained to compute optimal weights for every forest taking into account the consensus principle. The numerical experiments illustrate the proposed distance metric method.

Keywords: Classification · Random forest · Decision tree
Transfer learning · Quadratic optimization

1 Introduction

Transfer learning can be regarded as a very promising approach to address the problem when data may be too few to build a good classifier and the distribution of the training data from the source domain is different from that of the target domain [15]. It is one of the active topics in current machine learning research. Sun et al. [17] point out that developments of transfer learning or multitask learning have shown that knowledge learned in one or more source tasks can be transfer to a related target task to significantly improve learning. Transfer learning or domain adaptation aims to extract common knowledge across domains such that a model trained on one domain can be adapted effectively to other domains [15]. This aim is due to an assumption that although distributions between source and target domain are different, there are some common knowledge structures across domains. A huge amount of papers are devoted to various modifications of classification algorithms in order to solve the transfer learning problems, including SVM, boosting, deep neural networks, etc. [1,2,9–11]. Many transfer learning methods refer to the training domain where labeled data is abundant as the source domain, and the test domain where labeled data is not available or very little as the target domain. In other words, domain adaptation generalizes a classifier that is trained

© Springer International Publishing AG 2018
A. Filchenkov et al. (Eds.): AINL 2017, CCIS 789, pp. 194–208, 2018.
https://doi.org/10.1007/978-3-319-71746-3_17

on a source domain, for which a large amount of training data is available, to a target domain, for which data is scarce [2,3,23]. Comprehensive review papers devoted to various transfer learning tasks are provided by several authors, for example, [13,15,20]. It follows from the reviews that most algorithms of transfer learning are focused on learning weights for different domains based on the similarities between each source domain and the target domain or learning more precise classifiers from the source domain data jointly by maximizing their consensus of predictions on the target domain data [26].

Formally, one of the transfer learning problem statements can be formulated as follows. Suppose that there is a *source domain* of labeled data denoted as $\mathcal{P}^S$, defined by a feature space $\mathcal{X}$ and a marginal probability distribution $P(X)$, i.e., $\mathcal{P}^S = \{\mathcal{X}^S, P(X)\}$, where $X = \{\mathbf{x}_1, ..., \mathbf{x}_n\} \in \mathcal{X}^S$. The data set from the source domain consists of n_S training instances and is represented by the *source domain data* $\mathcal{D}^S = \{(\mathbf{x}_j, y_j), \ j = 1, ..., n_S\}$. Here $y_j \in \{1, 2, ..., C\}$ is the class label of $\mathbf{x}_j$. We assume that there are C classes of data. Similarly, the unlabeled *target domain data* can be defined and denoted as $\mathcal{D}^T = \{(\mathbf{z}_j), \ j = 1, ..., n_T\}$, where $\mathbf{z}_j$ is the j-th instance from the target domain data. Here n_T is the number of target data. This is the statement of the transductive transfer learning when it is assumed that the source and target tasks are the same, while the source and target domains are different [15]. We consider a case when the feature spaces between the source domain and the target domain are the same, but the marginal probability distributions of the input data are different. We aim to train a classifier to make precise predictions on the target domain data $\mathcal{D}^T$.

One of the very promising classification methods is the deep forest. It has been proposed by Zhou and Feng [25] as an alternative to deep neural networks and has been called the gcForest. The classification method uses a multi-layer structure where each layer contains many random forests. Such the structure can be regarded as an ensemble of decision tree ensembles. Zhou and Feng [25] point out that their approach is highly competitive to deep neural networks. In contrast to deep neural networks which require great effort in hyperparameter tuning and large-scale training data, gcForest is much easier to train and can perfectly work when there are only small-scale training data. Therefore, by taking into account its advantages, it is important to modify it in order to develop a structure solving the transfer learning problem. We propose the so-called Transfer Learning Deep Forest (TLDF) which is a modification of the gcForest [25].

A large part of multi-view classification algorithms considering the relationships between multiple views are based on the so-called consensus principle which aims to maximize the agreement on multiple distinct views [5,22]. The algorithms adapt their classification parameters in order to achieve the highest agreement between source and target domains. The same idea is applied to the TLDF. We introduce weights of trees in forests as the parameters to control a consensus measure.

Our contributions are as follows:

1. The main contribution of this work is a method for transfer learning on the basis of the deep forest or the gcForest [25]. Its main idea is to introduce weights of trees in forests which impact on the forest class probability distributions and can be regarded as training parameters of the deep forest. The class distributions in the deep forest are viewed as the weighted sum of the tree class probabilities where the weights are determined in order to maximize the agreement on domains.
2. We propose a way for constructing the convex quadratic optimization problems with linear constraints to compute optimal weights for every forest taking into account the consensus principle. The obtained optimization problems can be solved by means of standard efficient optimization algorithms.
3. We demonstrate performance of the proposed TLDF on several representation learning benchmarks.

It should be noted that the idea to introduce weights of trees in the metric learning problems on the basis of the gcForest has been proposed by Utkin and Ryabinin [18,19].

2   Deep Forest

Before considering the TLDF, we briefly introduce the gcForest proposed by Zhou and Feng [25]. The gcForest can be divided into two parts. The first part is the so-called Multi-Grained Scanning structure which uses sliding windows to scan the raw features. Its output is a set of feature vectors produced by sliding windows of multiple sizes. The second part of the gcForest is a cascade forest structure where each level of a cascade receives feature information processed by its preceding level, and outputs its processing result to the next level [25].

One of the important ideas underlying the cascade forest structure is a class distribution produced by every tree for each input instance. The distribution is computed by counting the percentage of different classes of instances at the leaf node where the concerned instance falls into. It produces a class vector by means of averaging class distributions across all trees in the same forest. The class vector is then concatenated with the original vector to be input to the next level of the cascade.

The use of the class vector as a result of the random forest classification is very similar to the idea underlying the stacking method [21]. The stacking algorithm trains the first-level learners using the original training data set. Then it generates a new data set for training the second-level learner (meta-learner) such that the outputs of the first-level learners are regarded as input features for the second-level learner while the original labels are still regarded as labels of the new training data. In fact, the class vectors in the gcForest can be viewed as the meta-learners. In contrast to the stacking algorithm, the gcForest simultaneously uses the original vector and the class vectors (meta-learners) at the next cascade level by means of their concatenation. This implies that the feature vector is enlarged and enlarged after every cascade level. The architecture of the cascade proposed by Zhou and Feng [25] is shown in Fig. 1. It can be seen from the figure that each level of the cascade consists of two different pairs of random forests which generate 3-dimensional class vectors concatenated each other and with the original input. It should be noted that this structure of forests can be modified in order to improve the gcForest for a certain application. After the last level, we have the feature representation of the input feature vector, which can be classified in order to get the final prediction.

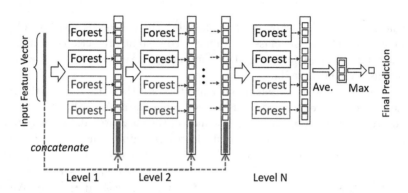

Fig. 1. The architecture of the cascade forest [25]

Let us introduce notations for indices corresponding to different deep forest components. The indices and their sets of values are shown in Table 1. One can see from Table 1, that there are Q levels of the deep forest or the cascade, every level contains M_q forests such that every forest consists of $T_{k,q}$ trees.

Table 1. Notations for indices

Type	Index
Cascade level	$q = 1, ..., Q$
Forest	$k = 1, ..., M_q$
Tree	$t = 1, ..., T_{k,q}$
Class	$c = 1, ..., C$

3   Consensus Measures and Training the TLDF

Many algorithms of transfer learning are based on the similarities between source domains and the target domain and maximize the consensus of predictions on the target domain data [26]. A common idea underlying the consensus transfer learning is to train a classifier on the source domain data by computing its parameters Θ such that the consensus measure of the predictions of the classifier on the target domain data $\mathbf{z} \in \mathcal{D}^T$ has to be maximal. The consensus measure denoted as $C(\mathbf{p})$ is defined in [27] through the Shannon entropy $E(\mathbf{p})$ as $C(\mathbf{p}) = -E(\mathbf{p})$, where $\mathbf{p} = (p_1, ..., p_C)$ is the vector of predicted probabilities of classes and the Shannon entropy is defined as follows:

$$E(\mathbf{p}) = -\sum_{c=1}^{C} p_c \log p_c. \tag{1}$$

Zhuang et al. [26,27] show that maximizing the consensus measure $C(\mathbf{p})$ is equivalent to enforcing the classifier to make consistent predictions on instances from the target domain data as well as minimizing the entropy of the predictions of each classifier on these data.

We apply the above consensus measure to train the q-th level of the forest cascade. All forests are trained by using the source data. According to [25], each forest of a cascade level produces an estimate of the class probability distribution by counting the percentage of different classes of training instances at the leaf node where the concerned instance falls into, and then averaging across all trees in the same forest. Suppose $p_c^{(t,k)}(\mathbf{x})$ is the probability of class c for $\mathbf{x} \in \mathcal{D}^S$ produced by the t-th tree from the k-th forest at the cascade level q. We will omit the index q in order to reduce the number of indices because all derivations will concern only level q. Then we can write a class probability distribution as

$$\mathbf{p}^{(t,k)}(\mathbf{x}) = (p_1^{(t,k)}(\mathbf{x}), ..., p_C^{(t,k)}(\mathbf{x})). \tag{2}$$

Suppose we have trained all trees in the TLDF by using the source data. According to [25], the k-th forest class distribution forms a class vector $V^{(k)}(\mathbf{x}) = (v_1^{(k)}(\mathbf{x}), ..., v_C^{(k)}(\mathbf{x}))$ which is then concatenated with the original vector $\mathbf{x}$ to be input to the next level of the cascade. Suppose an origin vector is $\mathbf{x} \in \mathcal{D}^S$, and the $p_c^{(t,k)}(\mathbf{x})$ is the probability of class c for $\mathbf{x}$ produced by the t-th tree from the k-th forest at the cascade level q. Following the results given in [25], the element $v_c^{(k)}(\mathbf{x})$ of the class vector corresponding to class c and produced by the k-th forest in the gcForest is determined as

$$v_c^{(k)}(\mathbf{x}) = T_k^{-1} \sum_{t=1}^{T_k} p_c^{(t,k)}(\mathbf{x}). \tag{3}$$

Then the concatenated vector $\mathbf{x}$ after a current level of the cascade is $\mathbf{x} \leftarrow (\mathbf{x}, V^{(1)}(\mathbf{x}), ..., V^{(M)}(\mathbf{x}))$. It is composed of the original vector $\mathbf{x}$ and class vectors obtained from M forests at the current level.

Classifying an instance $\mathbf{z}$ from the target data by means of the t-th tree from the k-th forest trained on the source data, we get another class distribution

$$\mathbf{p}^{(t,k)}(\mathbf{z}) = (p_1^{(t,k)}(\mathbf{z}), ..., p_C^{(t,k)}(\mathbf{z})). \tag{4}$$

It is important to note that the probability distribution $\mathbf{p}^{(t,k)}(\mathbf{z})$ is obtained by using the source data, i.e., it has been determined by counting the percentage of different classes of *source* instances at the leaf node where the *target* instance $\mathbf{z} \in \mathcal{D}^T$ falls into. As a result, we get probability distributions for every $\mathbf{z} \in \mathcal{D}^T$ and every tree $t = 1, ..., T_k$, from the k-th forest.

3.1   Weighted Average of Class Probabilities

In order to solve the transfer learning problem, we propose a method whose idea is to define the forest class distribution as a weighted sum of the tree class probabilities. In other words, we assign weights to every tree. At that, the weights are assigned in an optimal way in order to minimize the entropy of the predictions of each classifier on the target data. Suppose that we know optimal weights. Then an element of a class vector produced by the k-th forest consists of C probabilities is of the form:

$$v_c^{(k)}(\mathbf{z}, \mathbf{w}^{(k)}) = \sum_{t=1}^{T_k} p_c^{(t,k)}(\mathbf{z}) w^{(t,k)} = \mathbf{p}_c^{(k)}(\mathbf{z}) \cdot \mathbf{w}^{(k)}, \ c = 1, ..., C. \qquad (5)$$

where $w^{(t,k)}$ is the weight of the t-th tree from the k-th forest, which will be viewed as a parameter of the proposed transfer learning model; $\mathbf{w}^{(k)} = (w^{(1,k)}, ..., w^{(T_k,k)})^{\mathrm{T}}$ is the vector of weights of all trees from the k-th forest; $\mathbf{p}_c^{(k)}(\mathbf{z}) = (p_c^{(1,k)}(\mathbf{z}), ..., p_c^{(T_k,k)}(\mathbf{z}))$ is the vector of probabilities for all trees from the k-th forest corresponding to the c-th class.

It should be noted that the vectors of weights do not depend on the class and on $\mathbf{z}$. An illustration of the weighted averaging is shown in Fig. 2, where we partly modify the original picture from [25] and pictures from [18,19] in order to show how elements of the class vector are derived as a simple weighted sum. One can see from Fig. 2 that two-class distribution is estimated by counting the percentage of different classes of a new training instances $\mathbf{x} \in \mathcal{D}^S$ or $\mathbf{z} \in \mathcal{D}^T$ at the leaf nodes where the concerned instances $\mathbf{x}$ and $\mathbf{z}$ fall into, respectively. Figure 2 shows the class probability distributions obtained for the vector $\mathbf{z}$ or the vector $\mathbf{x}$ under condition of two classes. They are $(0.4; 0.6)$, $(0.2; 0.8)$, $(1.0; 0.0)$. Of course, they should be different for every instance, but we illustrate the same probability distribution for $\mathbf{z}$ as well as for $\mathbf{x}$ for short. By using the instances from the source domain data, we train all trees of the q-th level of the cascade and get the augmented features $v_c^{(k)}(\mathbf{x})$, $c = 1, 2$, by averaging the class probabilities of all trees in the k-th forest without using weights. The corresponding class vector is $(0.53; 0.47)$ as it is shown in Fig. 2. After concatenation of the original vector $\mathbf{x}$ with the augmented features $v_c^{(k)}(\mathbf{x})$, we get a new vector $\mathbf{x}$ for the next level of the forest cascade. By using the trees trained by means of the source domain data, we classify instances $\mathbf{z}$ from the target domain data. Then the class vector of $\mathbf{z}$ is computed as the weighted average. As a result, we get the augmented features $v_c^{(k)}(\mathbf{z}, \mathbf{w}^{(k)})$, $c = 1, 2$, corresponding to the q-th forest and obtained as weighted sums, i.e., there hold

$$v_1^{(k)}(\mathbf{z}, \mathbf{w}^{(k)}) = 0.4w^{(1,k)} + 0.2w^{(2,k)} + 1.0w^{(3,k)}, \tag{6}$$

$$v_2^{(k)}(\mathbf{z}, \mathbf{w}^{(k)}) = 0.6w^{(1,k)} + 0.8w^{(2,k)} + 0.0w^{(3,k)}. \tag{7}$$

After concatenation of the original vector $\mathbf{z}$ with the augmented features $v_c^{(k)}(\mathbf{z}, \mathbf{w}^{(k)})$, we get a new vector $\mathbf{z}$ for the next level of the forest cascade. We apply the greedy algorithm for training the TLDF, i.e., we train separately every level starting from the first level such that every next level uses results of training obtained at the previous level.

By having the weighted averages for every forest, where the weights are trained parameters, the next task is to develop an algorithm for training the TLDF, in particular, for computing the weights for every forest and for every cascade level.

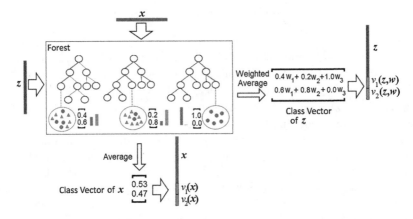

Fig. 2. An illustration of the class vector generation taking into account the weights

The elements $v_c^{(k)}(\mathbf{z}, \mathbf{w}^{(k)})$ form the k-th forest class probability distribution

$$V^{(k)}(\mathbf{z}, \mathbf{w}^{(k)}) = (v_1^{(k)}(\mathbf{z}, \mathbf{w}^{(k)}), ..., v_C^{(k)}(\mathbf{z}, \mathbf{w}^{(k)})). \tag{8}$$

In the same way, the concatenated vector $\mathbf{z}$ after the current level of the cascade is

$$\mathbf{z} \leftarrow \left(\mathbf{z}, V^{(1)}(\mathbf{z}, \mathbf{w}^{(1)}), ..., V^{(M)}(\mathbf{z}, \mathbf{w}^{(M)})\right). \tag{9}$$

3.2   The Shannon Entropy as a Consensus Measure

One of the ways for constructing the transfer learning model is to maximize the consensus measure or to minimize the Shannon entropy $H(V^{(k)}(\mathbf{z}))$, which is defined for the instance $\mathbf{z}$ as

$$H(V^{(k)}(\mathbf{z}, \mathbf{w}^{(k)})) = -\sum_{c=1}^{C} v_c^{(k)}(\mathbf{z}, \mathbf{w}^{(k)}) \log v_c^{(k)}(\mathbf{z}, \mathbf{w}^{(k)})$$

$$= -\sum_{c=1}^{C} \left(\mathbf{p}_c^{(k)}(\mathbf{z})\mathbf{w}^{(k)} \right) \log \left(\mathbf{p}_c^{(k)}(\mathbf{z})\mathbf{w}^{(k)} \right). \tag{10}$$

Hence, the mean entropy over all M forests and all instances from target data is defined as

$$H = \sum_{\mathbf{z} \in \mathcal{D}^T} \sum_{k=1}^{M} H(V^{(k)}(\mathbf{z}, \mathbf{w}^{(k)})). \tag{11}$$

In order to find optimal values of weights $\mathbf{w}^{(k)}$, we have to solve the following optimization problem:

$$\min_{\mathbf{w}^{(k)}, k=1,\ldots,M} \sum_{\mathbf{z} \in \mathcal{D}^T} \sum_{k=1}^{M} H(V^{(k)}(\mathbf{z}, \mathbf{w}^{(k)})) + \lambda R(\mathbf{w}), \tag{12}$$

subject to

$$\sum_{t=1}^{T_k} w^{(t,k)} = \mathbf{w}^{(k)} \mathbf{1}_k^{\mathrm{T}} = 1, \ w^{(t,k)} \geq 0, \ \forall t = 1, \ldots, T_k, \ k = 1, \ldots, M. \tag{13}$$

Here $R(\mathbf{w})$ is a regularization term; λ is a hyper-parameter which controls the strength of the regularization; $\mathbf{w} = (\mathbf{w}^{(1)}, \ldots, \mathbf{w}^{(M)})$ is the vector of all weights of a level; $\mathbf{1}_k$ is a unit vector having T_k elements. We define the regularization term as $R(\mathbf{w}) = \|\mathbf{w}\|^2$. The objective function (12) can be written as

$$\min_{\mathbf{w}^{(k)}, k=1,\ldots,M} \sum_{k=1}^{M} \sum_{\mathbf{z} \in \mathcal{D}^T} H(V^{(k)}(\mathbf{z}, \mathbf{w}^{(k)})) + \lambda R(\mathbf{w}). \tag{14}$$

Note that the vectors $\mathbf{w}^{(1)}, \ldots, \mathbf{w}^{(M)}$ have separate constraints. This implies that the optimization problem (12)–(13) can be decomposed into M problems of the form:

$$\min_{\mathbf{w}^{(k)}} \sum_{\mathbf{z} \in \mathcal{D}^T} H(V^{(k)}(\mathbf{z}, \mathbf{w}^{(k)})) + \lambda R(\mathbf{w}^{(k)}), \tag{15}$$

subject to

$$\mathbf{w}^{(k)} \mathbf{1}_k^{\mathrm{T}} = 1, \ w^{(t,k)} \geq 0, \ \forall t = 1, \ldots, T_k. \tag{16}$$

If to substitute (10) into (15)–(16), then we get an opimization problem which may be non-convex and NP-hard. Indeed, the function (10) is concave with respect to the variables $\mathbf{w}^{(k)}$. For simplifying the above optimization problem, we replace the entropy with a quite different convex function which can be regarded as a measure of the transfer learning consistence between source and target domains.

4   Convex Measure of the Transfer Learning Consistence

Before considering the objective function (15) replacement, we analyze the Shannon entropy (10). The entropy can be derived through the well-known Kullback-Leibler divergence $D_{KL}(\mathbf{p}\|\mathbf{q})$ as a distance between probability distribution $\mathbf{p}$ and the uniform distribution $\mathbf{q} = (C^{-1}, ..., C^{-1})$, i.e., we can write

$$D_{KL}(\mathbf{p}\|\mathbf{q}) = H(\mathbf{p}, \mathbf{q}) - H(\mathbf{p}) = \log C - H(\mathbf{p}). \tag{17}$$

Here $H(\mathbf{p}, \mathbf{q})$ is the cross entropy of $\mathbf{p}$ and $\mathbf{q}$. By minimizing $H(\mathbf{p})$, we maximize the distance $D_{KL}(\mathbf{p}\|\mathbf{q})$. Note that the entropy achieves its maximum at $\mathbf{p} = \mathbf{q}$, i.e., when $\mathbf{p}$ is uniform. This means that we aim to avoid the case when $\mathbf{p}$ is close to the uniform distribution. In other words, we search for the probability distribution $\mathbf{p}$ which is far from the uniform distribution.

It is obvious that the optimal class distribution is close to one of the vertices of the unit simplex having C vertices of the form $S_j = (0, ..., 0, 1_j, 0, ..., 0)$. Then we have to minimize the distances $d(\mathbf{p}, S_j)$ between the class distribution $\mathbf{p} = V^{(k)}(\mathbf{z}, \mathbf{w}^{(k)})$ and the distributions S_j, $j = 1, ..., C$. Since there are C distances, then we apply one of the pessimistic decision strategies for deriving the objective function. According to this strategy, we minimize the largest distance among distances $d(\mathbf{p}, S_j)$, $j = 1, ..., C$. As a result, the objective function is of the form:

$$\min_{\mathbf{w}^{(k)}} \sum_{\mathbf{z} \in \mathcal{D}^T} \max_{j=1,...,C} d(V^{(k)}(\mathbf{z}, \mathbf{w}^{(k)}), S_j) + \lambda R(\mathbf{w}^{(k)}). \tag{18}$$

Let us introduce a new variable $\beta(\mathbf{z})$, $\mathbf{z} \in \mathcal{D}^T$, which is defined for every $\mathbf{z}$ as $\beta(\mathbf{z}) = \max_{j=1,...,C} d(V^{(k)}(\mathbf{z}, \mathbf{w}^{(k)}), S_j)$. Then we can write the following optimization problem:

$$\min_{\mathbf{w}^{(k)}} \sum_{\mathbf{z} \in \mathcal{D}^T} \beta(\mathbf{z}) + \lambda R(\mathbf{w}^{(k)}), \tag{19}$$

subject to (16) and

$$\beta(\mathbf{z}) \geq d(V^{(k)}(\mathbf{z}, \mathbf{w}^{(k)}), S_j), \ \forall j \in \{1, ..., C\}, \ \mathbf{z} \in \mathcal{D}^T. \tag{20}$$

In order to get the quadratic optimization problem, we apply the city block L_1 distance, i.e.,

$$d(V^{(k)}(\mathbf{z}, \mathbf{w}^{(k)}), S_j) = \sum_{c=1}^{C} \left| \mathbf{p}_c^{(k)}(\mathbf{z})\mathbf{w}^{(k)} - S_{j,c} \right|, \tag{21}$$

where $S_{j,c}$ is the c-th element of the probability distribution S_j. We can simplify the above expression as follows:

$$
\begin{aligned}
d(V^{(k)}(\mathbf{z}, \mathbf{w}^{(k)}), S_j) &= \sum_{c=1, c \neq j}^{C} \left| \mathbf{p}_c^{(k)}(\mathbf{z})\mathbf{w}^{(k)} - 0 \right| + \left| \mathbf{p}_j^{(k)}(\mathbf{z})\mathbf{w}^{(k)} - 1 \right| \\
&= \left(1 - \mathbf{p}_j^{(k)}(\mathbf{z}) \right) \mathbf{w}^{(k)} + (1 - \mathbf{p}_j^{(k)}(\mathbf{z})\mathbf{w}^{(k)}) \\
&= 1 - (2\mathbf{p}_j^{(k)}(\mathbf{z}) - 1)\mathbf{w}^{(k)} = 2(1 - \mathbf{p}_j^{(k)}(\mathbf{z})\mathbf{w}^{(k)}).
\end{aligned}
\tag{22}
$$

Finally, we rewrite constraints (20) as

$$
\beta(\mathbf{z}) + 2\mathbf{p}_j^{(k)}(\mathbf{z})\mathbf{w}^{(k)} \geq 2, \ \forall j \in \{1, ..., C\}.
\tag{23}
$$

We have obtained the standard quadratic optimization problem with objective function (19) and linear constraints (16), (23).

5   An Algorithm for the TLDF Training

In sum, we can write a general algorithm for training the TLDF (see Algorithm 1). Its complexity mainly depends on the number of levels and other parameters, for instance, the number of trees in every forest.

Algorithm 1. A general algorithm for training the TLDF

Require: Source domain data $\mathcal{D}^S = \{(\mathbf{x}_i, y_i), \ i = 1, ..., n_S\}$; target domain data $\mathcal{D}^T = \{(\mathbf{x}_j), \ j = 1, ..., n_T\}$; number of cascade levels Q

Ensure: $\mathbf{w}$ for every $q = 1, ..., Q$

1: **for** $q = 1, q \leq Q$ **do**
2:     Train all trees at the q-th level by using the source domain data $\mathcal{D}^S$
3:     Get probability distributions of classes for every tree
4:     Classify every instance $\mathbf{z}$ from the target domain data $\mathcal{D}^T$ by using trees trained by means of $\mathcal{D}^S$ and find the probability distribution of classes $\mathbf{p}_c^{(k)}(\mathbf{z})$
5:     **for** $k = 1, k \leq M_q$ **do**
6:         Compute weights $\mathbf{w}^{(k)}$ at the q-th level from the k-th quadratic optimization problem with the objective function (19) and constraints (16), (23)
7:     **end for**
8:     For every $\mathbf{z}$ from $\mathcal{D}^T$, compute $V^{(k)}(\mathbf{z}, \mathbf{w}^{(k)}) = (v_1^{(k)}(\mathbf{z}, \mathbf{w}^{(k)}), ..., v_C^{(k)}(\mathbf{z}, \mathbf{w}^{(k)}))$ at the q-th level by using (5)-(8), $k = 1, ..., M$
9:     For every $\mathbf{z}$ from $\mathcal{D}^T$, form the concatenated vector $\mathbf{z} \leftarrow (\mathbf{z}, V^{(1)}(\mathbf{z}, \mathbf{w}^{(1)}), ..., V^{(M)}(\mathbf{z}, \mathbf{w}^{(M)}))$ for the next level
10: **end for**

---

6   Numerical Experiments

In order to evaluate the TLDF, we use the same cascade structure as the standard gcForest described in [25]. Each level of the cascade structure consists of 2 complete-random tree forests and 2 random forests. Three-fold cross-validation is used for the class vector generation. The number of cascade levels is automatically determined. A software in Python implementing the gcForest is available at https://github.com/leopiney/deep-forest. We modify this software in order to implement the procedure for computing optimal weights and weighted averages $v_{ij,c}^{(k)}$. A version of the modified software in Python is available at http://pml.spbstu.ru/wp-content/uploads/2017/02/TLDF-1.zip.

The accuracy measure Acc is defined as

$$Acc = \frac{\left|\mathbf{z} : \mathbf{z} \in \mathcal{D}^T \wedge y^*(\mathbf{z}) = y(\mathbf{z})\right|}{\left|\mathbf{z} : \mathbf{z} \in \mathcal{D}^T\right|} \cdot 100\%, \tag{24}$$

where $y^*(\mathbf{z})$ is the predicted label and $y(\mathbf{z})$ is the ground truth label for a test data $\mathbf{z}$.

To evaluate the average accuracy, we perform a cross-validation with 100 repetitions, where in each run, we randomly select N training data and $N_{\text{test}} = 2N/3$ test data. Different values for the regularization hyper-parameter λ have been tested, choosing those leading to the best results.

In order to evaluate the TLDF, we consider a numerical example given in [12] where a combination of two well-known public datasets USPS and MNIST is used for getting the source and target data. USPS[1] dataset contains handwritten digits from US post office and consists of 7,291 training images and 2,007 test images of size 16×16. Moreover, test data comes from totally different distribution than training data. MNIST[2] dataset has a training set of 60,000 instances and a test set of 10,000 instances of size 28×28. USPS and MNIST datasets follow very different distributions but they share 10 classes of digits. Long et al. [12] propose to construct one dataset USPS vs MNIST by randomly sampling 1,800 images in USPS to form the source domain, and sampling 2,000 images in MNIST to form the target domain. Then the source/target pair is switched to get another dataset MNIST vs USPS. The images are rescaled to 16×16 pixels, and each represented by a feature vector encoding the gray-scale pixel values. Hence the source and target data can share the same feature space [12].

We also use numerical results represented by Farajidavar et al. [7] with the USPS vs MNIST and the MNIST vs USPS datasets obtained by means of the adaptive transductive transfer machines (ATTM). We also use results provided by Epstein et al. [6] and obtained by means of joint auto-encoders (JAE). Moreover, we use the results given by Luo et al. [14] obtained by the close yet distinctive domain adaptation method (CDDA-a) and its modification (CDDA-b).

---

[1] https://www.csie.ntu.edu.tw/~cjlin/libsvmtools/datasets/multiclass.html.

[2] http://yann.lecun.com/exdb/mnist.

Table 2. Accuracy measures for the MNIST and USPS datasets by several transfer learning methods

	ATTM	JAE	CDDA-a	CDDA-b	**TLDF**
MNIST vs USPS	77.94	87.6	76.22	82.33	79.1
USPS vs MNIST	61.15	86.9	62.05	70.75	73.4

Table 3. Accuracy measures for the Office-Caltech dataset by several transfer learning methods

	DLRC	DTN	CDDA-a	CDDA-b	**TLDF**
A/W	42.80	43.00	41.69	38.31	37.70
A/D	41.80	56.00	37.58	38.22	40.90
A/C	42.70	42.90	42.12	41.32	42.40
W/A	38.50	36.89	37.27	41.75	41.30
W/D	94.30	84.00	87.90	89.81	86.40
W/C	33.80	34.18	31.97	33.30	35.10
C/A	49.70	54.00	48.33	52.09	51.70
C/W	41.70	58.50	44.75	47.12	53.90
C/D	47.50	56.00	48.41	45.86	49.20

The numerical results are represented in Table 2. It follows from Table 2 that we achieve results comparable or superior to existing methods.

Another publicly available dataset used for evaluating the proposed TLDF is the Office-Caltech dataset[3]. It consists of the Office dataset and the Caltech-256 dataset [8]. Office dataset contains three domains: Amazon (A), DSLR (D) and Webcam (W). The Caltech-256 dataset (C) has 30,607 images in 256 categories. The four domains A, D, W, C share 10 object categories in total. Every category consists of 958, 157, 295, 1123 images, respectively. By randomly selecting two different domains out of four domains, we form the source and the target datasets denoted as A/W, A/D, ..., C/D. Here the first letter corresponds to the source data, the second letter corresponds to the target data.

We use numerical results represented by Ding et al. [4] obtained by means of the Deep Low-Rank Coding (DLRC) method. Another set of results is taken from [24], where a domain adaptation framework named Deep Transfer Network (DTN) is proposed. Moreover, we use the results given by Luo et al. [14] obtained by the CDDA-a and CDDA-b methods. The numerical results are represented in Table 3. It follows from Table 3 that we achieve results comparable with existing methods. It should be also noted that some results are inferior to many existing methods. However, Zhou and Feng [25] have mentioned: "The performance of gcForest can be further improved via task-specific tuning." An additional study

---

[3] https://people.eecs.berkeley.edu/~jhoffman/domainadapt/#datasets_code.

is required in order to find the best architecture of the TLDF, which provides the outperforming results.

7 Conclusion

An application of the deep forest method proposed by Zhou and Feng [25] to the transfer learning problem has been considered in the paper. Two main ideas underlying the proposed application have been used. The first one is to introduce weights of decision trees as training parameters. The second idea is to train the weights in order to maximize the consensus measure. An efficient algorithm has been proposed for training the weights and for solving the transfer learning problem. It consists of standard convex quadratic programming problems whose solution does not meet any difficulties. It should be noted that the convexity is a useful condition for simplifying the optimization problem solution. However, different criteria and non-convex objective functions can be studied. In this case, we can use special optimization algorithms, for example, a non-convex modification of the well-known Frank-Wolfe algorithm proposed by Reddi et al. [16]. This is a direction for further research. Another direction for further research is to take into account a set of source domains. The consensus measure can be a direct way for studying this case. Another direction for research is to use the unit norm as the regularization term. In this case, we get a linear programming problem which is rather simple and provides a sparse solution. This direction could also lead to interesting results.

The main disadvantage of the proposed TLDF is that the transfer learning is based only on the augmented features and does not take into account the concatenated original vectors. In order to overcome this difficulty, we can apply a self-labeling method which includes unlabeled target domain samples in the training process by means of initializing their labels and then iteratively refining the labels. The refining is carried out at every level of the forest cascade. The labels are initially assigned to target instances by comparing distances between every instance from the target data and centers of classes of the source data. A combination of the source and target data can be implemented by applying the self-labeling method.

Another way for improving the TLDF is to introduce weights of the source data, which can be defined through distances between a center of the target data and every instance from the source data. The small distance leads to the large weight of the source instance. The decision tree training in this case is carried out by using the well-known reweighting or resampling procedures.

Acknowledgement. The authors would like to express their appreciation to the anonymous referees whose very valuable comments have improved the paper.

The reported study was partially supported by RFBR, research project No. 17-01-00118.

References

1. Arnold, A., Nallapati, R., Cohen, W.: A comparative study of methods for transductive transfer learning. In: Proceedings of the 7th IEEE International Conference on Data Mining Workshops, pp. 77–82. IEEE Computer Society, Washington (2007)
2. Ben-David, S., Blitzer, J., Crammer, K., Pereira, F.: Analysis of representations for domain adaptation. Adv. Neural Inf. Process. Syst. **19**, 137–144 (2007)
3. Chen, M., Blitzer, J., Weinberger, K.: Co-training for domain adaptation. Adv. Neural Inf. Process. Syst. **24**, 2456–2464 (2011)
4. Ding, Z., Shao, M., Fu, Y.: Deep low-rank coding for transfer learning. In: Proceedings of the 24th International Conference on Artificial Intelligence (IJCAI 2015), pp. 3453–3459. AAAI Press (2015)
5. Duan, L., Tsang, I., Xu, D., Chua, T.S.: Domain adaptation from multiple sources via auxiliary classifiers. In: Proceedings of the 26th Annual International Conference on Machine Learning, pp. 289–296. ACM (2009)
6. Epstein, B., Meir, R., Michaeli, T.: Joint auto-encoders: a flexible multi-task learning framework, 30 May 2017. arXiv:1705.10494v1
7. Farajidavar, N., deCampos, T., Kittler, J.: Adaptive transductive transfer machines. In: British Machine Vision Conference, vol. 25, pp. 1–12. BMVA Press, Nottingham, September 2014
8. Gong, B., Shi, Y., Sha, F., Grauman, K.: Geodesic flow kernel for unsupervised domain adaptation. In: 2012 IEEE Conference on Computer Vision and Pattern Recognition (CVPR), p. 20662073. IEEE (2012)
9. Habrard, A., Peyrache, J., Sebban, M.: A new boosting algorithm for provably accurate unsupervised domain adaptation. Knowl. Inf. Syst. **47**(1), 45–73 (2016)
10. Hu, J., Lu, J., Tan, Y.P.: Deep transfer metric learning. In: The IEEE Conference on Computer Vision and Pattern Recognition (CVPR), pp. 325–333. IEEE (2015)
11. Joachims, T.: Transductive inference for text classification using support vector machines. In: Proceedings of the Sixteenth International Conference on Machine Learning, San Francisco, CA, USA, pp. 200–209 (1999)
12. Long, M., Wang, J., Ding, G., Pan, S., Yu, P.: Adaptation regularization: a general framework for transfer learning. IEEE Trans. Knowl. Data Eng. **26**(5), 1076–1089 (2014)
13. Lu, J., Xue, S., Zhang, X., Han, Y.: A neural network-based interval pattern matcher. Information **6**, 388–398 (2015)
14. Luo, L., Wang, X., Hu, S., Chen, L.: Robust data geometric structure aligned close yet discriminative domain adaptation, 24 May 2017. arXiv:1705.08620v1
15. Pan, S., Yang, Q.: A survey on transfer learning. IEEE Trans. Knowl. Data Eng. **22**(10), 1345–1359 (2010)
16. Reddi, S., Sra, S., Poczos, B., Smola, A.: Stochastic frank-wolfe methods for nonconvex optimization, July 2016. arXiv:1607.08254v2
17. Sun, S.: A survey of multi-view machine learning. Neural Comput. Appl. **23**(7), 2031–2038 (2013)
18. Utkin, L., Ryabinin, M.: Discriminative metric learning with deep forest, May 2017. arXiv:1705.09620v1
19. Utkin, L., Ryabinin, M.: A Siamese deep forest, April 2017. arXiv:1704.08715v1
20. Weiss, K., Khoshgoftaar, T., Wang, D.: A survey of transfer learning. J. Big Data **3**(1), 1–40 (2016)
21. Wolpert, D.: Stacked generalization. Neural Netw. **5**(2), 241–259 (1992)

22. Xu, C., Tao, D., Xu, C.: A survey on multi-view learning, April 2013. arXiv:1304.5634v1

23. Xu, Z., Sun, S.: Multi-source transfer learning with multi-view Adaboost. In: Huang, T., Zeng, Z., Li, C., Leung, C.S. (eds.) ICONIP 2012. LNCS, vol. 7665, pp. 332–339. Springer, Heidelberg (2012). https://doi.org/10.1007/978-3-642-34487-9_41

24. Zhang, X., Wang, S., Yu, F., Chang, S.F.: Deep transfer network: unsupervised domain adaptation, March 2015. arXiv:1503.00591v1

25. Zhou, Z.H., Feng, J.: Deep forest: towards an alternative to deep neural networks, May 2017. arXiv:1702.08835v2

26. Zhuang, F., Cheng, X., Pan, S.J., Yu, W., He, Q., Shi, Z.: Transfer learning with multiple sources via consensus regularized autoencoders. In: Calders, T., Esposito, F., Hüllermeier, E., Meo, R. (eds.) ECML PKDD 2014. LNCS (LNAI), vol. 8726, pp. 417–431. Springer, Heidelberg (2014). https://doi.org/10.1007/978-3-662-44845-8_27

27. Zhuang, F., Luo, P., Xiong, H., Xiong, Y., He, Q., Shi, Z.: Cross-domain learning from multiple sources: a consensus regularization perspective. IEEE Trans. Knowl. Data Eng. **22**(12), 1664–1678 (2010)

Russian Paraphrase Detection Shared Task

ParaPhraser: Russian Paraphrase Corpus and Shared Task

Lidia Pivovarova[1(✉)], Ekaterina Pronoza[2], Elena Yagunova[2], and Anton Pronoza[3]

[1] University of Helsinki, Helsinki, Finland
lidia.pivovarova@helsinki.fi
[2] St.-Petersburg State University, St.-Petersburg, Russian Federation
katpronoza@gmail.com, iagounova.elena@gmail.com
[3] Institute for Informatics and Automation of the Russian Academy of Sciences,
St.-Petersburg, Russian Federation
antpro@list.ru

Abstract. The paper describes the results of the First Russian Paraphrase Detection Shared Task held in St.-Petersburg, Russia, in October 2016. Research in the area of paraphrase extraction, detection and generation has been successfully developing for a long time while there has been only a recent surge of interest towards the problem in the Russian community of computational linguistics. We try to overcome this gap by introducing the project ParaPhraser.ru dedicated to the collection of Russian paraphrase corpus and organizing a Paraphrase Detection Shared Task, which uses the corpus as the training data. The participants of the task applied a wide variety of techniques to the problem of paraphrase detection, from rule-based approaches to deep learning, and results of the task reflect the following tendencies: the best scores are obtained by the strategy of using traditional classifiers combined with fine-grained linguistic features, however, complex neural networks, shallow methods and purely technical methods also demonstrate competitive results.

Keywords: Shared task · Russian paraphrase · Paraphrase detection
Paraphrase corpus

1 Introduction

Paraphrase is one of the most problematic concepts in computational linguistics. It has been shown that a narrow definition – "paraphrases must be exactly logically equivalent" – does not cover many cases that are usually considered as paraphrase or quasi-paraphrase (Bhagat and Hovy 2013). In most practical cases a more relaxed definition of paraphrases is used, e.g. "alternative expressions of the same (or similar) meaning" (Agirre et al. 2015). This notion of similar meaning encompasses a variety of linguistic phenomena, which have a "broad and multi-faceted nature" (Vila et al. 2014). Moreover, in some cases it is hard to distinguish paraphrase and textual entailment, i.e. the implication relation between sentences.

© Springer International Publishing AG 2018
A. Filchenkov et al. (Eds.): AINL 2017, CCIS 789, pp. 211–225, 2018.
https://doi.org/10.1007/978-3-319-71746-3_18

Since it is difficult to work out an exact definition of paraphrase, a data-driven approach might be a reasonable choice. In this case we do not try to give a formal definition of paraphrase but instead lean on native speakers and their judgments whether a particular pair of sentences is a paraphrase or not. In practice, this data-driven approach requires a construction of large paraphrase corpora with manual or semi-automatic paraphrase annotation, which is obviously a time-consuming task that should be done anew for any given language. On the other hand, recent growth of machine-learning techniques in language processing turns such corpora into valuable resources that can be used to build automatic paraphrase detection systems.

In this paper we present a ParaPhraser project (http://www.paraphraser.ru/) aimed at building of Russian paraphrase corpus, studying of paraphrase phenomena in Russian news and development of automatic paraphrase detection and generation methods (Pronoza and Yagunova 2015a), (Pronoza and Yagunova 2015b), (Pronoza et al. 2015), (Pronoza et al. 2017). The project was launched in 2014 in St.-Petersburg State University; by the beginning of 2016 we have collected 11 thousand pairs of Russian news titles, which were manually collected as either paraphrase, partial paraphrase or non-paraphrase. The corpus construction process is a combination of automatic paraphrase candidates extraction and manual post-processing of candidate pairs using crowd-sourcing. As far as we aware this is the first sentential corpus of Russian paraphrase. From the very beginning the corpus has been publicly available. The current stage of the corpus allowed to perform various research, including linguistic study of paraphrase and study of information flow in news. It also can be used to train automatic paraphrase detection systems, including shared task organized in Fall 2016.

The rest of the paper is organized as follows: in Sect. 2 we briefly overview related research, including general paraphrase studies, paraphrase corpora and shared tasks; in Sect. 3 we present the ParaPhraser project and describe the corpus construction process; in Sect. 4 we present the shared task and its results.

2 Background

2.1 Paraphrase Extraction and Recognition

The detailed survey of paraphrase and textual entailment studies can be found in (Androutsopoulos and Malakasiotis 2010). We use their exhaustive work as a frame for this section; at the same time, we would like to point out some major changes introduced in the area during the most recent years.

According to (Androutsopoulos and Malakasiotis 2010), all the tasks related to paraphrases are broken into three main groups: extraction, recognition and generation. Paraphrase extraction is a processing of large corpora aiming at finding paraphrastic sentences or phrases; this is the task we had to solve in the initial step of ParaPhraser corpus generation (see Sect. 3). Paraphrase recognition means that for a given sentence pair a system should determine whether this is a paraphrase or not; we believe that this task can be solved using ParaPhraser corpus as training data; one of the goals of the shared task, described in Sect. 4, is to test this assumption. Paraphrase generation, that is a producing of artificial paraphrase for a given sentence, is beyond the scope of this paper, though we are working on this problem as the part of the ParaPhraser project.

Our paraphrase extraction method is based on approach introduced in (Fernando and Stevenson 2008). They proposed a matrix similarity metric that measures a distance between two sentences based on their word similarity in WordNet. Since a comprehensive Russian WordNet is not currently available we used a synonym dictionary instead of WordNet; we also introduced several modifications into Fernando and Stevenson similarity metric (Pronoza and Yagunova 2015b).

(Androutsopoulos and Malakasiotis 2010) listed several methods for paraphrase recognition, including logic-based methods, vector-based methods, those based on surface string similarity, based on syntactic similarity, based on symbolic meaning representation, machine learning methods, and decoding-based methods. Though they mention machine learning as only one method among others, which can be used to combine various features, machine learning methods has become dominating in paraphrase detection area over last years. This does not mean that other methods do not appear in literature; e.g., (Pham et al. 2013) used distributional semantics approach to paraphrase detection. Moreover, it is hard to place a certain approach into single class of the classification. E.g. (Madnani et al. 2012) demonstrated that machine-translation evaluation metrics, such as BLEU, can be effectively used in paraphrase recognition task; most of these metrics utilize surface-string similarity but SVM classifier is used on top of it.

Recent boost in deep learning methods has also affected paraphrase detection studies. Already in 2011, a recursive autoencoder was trained that outperformed state of the art in paraphrase detection task (Socher et al. 2011). An attention-based long short-term memory architecture was used to automatically align pair of sentences and thus measure their similarity (Rocktäschel et al. 2015). A convolutional neural network achieved competitive performance in paraphrase detection task (He et al. 2015).

In the survey conducted by (Androutsopoulos and Malakasiotis 2010) several natural language processing tasks are mentioned where paraphrase methods can be applied, including question answering, text summarization, information extraction, machine translation, and natural language generation. More recently, even more directions of paraphrase applications have appeared in literature. (Barrón-Cedeño et al. 2013) the authors have demonstrated the importance of paraphrase for plagiarism detection and annotated a plagiarism corpus with paraphrase types. In (Petrović et al. 2012) paraphrase was used for first entity detection task, i.e. to find out the first document that describes a certain news event; they argued that lexical variation is a major obstacle for this task, as well as in number of other tasks, which can be overcome using paraphrase detection techniques. In (Pavlick and Nenkova 2015) importance of stylistic shifts in paraphrase for genre identification was demonstrated. In (Wieting et al. 2015) the authors used paraphrase corpus to train word embeddings and this improved performance in lexical similarity task. In (Hintz 2016) it was claimed that paraphrase can be used for stylistic harmonization in multi-document text summarization systems.

Even though the majority of work is done on English data, there is a certain interest in paraphrase research for other languages. For example, in (Eshkol-Taravella and

Grabar 2014) paraphrastic reformulations in French spoken corpora are studied. In (Nevěřilová 2014) a paraphrase generation system for Czech was proposed.

There are several publications on paraphrase detection and text reuse for the Russian language, e.g. (Bakhteev et al. 2015), (Khritankov et al. 2015), (Malykh 2016), however, the amount of research is rather small compared to other languages and to other natural language processing tasks for Russian. One of the missions of the ParaPhraser project is to overcome this gap.

A number of shared tasks on semantic textual similarity have been organized during the last five years as a part of SemEval conferences (Agirre et al. 2012, 2013). The paraphrase detection is very similar to this task, the only difference is that in our task the classification is discrete (paraphrase – non-paraphrase) while in textual similarity the task is to compute a semantic distance using continuous scale. SemEval shared tasks used English and Spanish data (Agirre et al. 2014, 2015). In the most recent shared task there was a sub-task on cross-lingual paraphrase detection (Agirre et al. 2016). There has been organized a special task on semantic similarity in Twitter (Xu et al. 2015). The shared task for paraphrased plagiarism detection has been organized as a part of Russian plagiarism detection shared task (Smirnov et al. 2017) though only one response has been submitted (Zubarev and Sochenkov 2017). Thus, we can claim that this is a first successful attempt to organize a shared task on Russian paraphrase detection.

2.2 Paraphrase Corpora

There exist a number of available paraphrase corpora. Microsoft Research Paraphrase Corpus (MSRP) (Dolan et al. 2004) is the most known of them. It consists of 5801 pairs of sentences (3900 of them being paraphrases) collected from news clusters. Although it is noted for its loose definition of a paraphrase, its 2-way annotation and high lexical overlap between the sentences (see, for example, Rus et al. 2016, Triantafillou et al. 2016, Liang et al. 2016), it is widely used in paraphrase detection task, and it is the corpus which inspired the development of other paraphrase resources (including our ParaPhraser corpus). MSRP is used as a dataset to monitor state-of-the-art result for paraphrase identification.

Other paraphrase corpora can be classified into several groups depending on the level of paraphrase they cover. Some corpora are purely sentential, while others have additional phrase- or word-level markup. There are also resources which only contain phrasal and word-level paraphrases.

Based on the source of paraphrases, paraphrase corpora can be classified as constructed automatically or manually. The former include parallel multilingual corpora and comparable monolingual corpora, suach as different translations of the same texts, news texts, texts on similar topics, e.g., from the social networks or students' answers to the questions, social media, Wikipedia, different descriptions of the same videos.

Sentential Corpora. One of the oldest sentential corpora known to us is the KMC corpus (Knight and Marcu 2002) collected from pairs of texts and their summaries.

User Language Paraphrase Corpus (McCarthy and McNamara 2008) is collected from student paraphrases of biology textbook sentences. Question Paraphrase Corpus

(Bernhard and Gurevych 2008) includes sentences pairs derived from WikiAnswers and annotated by social media users. Microsoft Research Video Description Corpus (Chen and Dolan 2011) is collected from short descriptions of videos annotated on the Amazon Mechanical Turk crowdsourcing platform.

Regneri and Wang corpus (Regneri et al. 2014) is collected from summaries of TV show episodes. Twitter Paraphrase Corpus (Xu et al. 2013) is derived from tweets corresponding to the same events (referring to the same date and mentioning the same named entity). Student Response Analysis Corpus (Dzikovska et al. 2013), is collected from students' answers to explanation and definition question. Semantic Textual Similarity Corpus (Agirre et al. 2013) is collected from several sources including news texts, Framenet-WordNet glosses and OntoNotes-WordNet glosses.

Non-English sentential paraphrase corpora known to us are Japanese Paraphrase Corpus for Speech Translation (Shimohata et al. 2004), consisting of sentences derived from travel conversation and versions of them paraphrased by humans, and Turkish Paraphrase Corpus (Demir et al. 2012), covering both sentence- and word- and phrase-level paraphrases, and derived from several sources: translations of a famous novel, subtitles, translations from an English-Turkish parallel corpus, and articles from a news website. More recently, another Turkish paraphrase corpus has been compelled.

Phrasal Corpora. The corpus compiled by (Cohn et al. 2008) is derived from three different sources: the multi-translation Chinese corpus (mtc), Jules Verne's "20,000 leagues under the sea" novels and MSRP (with non-paraphases).

WiCoPaCo (Max and Wisnewski 2010) is a corpus of French paraphrases collected from Wikipedia's revision history. WRPA (Vila et al. 2010) is another corpus based on Wikipedia and taking advantage of its structure. Unlike WiCoPaCo it captures only paraphrases of specific relationions (authorship, person-date of birth relation, etc.). The SEMILAR Corpus (The SEMantic SimILARity Corpus, (Rus et al. 2012)) is based solely on MSRP, enriched with word level similarity and alignments.

The Paraphrase Database developed by (Ganitkevitch and Callison-Burch 2014) is a rich paraphrase resource, which includes billions of paraphrase pairs. It is collected for more than 20 languages, including Russian, from bilingual parallel corpora. The authors use a language independent method to extract paraphrases from parallel bilingual texts: paraphrases are found in a single language by "pivoting" over a shared translation in another language. This approach was introduced by (Bannard and Callison-Burch 2005) and has been successfully applied by many researchers. PPDB includes lexical, phrasal and syntactic paraphrases, all of which are annotated with metrics from machine translation.

3   The ParaPhraser Project

There have been no publicly available paraphrase resources for the Russian language known to us, with the only exception of the dataset published by (Ganitkevitch and Callison-Burch 2014) as part of The Paraphrase Database project. The latter includes paraphrases on the word-, phrase- and syntactic levels, but it lacks information on the context of paraphrases. That is why we have constructed a sentential paraphrase corpus as part of our ParaPhraser project. The project is aimed at studying paraphrase phenomenon in Russian, including paraphrase extraction, paraphrase corpora construction and building paraphrase identification and generation models. Our results of solving paraphrase generation problem are available in the form of RESTful API service (https://paraphraser.ru/api/form), and the collected paraphrase corpus is also freely available on our website (http://paraphraser.ru/download). The corpus is not intended to be a general-purpose one. It consists of sentential paraphrases, extracted from news headlines, since news analysis is our primary interest, with the focus on such practical tasks as information extraction and text summarization.

To build the corpus we use a two-step procedure: first, automatic collection of candidate pairs and then manual annotation using crowdsourcing. We now describe both stages in more details.

3.1 The Construction Process

In the ParaPhraser project, we collect sentence pairs in real time. We parse web sites of several Russian news agencies and extract headlines of the articles. The headlines, as in the strategy proposed by (Wubben et al. 2009), are compared to each other, and paraphrase candidates are extracted using a similarity metric which extends the unsupervised matrix similarity metric proposed by (Fernando and Stevenson 2008) and is also a variant of soft cosine measure (Sidorov et al. 2014). A detailed description of metric calculation can be found in (Pronoza and Yagunova 2015). We include in the corpus pairs of sentences with the similarity metric value larger than a certain threshold. To provide more negative instances, we also include in the corpus a small random portion of sentence pairs with similarity metric value below the threshold.

3.2 Crowdsourcing

Potential paraphrases are annotated via our online interface[1]. The annotators are native speakers of Russian. Most of them are naïve speakers but there are also expert linguists and students of linguistics. Two sentences at a time are shown to an annotator and she/he decides whether the sentences convey the same meaning (1), similar meanings (0) or different meanings (-1). There are no specific instructions; instead, we let them use their own judgment and intuition. We introduce an entertainment element into the tedious annotation process: the annotators are shown funny pictures and/or facts at random intervals and are encouraged to work further. Inter-annotator agreement, calculated as Kohen's Cappa for all pairs of annotators, does not exceed 0.6.

When calculating resulting paraphrase classes, we only consider sentence pairs annotated by at least 3 users. We discard sentence pairs with opposite judgments (-1 and 1). Paraphrase class of a pair of sentences in the corpus is calculated as the median of all the scores given to this pair by the annotators (in case of ties the values are round down to the previous integers (0.5 to 0 and -0.5 to -1).

---

[1] http://paraphraser.ru/scorer.

3.3   Evaluation

Due to our paraphrase construction method only small subset of instances classified as negative by the algorithm is selected for manual assignment, which is not sufficient to compute recall. Thus we use precision to evaluate the quality of the unsupervised similarity metric for corpus construction. Precision of the metric on the current corpus, i.e. the training dataset used for the Shared Task, is 79.92%. Previously we evaluated our metric used for corpus construction (Pronoza et al. 2015) when the corpus consisted of about 5 thousand sentence pairs, and metric precision was 80.24%. Such results are quite promising compared with the original metric by Fernando and Stevenson that achieved 75.2% against MSRP.

4   Shared Task

4.1   The Task

The task input was a set of sentence pairs collected from news headlines and manually annotated by three native speakers as precise paraphrase, near paraphrase and non-paraphrase, as it is described in the previous section. We used approximately 7 and 2 thousand pairs for the training and test sets respectively. Both training and test sets are freely available[2].

The ParaPhraser corpus has been freely available from the very beginning, which means that all manually annotated data immediately became public. Only when we decided to organize the shared task we stopped publishing data to collect a test set. Thus, the training and the test sets are collected during different time periods and potentially annotated by different people. Some participants of the shared task noticed that cross-validation results were slightly better than results obtained on the test set, which can be explained by the fact that the test set was not a random sample from the data.

The shared task consisted of two sub-tasks:

Task 1. Three-class classification: given a pair of sentences, to predict whether they are precise paraphrases, near paraphrases or non-paraphrases.

Task 2. Binary classification: given a pair of sentences, to predict whether they are paraphrases (whether precise or near paraphrases) or non-paraphrases.

For each task we allowed standard and non-standard runs. In standard runs participants could not use any corpora but ParaPhraser or any derivatives from these external corpora (such as embeddings). However, we allowed to use any language processing tools or manually compiled dictionaries in standard runs. Any resources were allowed for non-standard runs.

Submissions of the participants were evaluated using accuracy and F1-score (F1-micro and F1-macro for three-class classification task).

---

[2] http://www.paraphraser.ru/download/.

4.2 Baselines

We provided two baselines for both tasks (2-class and 3-class classification). The first baseline assigns *random* class to each pair of sentences. The second baseline (*baseline2*) is a bit more complicated and consists of two steps. First, we conduct stemming of all words consisting of more than two characters by cutting off two characters from the end of a word. Then we compute a number of the overlapping words. For two-way classification a pair is classified as a paraphrase if more than a half of words from the longer sentence are mentioned in the shorter one. For 3-way classification we consider that the pair is a near-paraphrase if overlap of words is between 33% and 50% and precise paraphrase pair in case the overlap is more than 50%. Despite the simplicity of the technique the results appeared not to be the worst.

4.3 Results

For each task each participant might submit 20 standard and 20 non-standard runs. Since none of the participants made that many submissions we can assume that all participants submitted as many different responses as they wanted.

In total, 16 teams registered to the shared task, 11 submitted at least one result. The organizers submitted baseline results and an additional algorithm, described in the next section. The final results are presented in Tables 1, 2, 3 and 4[3]. For each team we present only the best result.

As can be seen from the tables, three-way classification is a more difficult task than two-way classification, for those systems that participated in both tasks the difference is up to 15% in accuracy and up to 30% in F-measure. The difference between standard and non-standard runs is not that high, which might be explained by the nature (and rather small amount) of our data: the sentences in the ParaPhraser corpus are highly

Table 1. Results: 3-way classification, standard run

Team	Accuracy	F1 (macro)	Method
Team3448	0.5901	0.5692	Classifier + linguistic features
AsoBek	0.5732	0.5557	Classifier + surface features
Penguins	0.5721	0.4443	Textula similarity based on word embeddings
MLforNLP	0.5695	0.5437	Technological approach (including translation into English)
True positive	0.5631	0.5382	Classifier + linguistic features
Dups	0.5478	0.5175	Neural networks
Baseline2	0.5325	0.5096	
DHL	0.4881	0.4483	Neural networks?
PhraseAnalog	0.4522	0.4344	Rule-based system
Team	0.4068	0.3699	
Random	0.3439	0.3341	

---

[3] In some cases we don't know, what method was used.

overlapping, which is common for corpora constructed from news texts, and simple shallow methods are usually more successful when tried against such data.

Table 2. Results: 2-way classification, standard run

Team	Accuracy	F1	Method
Dups	0.7459	0.8044	Neural networks
Team3448	0.7448	0.8078	Classifier + linguistic features
NLX	0.7274	0.7880	Neural networks
AsoBek	0.7211	0.7873	Classifier + surface features
True positive	0.7179	0.7656	Classifier + linguistic features
MLforNLP	0.7153	0.7853	Technological approach (including translation into English)
DHL	0.6292	0.7325	Neural networks
Baseline2	0.5858	0.5094	
Random	0.4966	0.5403	
Penguins	0.4702	0.2170	Textual similarity based on word embeddings

Table 3. Results: 3-way classification, non-standard run

Team	Accuracy	F1 (macro)	Method
True positive	0.6181	0.5838	Classifier + linguistic features
Dups	0.5969	0.5680	Neural networks
Team3448	0.5853	0.5642	Classifier + linguistic features
L533	0.5832	0.5567	
DHL	0.4099	0.3576	Neural networks?

Table 4. Results: 2-way classification, non-standard run

Team	Accuracy	F1	Method
True positive	0.7739	0.8110	Classifier + linguistic features
Dups	0.7665	0.7982	Neural networks
Team3448	0.7343	0.7827	Classifier + linguistic features
L533	0.6926	0.7794	
DHL	0.5605	0.6916	Neural networks

The participants of the task used a wide variety of techniques, from rule-based approaches to deep learning, and results of the task reflect the following tendencies: the best scores are obtained by the strategy of using traditional classifiers combined with fine-grained linguistic features, however, complex neural networks, shallow methods and purely machine learning methods also demonstrate competitive results[4]. The best results for two-way classification are slightly lower than English state of the art: the best result reported on ACL Anthology wiki page[5] yields accuracy 80.4% and F1-measure 85.9% though it is hard to compare results obtained on different corpora.

The papers, submitted to this volume present a variety of methods:

- Rule-based semantic parser (Boyarsky and Kanevsky 2017)
- SVM or Random Forest classifiers on top of thesaurus-based similarity features (Loukachevitch et al. 2017)
- SVM classifier on top of word and character unigrams, bigrams and trigrams (Eyecioglu and Keller 2017)
- Gradient Boosting classifier on top of features obtained from existing toolkits, including machine translation and similarity detection tools for the English language (Kravchenko 2017)
- Convolutional neural networks (Maraev et al. 2017)

4.4 Experiments

The task organizers submitted runs for both tasks as Team3448. Our approach towards the problem of paraphrase detection is based on the use of three types of sentence similarity measures as features in the paraphrase classification task (Pronoza and Yagunova 2015a): (1) surface, or shallow, similarity measures based on the overlap of n-grams, words and characters in the sentences; (2) semantic similarity measures that cover synonymy relations and derivation morphology; (3) distributional measures that use vector representations of words and phrases.

In total, we use 24 shallow features, 11 semantic features and 45 distributional features. Most of the features are described in (Pronoza and Yagunova 2015a), the others are distributional features with *phrase* embeddings with discriminative weights and 3-nearest neighbours smoothing for unknown words.

We submitted both standard and non-standard runs. Our models for the non-standard runs were built using all the described features. In the standard setting we cut off distributional features since they used external corpora. We tried two classifiers: SVM and Gradient Tree Boosting. Parameters of SVM and GTB were optimized on the development set (20% of the training set).

This approach achieved quite competitive results and even obtained the 1st place in the standard run of Task 1. Surprisingly, results of our standard runs are better than those of non-standard runs (the former use external resources and richer feature sets than the latter ones). This is similar to a general tendency, presented in Tables 1, 2, 3 and 4, where non-standard runs gain only little improvement.

---

[4] These are observations done during the shared task workshop at the AINL 2016 conference. Unfortunately, not all participants submitted a paper though some presentations are available on the conference webpage: http://ainlconf.ru/2016/materials.

[5] https://aclweb.org/aclwiki/index.php?title=Paraphrase_Identification_(State_of_the_art)).

5   Conclusion

We presented a freely available ParaPhraser corpus and the first shared task on Russian Paraphrase detection. We demonstrate that the corpus can be used for such task, which means that it is potentially useful for practical applications that require paraphrase identification step, such as cross-document text summarization or information extraction. The shared task results demonstrate that paraphrase detection methods developed for other languages may be applied to Russian and yield results only little worse than the English state of the art.

According to the results of the Shared task, various methods, from rule-based systems to deep learning, can be used for paraphrase detection, and most of them are quite successful at the task in question. As our dataset is not large, we expected a traditional (classifier + fine-grained features) approach to achieve best scores, and the results of the task met our expectations. However, deep learning approach also obtained high results (and the first place in one of the subtasks), and other methods, both surface and complex ones, appeared to be competitive.

We continue collecting data for the corpus. In total, we have already collected about 11 thousand pairs of sentences, which is 2 thousand more than we had during the shared task evaluation campaign. These 2 thousand are not yet publicly available since we plan to use part of it as a test set in the next shared task.

Though the shared task was quite successful there are also lessons learned, that we will use in the next shared tasks. First, we should have asked all participants to submit a short description of their method, so that we knew which approaches were tried even if the team decided not to submit the paper. Second, we should try to balance training and test set, so that training set contains some sentence pairs annotated by the same annotators as the test set and during the same period of time.

Another idea is to use much larger test set, where some pairs would be manually annotated and used to compute the evaluation measures and some pairs would be only automatically collected. These would serve for two goals: make it more difficult to optimize systems to a particular test set and reduce human efforts in annotation since the pairs on which all participating systems agree might be added to the corpus without human annotation.

References

Agirre, E., Banea, C., Cardie, C., Cer, D., Diab, M., Gonzalez-Agirre, A., Guo, W., Lopez-Gazpio, I., Maritxalar, M., Mihalcea, R., Rigau, G., Wiebe, J.: SemEval-2014 task 10: multilingual semantic textual similarity. In: Proceedings of SemEval 2014 (2014)

Agirre, E., Banea, C., Cardie, C., Cer, D., Diab, M., Gonzalez-Agirre, A., Guo, W., Lopez-Gazpio, I., Maritxalar, M., Mihalcea, R., Rigau, G.; Uria, L., Wiebe, J.: SemEval-2015 task 2: semantic textual similarity, English, Spanish and pilot on interpretability. In: Proceedings of SemEval 2015 (2015)

Agirre, E., Banea, C., Cer, D., Diab, M., Gonzalez-Agirre, A., Mihalcea, R.; Rigau, G., Wiebe, J.: Semeval-2016 task 1: semantic textual similarity, monolingual and cross-lingual evaluation. In: Proceedings of SemEval 2016 (2016)

Agirre, E., Cer, D., Diab, M., Gonzalez-Agirre, A.: SemEval-2012 task 6: a pilot on semantic textual similarity. In: Proceedings of SemEval 2012 (2012)

Agirre, E., Cer, D., Diab, M., Gonzalez-Agirre, A., Guo. W.: *SEM 2013 shared task: semantic textual similarity. In: Proceedings of *SEM 2013 (2013)

Androutsopoulos, I., Prodromos Malakasiotis, P.: A survey of paraphrasing and textual entailment methods. J. Artif. Intell. Res. **38**, 135–187 (2010)

Bakhteev, O., Kuznetsova, R., Romanov, A., Khritankov, A.: A monolingual approach to detection of text reuse in Russian-English collection. In: Artificial Intelligence and Natural Language and Information Extraction, Social Media and Web Search FRUCT Conference (AINL-ISMW FRUCT), pp. 3–10. IEEE (2015)

Bannard, C., Callison-Burch, C.: Paraphrasing with bilingual parallel corpora. In: Proceedings of the 43rd Annual Meeting of the ACL, pp. 597–604 (2005)

Barrón-Cedeño, A., Vila, M., Martí, M.A., Rosso, P.: Plagiarism meets paraphrasing: insights for the next generation in automatic plagiarism detection. Comput. Linguist. **39**(4), 917–947 (2013)

Bernhard, D., Gurevych, I.: Answering learners' questions by retrieving question paraphrases from social Q&A sites. In: Proceedings of the ACL 2008 3rd Workshop on Innovative Use of NLP for Building Educational Applications, pp. 44–52 (2008)

Bhagat, R., Hovy, E.: What is a paraphrase? Comput. Linguist. **39**(3), 463–472 (2013)

Boyarsky, K., Kanevsky, E.: Effect of semantic parsing depth on the identification of paraphrases in Russian texts. In: Filchenkov, A. et al. (eds.) AINL 2017. CCIS, vol. 789, pp.226–241. Springer, Cham (2017)

Chen, D.L., Dolan, W.B.: Collecting highly parallel data for paraphrase evaluation. In: Proceedings of the 49th Annual Meeting of the Association for Computational Linguistics, Portland, Oregon, USA, pp. 190–200 (2011)

Cohn, T., Callison-Burch, C., Lapata, M.: Constructing corpora for the development and evaluation of paraphrase systems. Comput. Linguist. **34**(4), 597–614 (2008)

Demir, S., El-Kahlout, I.D., Unal, E., Kaya, H.: Turkish paraphrase corpus. In: proceedings of LREC 2012, pp. 4081–4091 (2012)

Dolan, B., Quirk, C., Brockett, C.: Unsupervised construction of large paraphrase corpora: exploiting massively parallel news sources. In: Proceedings of the 20th International Conference on Computational Linguistics, p. 350. Association for Computational Linguistics (2004)

Dzikovska, M.O., Nielsen, R., Brew, C., Leacock, C., Giampiccolo, D., Bentivogli, L., Clark, P., Dagan, I., Dang, H.T.: SemEval - 2013 task 7: the joint student response analysis and 8th recognizing textual entailment challenge. In: Proceedings of the 7th International Workshop on Semantic Evaluation (SemEval 2013), Atlanta, Georgia, USA (2013)

Eshkol-Taravella, I., Grabar, N.: Paraphrastic reformulations in spoken corpora. In: Przepiórkowski, A., Ogrodniczuk, M. (eds.) NLP 2014. LNCS (LNAI), vol. 8686, pp. 425–437. Springer, Cham (2014). https://doi.org/10.1007/978-3-319-10888-9_42

Eyecioglu, A., Keller, B.: Knowledge-lean paraphrase identification using character-based features. In: Filchenkov, A. et al. (eds.) AINL 2017. CCIS, vol. 789, pp. 257–276. Springer, Cham (2017)

Fernando, S., Stevenson, M.: A semantic similarity approach to paraphrase detection. In: Proceedings of the 11th Annual Research Colloquium of the UK Special Interest Group for Computational Linguistics, pp. 45–52 (2008)

Ganitkevitch, J., Callison-Burch, C.: The multilingual paraphrase database. In: Proceedings of the Ninth International Conference on Language Resources and Evaluation (LREC 2014). European Language Resources Association (ELRA), Reykjavik, Iceland (2014)

He, H., Gimpel, K., Lin, J.: Multi-perspective sentence similarity modeling with convolutional neural networks. In: Proceedings of the 2015 Conference on Empirical Methods in Natural Language Processing, pp. 1576–1586 (2015)

Hintz, G.: Data-driven paraphrasing and stylistic harmonization. In: Proceedings of NAACL-HLT, pp. 37–44 (2016)

Khritankov, A., Botov, P., Surovenko, N., Tsarkov, S., Viuchnov, D., Chekhovich, Y.: Discovering text reuse in large collections of documents: a study of theses in history sciences. In: Artificial Intelligence and Natural Language and Information Extraction, Social Media and Web Search FRUCT Conference (AINL-ISMW FRUCT), pp. 26–32. IEEE (2015)

Knight, K., Marcu, D.: Summarization beyond sentence extraction: a probabilistic approach to sentence compression. Artif. Intell. **139**(1), 91–107 (2002)

Kravchenko, D.: Paraphrase detection using machine translation and textual similarity algorithms. In: Filchenkov, A. et al. (eds.) AINL 2017. CCIS, vol. 789, pp. 277–292. Springer, Cham (2017)

Liang, C., Paritosh, P., Rajendran, V., Forbus, K.D.: Paraphrase identification with structural alignment. In: Proceedings of the 16th International Joint Conference on Artificial Intelligence, pp. 2859–2865 (2016)

Loukachevitch, N., Shevelev, A., Mozharova, V., Dobrov, B., Pavlov, A.: RuThes thesaurus in detecting Russian paraphrases. In: Filchenkov, A. et al. (eds.) AINL 2017. CCIS, vol. 789, pp. 242–256. Springer, Cham (2017)

Madnani, N., Tetreault, J., Chodorow, M.: Re-examining machine translation metrics for paraphrase identification. In: Proceedings of the 2012 Conference of the North American Chapter of the Association for Computational Linguistics: Human Language Technologies, pp. 182–190. Association for Computational Linguistics (2012)

Malykh, V.: Robust word vectors for Russian language. In: Proceedings of Artificial Intelligence and Natural Language AINL FRUCT 2016 Conference, Saint-Petersburg, Russia, 10–12 November 2016, pp. 95–98 (2016)

Maraev, V., Saedi, C., Rodrigues, J., Branco, A., Silva, J.: Character-level convolutional neural network for paraphrase detection and other experiments. In: Filchenkov, A. et al. (eds.) AINL 2017. CCIS, vol. 789, pp. 293–304. Springer, Cham (2017)

Max, A., Wisniewski, G.: Mining naturally-occurring corrections and paraphrases from Wikipedia's revision history. In: LREC 2010, Valetta, Malta (2010)

McCarthy, P.M., McNamara, D.S.: The user-language paraphrase corpus. In: Cross-Disciplinary Advances in Applied Natural Language Processing (2008)

Nevěřilová, Z.: Paraphrase and textual entailment generation in Czech. Computación y Sistemas **18**(3), 555–568 (2014)

Pavlick, E., Nenkova, A.: Inducing lexical style properties for paraphrase and genre differentiation. In: Proceedings of the 2015 Conference of the North American Chapter of the Association for Computational Linguistics: Human Language Technologies, pp. 218–224 (2015)

Petrović, S., Osborne, M., Lavrenko, V.: Using paraphrases for improving first story detection in news and Twitter. In: Proceedings of the 2012 Conference of the North American Chapter of the Association for Computational Linguistics: Human Language Technologies, pp. 338–346. Association for Computational Linguistics (2012)

Pham, N., Bernardi, R., Zhang, Y.Z., Baroni, M.: Sentence paraphrase detection: When determiners and word order make the difference. In: Proceedings of the Towards a Formal Distributional Semantics Workshop, IWCS 2013, pp. 21–29 (2013)

Pronoza, E., Yagunova, E., Kochetkova, N.: Sentence paraphrase graphs: classification based on predictive models or annotators' decisions? In: Sidorov, G., Herrera-Alcántara, O. (eds.) MICAI 2016. LNCS (LNAI), vol. 10061, pp. 41–52. Springer, Cham (2017). https://doi.org/10.1007/978-3-319-62434-1_4

Pronoza, E., Yagunova, E.: Comparison of sentence similarity measures for Russian paraphrase identification. In: Artificial Intelligence and Natural Language and Information Extraction, Social Media and Web Search FRUCT Conference (AINL-ISMW FRUCT), pp. 74–82. IEEE (2015a)

Pronoza, E., Yagunova, E.: Low-level features for paraphrase identification. In: Sidorov, G., Galicia-Haro, S.N. (eds.) MICAI 2015. LNCS (LNAI), vol. 9413, pp. 59–71. Springer, Cham (2015b). https://doi.org/10.1007/978-3-319-27060-9_5

Pronoza, E., Yagunova, E., Pronoza, A.: Construction of a Russian paraphrase corpus: unsupervised paraphrase extraction. In: Braslavski, P., Markov, I., Pardalos, P., Volkovich, Y., Ignatov, D.I., Koltsov, S., Koltsova, O. (eds.) RuSSIR 2015. CCIS, vol. 573, pp. 146–157. Springer, Cham (2016). https://doi.org/10.1007/978-3-319-41718-9_8

Regneri, M., Wangy, R., Pinkal, M.: Aligning predicate-argument structures for paraphrase fragment extraction. In: LREC 2014, pp. 4300–4307 (2014)

Rocktäschel, T., Grefenstette, E., Hermann, K.M., Kočiský, T., Blunsom, P.: Reasoning about entailment with neural attention. arXiv preprint arXiv:1509.06664 (2015)

Rus, V., Banjade, R., Lintean, M.: On paraphrase identification corpora. In: LREC 2014, pp. 2422–2429 (2016)

Rus, V., Lintean, M., Moldovan, C., Baggett, W., Niraula, N., Morgan, B.: The SEMILAR corpus: a resource to foster the qualitative understanding of semantic similarity of texts. In: Semantic Relations II: Enhancing Resources and Applications, The 8th Language Resources and Evaluation Conference (LREC 2012), May 23–25, Instanbul, Turkey (2012)

Shimohata, M., Sumita, E., Matsumoto, Y.: Building a paraphrase corpus for speech translation. In: Proceedings of 4th international conference on language resources and evaluation (LREC) (2004)

Sidorov, G., Gelbukh, A., Gómez-Adorno, H., Pinto, D.: Soft similarityand soft cosine measure: similarity of features in vector space model. Computación y Sistemas 18(3), 491–504 (2014)

Smirnov, I., Kuznetsova, R., Kopotev, M., Khazov, A., Lyashevskaya, O., Ivanova, L., Kutuzov, A.: Evaluation tracks on plagiarism detection algorithms for the russian language. In: Dialog 2017 (2017)

Socher, R., Huang, E. H., Pennin, J., Manning, C. D., Ng, A.Y.: Dynamic pooling and unfolding recursive autoencoders for paraphrase detection. In: Advances in Neural Information Processing Systems, pp. 801–809 (2011)

Triantafillou, E., Kiros, J.R., Urtasun, R., Zeme, R.: Towards generalizable sentence embeddings. In: Proceedings of the 1st Workshop on Representation Learning for NLP, pp. 239–248, Berlin, Germany (2016)

Vila, M., Martí, M.A., Rodríguez, H.: Is this a paraphrase? What kind? Paraphrase boundaries and typology. Open J. Modern Linguist. 4(01), 205 (2014)

Vila, M., Rodriguez, H., Marti, M.A.: WRPA: a system for relational paraphrase acquisition from Wikipedia. In: Procesamiento del Lenguaje Natural, Revista no. 45, septiembre 2010, pp. 11–19 (2010)

Wieting, J., Bansal, M., Gimpel, K., Livescu, K.: Trans. Assoc. Comput. Linguist. 3, 345–358 (2015)

Wubben, S., van den Bosch, A., Krahmer, E., Marsi, E.: Clustering and matching headlines for automatic paraphrase acquisition. In: Proceedings of the 12th European Workshop on Natural, Language Generation, pp. 122–125, Athens, Greece (2009)

Xu, W., Callison-Burch, C., Dolan, W.B.: SemEval-2015 task 1: paraphrase and semantic similarity in Twitter (PIT). In: Proceedings of SemEval (2015)

Xu, W., Ritter, A., Grishman, R.: Gathering and generating paraphrases from Twitter with application to normalization. In: Proceedings of the Sixth Workshop on Building and Using Comparable Corpora, August 2013, Sofia, Bulgaria, pp. 121–128 (2013)

Zubarev, D.V., Sochenkov, I.V.: Paraphrased plagiarism detection using sentence similarity. In: Dialog 2017 (2017)

Effect of Semantic Parsing Depth on the Identification of Paraphrases in Russian Texts

Kirill Boyarsky[✉] and Eugeni Kanevsky

ITMO University, St. Petersburg, Russia
boyarin9@yandex.ru

Abstract. As a tool to solve the problem of identification of paraphrases in Russian texts, the paper proposes the semantic-syntactic parser SemSin and a semantic classifier. Several alternative methods for evaluating the similarity of sentence pairs—by words, by lemmas, by classes, by semantically related concepts, by predicate groups—have been analyzed. Advantages and drawbacks of the methods are discussed. The paraphrase identification quality has been shown to rise with increasing depth of using the semantic information. Yet, complementing the analysis with predicate groups, identified by the dependency tree, may even cause the identification to degrade due to the growing number of false positive decisions.

Keywords: Russian texts · Paraphrases · Semantic dictionary · Lemmas Classifier · Classes · Semantic-syntactic parsing · Synonymy

1 Introduction

An intense interest of researchers in the field of information retrieval has been lately aroused by the paraphrase identification. The issue is the subject of rather many papers, but only few of them are concerned with the Russian language.

English-language publications include a great number of works on paraphrase identification, employing various lexical, syntactic, and semantic techniques [1, 2]. Most methods used learning, tokenization, POS tagging, and stemming for verbs and nouns only [3]. The authors assigned different weights to words with account of their grammatical role in sentences. Corley and Mihalcea [4] applied measurements of semantic similarity of texts with the help of WordNet [5]; the semantic similarity of words was assessed only for verbs and nouns, while adverbs, adjectives, and cardinal numerals were compared lexically. Such method was shown to be significantly more accurate than the simple lexical comparison. An allied technique, based on semantic information WordNet, featured in [6]. To identify paraphrases, Pershina [7] used additionally an idiom database. The best results of paraphrase identification in English texts yield F-measure about 82%.

The works on paraphrase identification in Russian texts are far from numerous [8]. Significant complexities arise due to specifics of the Russian language, which is notable for free word order and rich morphology. The present work aims to study the

© Springer International Publishing AG 2018
A. Filchenkov et al. (Eds.): AINL 2017, CCIS 789, pp. 226–241, 2018.
https://doi.org/10.1007/978-3-319-71746-3_19

efficiencies of alternative methods of paraphrase analysis in Russian texts and find the optimum degree of using the semantic information.

This work was performed within the framework of the contest on identification of Russian paraphrases [9] that were present in the corpus of paraphraser.ru [10]. Two alternative analyses of pairs of news headlines were offered for the contest, division into two groups (paraphrases and not paraphrases) and into three groups (with a group of loose paraphrases separated additionally). It turned out by the presented data that the first alternative analysis exhibits substantially better accuracy. To a large degree, this is related with the subjectivity of identifying the loose paraphrases. Therefore, division into two groups was selected.

To determine whether two sentences are paraphrases, i.e. their meanings are the same for native speakers, a numerical measure of similarity is to be introduced. As such measure, we used Jaccard index J: if two sentences A and B contain $n(A)$ and $n(B)$ lexical units,

$$J = \frac{n(A) \cap n(B)}{n(A) \cup n(B)} = \frac{n(A) \cap n(B)}{n(A) + n(B) - n(A) \cap n(B)}.$$

In compliance with the criterion, the similarity measure is defined as the ratio of the number of coincident units and the total number of different units.

Two issues should be resolved: what the unit to be compared is and what the comparison criterion is. We selected four alternatives for comparison.

Alternative 1. The comparison unit is taken to be a word or another continuous character sequence. The sequence is supposed long enough, so that one- and two-letter words are disregarded.

Advantages. No tools are needed for parsing the text. Identification of a coincident character group is very likely to signify the presence of cognate words, close in meaning. If such a group is a separate word, the word can possibly have one and the same function in both utterances.

Drawbacks. Similar fragments are highly probable to include functional words. Their coincidence is often of a random nature and fails to reflect the meaning of a text fragment (though the coincidence of, say, negative particle *не* (*no*) can be instrumental for finding the similarity measure). Besides, since Russian belongs to the synthetic type of languages, even a very slight rephrasing that has no impact whatsoever on the utterance meaning leads to changes in word forms. This degrades the analysis accuracy substantially.

Alternative 2. The comparison unit is taken to be the normalized word form (lemma) rather than a word form. Functional words are disregarded. The parser is adjusted to a subject matter area by excluding homonymic word forms that do not fall into the given area. For instance, the word form *белку* corresponds to two lemmas, *белок* (*protein*) и *белка* (*squirrel*). In a biochemistry text, only the former will be selected.

Advantages. The complexities related to the abundance of word forms of each lemma are eliminated. The analysis becomes more independent of the particular structure of a sentence (*человек, который смеется* vs *смеющийся человек*—*the man who laughs* vs. *the laughing man*).

Drawbacks. A tool is needed for morphological analysis. Often enough (roughly in 7% of cases), the identification of a lemma by a word form is ambiguous, i.e. the problem of homonymy arises. The issues related to functional words remain.

Alternative 3. The comparison takes account of the word semantics.

Advantages. The words are compared by meaning rather than spelling, which is, in principle, to raise the accuracy of paraphrase identification.

Drawbacks. Complexity and ambiguity of the semantic analysis. The need to use semantic dictionaries. If the function in English is fulfilled by the semantic web, the corresponding tools for the Russian language are underdeveloped and there is no semantic dictionary, adopted as a de-facto standard.

Alternative 4. The comparison involves the complete parsing tree.

Advantages. The possibility of analyzing the similarity of noncontiguous word groups and identifying conceptual blocks, describing the terms of the subject matter area.

Drawbacks. Raised probability of parser errors. Insufficiently explored issue of identifying contextual blocks. High sensitivity of the method to the replacement of one kind of constructions with another, e.g., a participle construction with a dependent clause.

Best known among Russian-language parsers, capable of in-depth sentence analysis, are ETAP-3 [11] and also those by Yandex [12], Abbyy [13], «Dictum» [14]. The first two parsers operate by a system of rules, the other two draw on in-house technologies. All parsers use dictionaries in some way or another.

The present work employed semantic-syntactic parser SemSin [15], with its main features examined below.

2 Parser

2.1 The SemSin System

SemSin is a system, comprising functions of lemmatizator, syntactic and semantic analyzer. The parser includes dictionaries, classifier, morphological analysis unit, and syntactic module driven by a set of production rules.

The main dictionary, built from modified Tuzov's dictionary [16], contains about 190,000 lexemes. Each lexeme has an indication of its morphological characteristics as well as the number (or numbers) of its semantic class and actants or valencies (to link dependent words). A lexeme can be in correspondence with several semantic homonyms, e.g., коса as a tress, коса as a foreland, коса as a scythe, which belong to different classes.

The dictionary of phrasemes contains more than 4900 items, consisting of two or more words, about 2000 of them being unmodifiable. These are complex prepositions or adverbs as well as metaphors or compound proper names. Unmodifiable idioms are joined together in a token prior to the syntactic analysis. Modifiable idioms are considered as separate lexemes, but are combined into a single unit with the appropriate semantic class.

The dictionary of prepositions includes more than 2240 items, each of them indicating the preposition proper, the case of the noun it requires, the semantic class of the noun, and the type of the resulting link (e.g., "*Where*", "*When*" etc.). One and the same preposition can produce different types of links, if usable with different cases or for nouns of various classes.

The classifier contains 1700 classes that form a tree, built on the semantic principle. A lexeme can fall into several classes. In greater detail, the system of classes and its use for paraphrase identification are discussed in Sect. 3.2.

The dictionary volume ensures recognition of about 96% of words at the transition to a more modern batch of news texts. In roughly half of cases, the words missing in the dictionary are proper names, which the system can recognize automatically. At the system training stage, the dictionary was supplemented with 430 words. Upon that, the number of unknown words in processing of the contest corpus did not exceed 0.5%. Note that appearance of words, unknown to the parser, is not too critical in the paraphrase identification problem, since the words are compared just contextually.

2.2 Syntactic Parsing and Semantic Analysis Using SemSin

The SemSin system analyzes the text by paragraphs. First, the text is divided into tokens, and each word is processed by the morphological analyzer. The parsing result is returned as one or several lemmas with indication of the corresponding actants. Upon that, the presyntactic module is run, which separates a paragraph into sentences, reviews spelling and morphological characteristics of some constructions (hyphenated words, compound and alphanumeric numerals), attempts resolving the problem with unknown words, and parses phrasemes [17].

Further on, the syntactic module is activated, which uses about 480 rules [18, 19]. While analyzing a sentence, grammatical and POS homonymy is lifted, the sentence is segmented, and the syntactic dependency tree is built. In many cases, the lexical homonymy is lifted, too. Dependency parsing is well suited for processing a morphologically rich, free-word-order language like Russian [20]. The result of the syntactic analysis is saved as an XML-file, each word given its lemma, POS, grammatical signs (animacy, gender, number, case, tense etc.), basic class number, parent node identifier and relation type as well as references to words in close semantic relation with given one (siblings) (Sect. 3.2).

The resulting tree contains as full as possible information on a sentence. The information can later underlie the solving of a variety of problems: term identification [21], text classification [22] etc. The present work discusses, which information exactly is useful for finding the similarity of sentence meanings, i.e. for paraphrase identification.

3 Text Analysis

The present work focused attention on the analysis of alternative ways of describing semantics and incorporating it into the paraphrase identification procedure. All examples and expert assessments were taken from the Russian paraphrase corpus [23].

3.1 Lemmatization

As shown in [24], the greatest power of differentiation in text clusterization problems in Russian (in contrast to English and French) belongs to nouns. Yet for paraphrase identification, extraction of nouns only is too coarse and cannot cover all nuances of meaning. So, kept for the comparison were nouns, adjectives, verbs, and verbal forms (participles, gerunds), and numerals. Examples of the effect of lemmatization on the Jaccard index value are given in Table 1.

Table 1. The lemmatization effect

Sentences		No lemmatization	With lemmatization
1	NI опубликовал список самого опасного вооружения флота России (NI published the list of the most dangerous weapons of the Russian Navy) В США опубликован топ-5 самых опасных вооружений ВМФ России (The top 5 most dangerous weapons of the Russian naval forces was published in the USA)	0.067	0.455
2	Путин впервые объявил минуту молчания на параде Победы (Putin first declared the minute's silence at the Victory parade) Пан Ги Мун поблагодарил Путина за организацию Парада Победы (Ban Ki-moon thanked Putin for organization of the Victory Day parade)	0.067	0.250

Note that the lemmatization increases the degree of cohesion of sentences both when they are paraphrases (Example 1) and when the coincidence of words is accidental and there is no coincidence of meanings (Example 2). Thus, lemmatization does unequivocally raise the accuracy of comparison of sentence vocabularies, yet is insufficient for comparing the meanings.

3.2 Semantics. Accounting for Classes

To compare sentences more accurately, a semantic classifier, containing about 1700 classes, was used. Its basis is Tuzov's classifier [16], targeted exactly on the computational analysis of texts. The tree of classes is built so that the semantic classes had certain syntactic properties. For instance, the list of activities, possible for a living creature, is other than that of actions of inanimate objects; different classes may have different attributes etc. In addition, the format of the classifier is suitable for machine use.

Our classifier differs, e.g., from that of Shvedova [25]. For instance, the word *жрец* (*priest*) is in both cases classified practically the same—as a certain type of man's profession. In the meantime, Shvedova places the word *желание* (*desire*) into the tree branch *духовный мир–чувства–...* (*spiritual world–feelings–...*), whereas our classifier treats psychic phenomena and feelings as human properties. The position of the word *priest* in classifier WordNet is on the whole in accord with the Russian classifiers, while that of the word *desire* is nearer to Shvedova's one.

When developing a classifier, it is very difficult, if possible at all, to ascertain, which tree tier some word is to be placed on, which attribute the subclasses are to be divided by, when the further branching is to be stopped. For instance, *коса as a scythe* belongs in WordNet to the class of cutting tools and in our and Shvedova's classifiers to that of agricultural implements. The paraphrase identification assumed the membership of different words of the first and second sentences in one class to mean the similarity of the meanings.

Often, a hyperonym of a particular word appears in its place in the text. To identify hyperonyms, a class was sought within accuracy ±1 hierarchy level. This ensures the coincidence of words highlighted with bold font in the following examples:

Лавров подарил Керри **помидоры**... vs. *Лавров подарил Керри* **овощи**... (*Lavrov presented Kerry with* **tomatoes**... vs. *Lavrov presented Kerry with* **vegetables**...)
Жертвами взрыва ... стали не менее трех **человек** vs. *Жертвой взрыва... стал* **гражданин** *Великобритании* (*At least three* **persons** ... *fell victims to an explosion* vs. *A Great Britain* **citizen** ... *became a victim to an explosion*)

Exceptions are proper names, since all city names, e.g., fall into one class, as do all last names of people.

The class comparison often overlays the synonymy relations. It is well to bear in mind, however, that the Russian language has significantly fewer coincidences of word forms of different parts of speech because of its morphological specifics. Thus, English *iron* is both a noun and an adjective, while Russian has the noun *железо*, and the adjective *железный*. When identifying the synonymy in Russian-language RusNet [26], the coincidence of parts of speech in synonymic words is meant. The comparison «by classes» is broader in this sense and allows conclude on meaning similarity even at a substantial rephrasing:

Турецкий сухогруз подвергся **обстрелу**... vs. *Турецкий сухогруз* **обстреляли** (*Turkish bulk carrier went under* **fire** vs. *Turkish bulk carrier was* **fired** at)

Undoubtedly, the issue of whether two nouns, belonging to one class, are synonyms is ambivalent. For instance, the words *веревка, бечевка* (*rope, twine*) are easily interchangeable in a text, whereas the words *помидор, огурец* (*tomato, cucumber*) are not. Nonetheless, the analysis suggests that the similarity of classes most frequently speaks of similarity of meanings.

3.3   Semantics. Synonymy and Additional Classes

In a number of cases, the comparison by classes proves insufficient. For instance, the «tennis problem» is known [27], which consists of the fact that words belonging to one subject matter area reside in quite different branches of the classifier. So, our dictionary was complemented with information on semantic similarity of words. We will refer to such words as **siblings**. Actually, the emergence of additional relations means transformation of the class tree into a semantic network. Note that the work is at its initial stage and the number of such relations is about 10,500 at the moment.

First of all, this concerns toponyms. As pointed out above, proper names are compared by lemmas only. Yet, one and the same country may pass by different names, and this is to be allowed for in paraphrase analysis. An abbreviation can often stand for a country name; so, the dictionary was complemented with the data on the identical synonymy of words and expressions.

*В **Британии** палата общин одобрила однополые браки* (The House of Commons approved same-sex marriages in **Britain**)
*Палата общин **Великобритании** одобрила однополые браки* (The House of Commons of **Great Britain** approved same-sex marriages)

*США просят **РФ** немедленно отменить запрет на ввоз мяса* (The US urges the RF to immediately lift the ban on meat imports)

*США призвали **Россию** немедленно снять запрет на импорт мяса* (The US urges Russia to immediately lift the ban on meat imports)

КНДР готовится нанести ракетный удар по США (**DPRK** is getting ready to deliver a missile strike on the US)

Северная Корея пригрозила ракетным ударом по США (**North Korea** threatened with a missile strike on the US)

*"Ночные волки": суд в **ФРГ** отпустил байкера* ("Night wolves": a court in **BRD** sets a biker free)

*Задержанного в **Германии** байкера «Ночных волков» отпустили* (The "Night wolves" biker detained in **Germany** was set free)

A kind of an analog for class-subclass relations for toponyms is the information, which region and country a populated place belongs to. This was added to the dictionary, though on a rather limited scale. For instance, it was indicated that *Дамаск* (*Damascus*) is the capital of *Сирии* (*Syria*) and city *Сент-Луис* (*St. Louis*) is in state *Миссури* (*Missouri*) of country *США* (the US):

*Неизвестный открыл огонь в бизнес-школе штата **Миссури*** (An unknown opened fire in a business school in state **Missouri**)
*В **Сент-Луисе** преступник открыл огонь в бизнес-школе* (In **St. Louis**, a criminal opened fire in a business school)

The next group of words with additional relations, introduced into the dictionary, is characterization of people by their ethnic origin and place of residence. One needs to know that *сибиряк* (*Siberian*) is a man's name by his place of residence and also that the place is exactly *Сибирь* (*Siberia*). Thus, each such word is «attached» to the corresponding country, region or city:

Американец выиграл в лотерею за 100 евро картину П. Пикассо (An **American** won a P. Picasso's picture in a lottery for 100 euros)
*Житель **Пенсильвании** выиграл в лотерею картину Пикассо* (A resident of **Pennsylvania** won a Picasso's picture in a lottery)

As siblings for some commonly occurring names of high-profile figures, the posts and countries of the latter are indicated:

*Синдзо **Абэ** в письме **президенту РФ** объяснил, почему не приедет на 9 мая (Sindzo **Abe** explained in his letter to the **RF President**, why he would not come for May, 9)*
*В письме **Путину** японский премьер объяснил причины отказа приехать в Москву 9 Мая (In his letter to **Putin**, **Japanese Prime Minister** explained the reasons of the refusal to come to Moscow on May, 9)*

***США** приостановили поставку истребителей в Египет (**The US** suspended the delivery of fighters to Egypt)*

***Обама** приостановил поставки истребителей F-16 в Египет (**Obama** suspended the delivery of F-16 fighters to Egypt)*

Another group of siblings comprises relations by membership of certain social strata, organizations etc. *Communist* names a man not just by membership in some civic association, but specifically in a communist party, while *hockey player* has to do with just ice hockey among all sports:

*Депутаты от **КПРФ** попросили Путина взять под защиту гималайского медведя (**Deputies** from the **CPRF** asked Putin to take Himalayan black bear under his protection)*
***Коммунисты** попросили Путина защитить гималайских медведей (**Communists** asked Putin to protect Himalayan black bears)*

*Сборная России по **хоккею** проиграла финнам и во втором матче Евротура (**Ice hockey** team Russia lost to Finns in the second match of Euro Hockey Tour, too)*

*Российские **хоккеисты** проиграли на Евротуре четыре матча подряд (Russian **hockey players** lost four Euro Hockey Tour matches in a row).*

A separate group of siblings is formed by relations of established terms and idiomatic expressions with their semantic analogs:

*В Красноярском крае **исчез** заместитель прокурора (Deputy Prosecutor **disappeared** in Krasnoyarsk Territory)*
*В Красноярском крае **пропал без вести** помощник прокурора района (Deputy District Prosecutor is **gone missing** in Krasnoyarsk Territory)*

***Ушел из жизни** Уго Чавес (Hugo Chavez **passed from this life**)*

***Умер** Уго Чавес (Hugo Chavez **died**)*

*Шезлонг с "Титаника" **продали** за 100 тысяч фунтов (Deck chair from Titanic was **sold** for a hundred thousand pounds)*

*Шезлонг с «Титаника» **ушел с молотка** за 100 тысяч фунтов (Deck chair from Titanic was **auctioned off** for a hundred thousand pounds)*

All lexemes in the dictionary are divided into basic and derivative, with the latter to be understood not only as the words, cognate with ones. Basic lexemes number somewhat less than a half (about 83,000). Roughly 20,000 of them form semantic families, including derivative words with similar meanings. So, the family of the basic word *чувство* (*feeling*) has more than 100 derivative words, found among them are nouns (*нечувствительность, аналгезия—insensitivity, analgesia*), adjectives (*чувствительный, душещипательный—sensitive, soulful*), verbs (*чувствовать, обуревать—feel, overwhelm*). The semantic class of a derivative word coincides in some cases with that of the basic one and differs in others. For instance, the basic word *сигнал* (*signal*) falls into a subclass of a branch of the semantic tree *информация* (*information*). Its derivatives *сигнальный, сигнализация, сигнализировать* (*signaling, signalization, signalize*) are in the same class, whereas the word *сигнальщик* (*signalman*) names a man of a certain occupation. Information on the membership of words in a common family contributes about additional 7,500 relations to the semantic network and was also allowed for in paraphrase identification.

Consider the example:

> *Оппозиция **ФРГ** угрожает правительству судом из-за шпионского скандала* (*The BRD opposition threatens the government to go to court because of the spy scandal*)
> ***Немецкая** оппозиция пригрозила правительству иском из-за скандала с BND* (*The **German** opposition menaced the government with a lawsuit because of the scandal with BND*)

If the sentences are compared by lemmas only, there are three coincidences (*оппозиция, правительство, скандал—opposition, government, scandal*) with Jaccard index $J = 0.27$. With classes accounted for, we get the coincidence for verbs *угрожать* (*threaten*) and *пригрозить* (*menace*), then $J = 0.40$. The name *ФРГ* (*BRD*) falls into the family of siblings for the word *Германия* (*Germany*), this very family contains the word *немецкий* (*German*), a derivative of the basic word *немец* (*a German*), which is in turn a sibling of *Германии* (*Germany*). We get $J = 0.55$. Finally, the lexemes *суд* (*court*) and *иск* (*lawsuit*) are siblings, too. Ultimately, we get $J = 0.75$, which reflects quite well the similarity of meanings of the sentences.

3.4   Dependency Tree

We attempted to use the parser-built dependency tree to refine the comparison of sentence meanings. The coincidence of the subject of action and the predicate proper was assumed to increase the probability of the sentences in question being paraphrases. In this case, the coincidence of lemmas was factored in the Jaccard index evaluation with weight 1.5.

However, the practice revealed that Jaccard index of sentence pairs with the said property is typically great as it is. Therefore, its slight growth moves such a pair from category «not paraphrase» to «paraphrase» comparatively seldom. But these rare transitions are also of certain interest. Table 2 lists the Jaccard index values with account of coincidences of lemmas, classes, and siblings (LCS) and additionally predicate pairs (LCSP).

The Table shows that allowing for subject–predicate coincidences makes finding paraphrases in some cases easier (Examples 1, 2). However, this can lead also to fallacious rise in the degree of similarity between sentences (Examples 3, 4). Occasionally, the decision whether the sentences in a pair are paraphrases is of borderline nature and can be interpreted by different experts differently (Example 5). Sometimes, plain expert's mistakes are possible (Example 6).

Generally, allowing for subject–predicate pair coincidences caused a slight lowering of the analysis quality with decrease of F-measure. Yet the change falls within the accuracy due to the subjectivity of the expert judgment on what a paraphrase is. A promising line of the work development, as we believe, is the further expansion of

Table 2. The effect of coincidence of subject-predicate pair on the Jaccard index value

Sentences		LCS	LCSP	Comment
1	МИД Чехии: дипломат получил выговор за высказывание о пожаре в Одессе (Czech Foreign Ministry: a diplomat was reprimanded for speaking about the fire in Odessa) Оправдавший сожжение людей в Одессе чешский дипломат получил выговор (The Czech diplomat who had justified the burning of people in Odessa was reprimanded)	0.33	0.50	The result is correct; the sentences are identical in meaning
2	Кобзон назвал результат России на Евровидении очень достойным (Kobzon called the result of Russia at the Eurovision very worthy) Иосиф Кобзон назвал второе место Полины Гагариной достойным (Iosif Kobzon called Polina Gagarina's second place worthy)	0.27	0.45	The result is correct; the sentences are identical in meaning
3	Лужков назвал свою ферму примером для российского правительства (Luzhkov called his farm an example for the Russian government) Лужков назвал российскую экономику "антинародной" (Luzhkov called the Russian economy "antipopular")	0.37	0.62	Erroneous increase in the degree of similarity, which is connected to disregarding the differences in direct objects
4	МВД насчитало 200 тыс. участников празднования Дня Победы в Севастополе (MIA has counted 200 thousand participants of the Victory Day celebration in Sevastopol) МВД насчитало 250 тыс. участников акции "Бессмертный полк" в Москве (MIA counted 250 thousand participants of the action "Immortal regiment" in Moscow)	0.28	0.48	Erroneous increase in the degree of similarity, which is connected to disregarding the differences in adverbs of place

(*continued*)

Table 2. (*continued*)

Sentences		LCS	LCSP	Comment
5	Суд оправдал Васильеву в хищении акций на 2 млрд. рублей (The court acquitted Vasilyeva in plunder of shares worth 2 billion rubles) Суд оправдал Васильеву в мошенничестве со зданием на Арбате (The court acquitted Vasilyeva of fraud with an Arbat building)	0.27	0.45	The sentences are not treated as paraphrases in the tagging standard. They are close in meaning; if considered as news headlines, there is some difference, but the reference is actually to the same event
6	Эксперты из России и Белоруссии направились с проверкой в Эстонию (Experts from Russia and Belarus went on inspection to Estonia) Российские военные эксперты направились с проверкой в Эстонию (Russian military experts went on inspection to Estonia)	0.71	0.86	The sentences are not treated as paraphrases in the tagging standard. The event is undoubtedly the same; it is a probable tagging error

the dependency tree analysis with the view of lowering the similarity coefficient for sentence pairs, differing in, e.g., place of action, and reducing the number of false positives.

4   Results

Below are the results of the semantic similarity analysis of sentence pairs for paraphrase identification, which was run according to the following schemes:

1. By words (W). The comparison involved all words, including the auxiliary parts of speech.
2. By lemmas (L). The comparison involved nouns, adjectives, verbs, numerals.
3. By classes (LC). The comparison additionally involved the coincidence of classes by the semantic tree.
4. By siblings (LCS). The comparison additionally involved the coincidence of classes by the semantic web.

Examples of the effect of the counting scheme on the Jaccard index value are given in Table 3.

Table 3. Jaccard indices for different coincidence counting schemes

No.	Sentences	W	L	LC	LCS
1	Крупный пожар вспыхнул на складе на северо-востоке Москвы (Large fire broke out at a warehouse in north-eastern Moscow) Крупный пожар в административном здании в центре Москвы потушен (Large fire in an office building was put out in downtown Moscow)	0.230	0.300	0.444	0.444
2	Продажи АвтоВАЗа в России в апреле сократились на 38,3% (AvtoVAZ April sales in Russia dropped by 38.3%) Продажи «АвтоВАЗа» в России рухнули на треть (AvtoVAZ Russian sales crashed by a third)	0.273	0.500	0.667	0.778
3	СМИ: поздравляя Вакарчука, Кличко ошибся с возрастом и именем юбиляра (Media: When congratulating Vakarchuk, Klichko made a mistake in the celebrant's age and name) Кличко перепутал в поздравлении имя и возраст солиста «Океана Эльзы» (Klichko confused the name and age of Okean Elzy's front man in his congratulation)	0.062	0.230	0.455	0.455
4	Выселен последний экс-депутат, незаконно занимавший жилье в Москве (The last ex-deputy who occupied a dwelling illegally was evicted in Moscow) В Москве выселили бывшего депутата Думы из служебной квартиры (Former Duma deputy was evicted out of official housing in Moscow)	0.071	0.181	0.300	0.600
5	Morgan Stanley взял на работу бывшего зампреда Банка России (Morgan Stanley employed the former deputy chairman of the Bank of Russia) Бывший глава ФСФР нашел работу в Morgan Stanley (Ex-head of FFMS landed a job with Morgan Stanley)	0.200	0.364	0.500	0.500

At the stage of system retraining by the test corpus, the optimum «cutoff level» was estimated: what the threshold Jaccard index value is to consider a sentence pair a paraphrase. The evaluation involved standard parameters, accuracy (A), precision (P), recall (R), and F-measure (F):

$$A = \frac{T_P + T_N}{T_P + T_N + F_P + F_N}$$

$$P = \frac{T_P}{T_P + F_P},$$

$$R = \frac{T_P}{T_P + F_N},$$

$$F = \frac{2PR}{P + R}.$$

Here T_P is the number of true positive decisions, i.e. both the expert and the machine categorized the pairs as paraphrases; T_N is the number of true negative decisions; F_P is the number of false positive decisions; F_N is the number of false negative decisions.

Typical distributions are shown on Fig. 1.

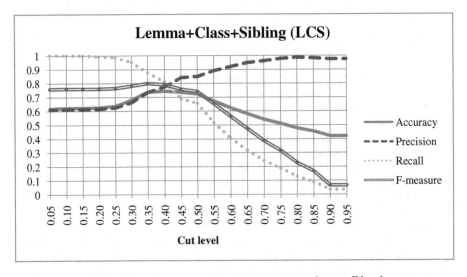

Fig. 1. Dependence of accuracy parameters on the cutoff level

The dependences for other counting schemes were of qualitatively similar nature. It is evident that lowering the cutoff level, i.e. classing almost all sentence pairs as paraphrases, can ensure an arbitrarily high value of recall, and raising it can do the same for accuracy. The analysis quality was assessed by the accuracy and F-measure parameters, for which the optimum cutoff level value fell into interval 0.35...0.40.

Table 4. Accuracy and F-measure at different comparison schemes

Scheme	W	L	LC	LCS	LCSP
A	0.6920	0.7185	0.7430	0.7435	0.7419
F	0.7625	0.7738	0.7836	0.800	0.7951

5 Conclusion

The results of the quality of paraphrase identification are given Table 4. The comparison proceeded in compliance to the Golden standard tagging [28]. Quoted for each considered comparison scheme are the best values of accuracy *A* and F-measure *F*.

Basing on the table, one can conclude that:

- In most cases, the news headlines, referring to one and the same event, are highly similar lexically and can be designated as paraphrases by any method.
- Application of semantic-syntactic comparison methods (LC) improves the results against not only simple character-at-a-time comparison but also that by lemmas.
- Increasing the semantic analysis «depth» by virtue of transition from the tree of classes to the semantic web (LCS) raises the analysis quality.
- Additional allowing for subject–predicate pair coincidences (LCSP) degrades the analysis quality insignificantly because of growing number of false positives.

Note that Table 5 in work [8] listed 15 pairs of sentences, particularly difficult for the paraphrase analysis. The three methods discussed in the paper yield 6, 6, and 7 errors, respectively. Our LCS method yields 2 errors.

It should be noted that achieved quality indicators are at the level of the best results of the contest on paraphrase identification by classification into two groups, paraphrases and not paraphrases, and compare well to those attained on English texts. The results for classification into three groups are substantially lower due to the complexity of separating the group of "loose-paraphrases".

An advantage of the method under discussion is that it operates on the dictionary information only, so that the system needs no retraining at a change of the subject matter area.

References

1. Barron-Cedeno, A., Vila, M., Marti, M.A., Rosso, P.: Plagiarism meets paraphrasing: insights for the next generation in automatic plagiarism detection. Comput. Linguis. **39**(4), 917–947 (2012)
2. Pham, N., Bernardi, R., Zhang, Y.Z., Baroni, M.: Sentence paraphrase detection: when determiners and word order make the difference. In: Proceedings of the Towards a Formal Distributional Semantics Workshop at IWCS, pp. 21–29 (2013)
3. Finch, A., Hwang, Y.S., Sumita, E.: Using machine translation evaluation techniques to determine sentence-level semantic equivalence. In: Proceedings of the 3rd International Workshop on Paraphrasing, pp. 17–24 (2005)

4. Corley, C., Mihalcea, R.: Measuring the semantic similarity of texts. In: Proceedings of the ACL Workshop on Empirical Modeling of Semantic Equivalence and Entailment, pp. 13–18. Association for Computational Linguistics, Stroudsburg (2005)
5. http://wordnet.princeton.edu/
6. Fernando, S., Stevenson, M.: A semantic similarity approach to paraphrase detection. In: 11th Annual Research Colloquium Proceedings of the Computational Linguistics UK (2008)
7. Pershina, M., He, Y., Grishman, R.: Idiom paraphrases: seventh heaven vs cloud nine. In: Proceedings of the EMNLP 2015 Workshop on Linking Models of Lexical, Sentential and Discourse-level Semantics, pp. 76–82 (2015)
8. Pronoza, E., Yagunova, E.: Comparison of sentence similarity measures for Russian paraphrase identification. In: Artificial Intelligence and Natural Language and Information Extraction, Social Media and Web Search FRUCT Conference, pp. 74–82 (2015)
9. Pivovarova, L., Pronoza, E., Yagunova, E., Pronoza, A.: ParaPhraser Russian paraphrase corpus and shared task. In: Filchenkov, A., et al. (eds.) AINL 2017. CCIS, vol. 789, pp. 211–225 (2018). https://doi.org/10.1007/978-3-319-71746-3_z
10. http://www.paraphraser.ru
11. Iomdin, L., Petrochenkov, V., Sizov, V., Tsinman, L.: ETAP parser: state of the art Komp'iuternaia Lingvistika i Intellektual'nye Tekhnologii: Trudy Mezhdunarodnoi Konferentsii "Dialog 2012". Bekasovo, pp. 119–131 (2012)
12. Antonova, A.A., Misyurev, A.V.: Russian dependency parser SyntAutom at the DIALOGUE-2012 parser evaluation task. Komp'iuternaia Lingvistika i Intellektual'nye Tekhnologii: Trudy Mezhdunarodnoi Konferentsii "Dialog 2012". Bekasovo, pp. 104–118 (2012)
13. Anisimovich, K.V., Druzhkin, K.J., Minlos, F.R., Petrova, M.A., Selegey, V.P., Zuev, K.A.: Syntactic and semantic parser based on ABBYY Compreno linguistic technologies. Komp'iuternaia Lingvistika i Intellektual'nye Tekhnologii: Trudy Mezhdunarodnoi Konferentsii "Dialog 2012". Bekasovo, pp. 91–103 (2012)
14. http://www.dictum.ru/ru/syntax-analysis/blog
15. Boyarsky, K.K., Kanevsky, E.A.: Semantiko-sintaksicheskiy parser SemSin // Nauchno-Tehnicheskii Vestnik Informatsionnykh Tekhnologii, Mekhaniki i Optiki, vol. 15, № 5, pp. 869–876 (2015). (in Russian)
16. Tuzov, V.A.: Komp'juternaja semantika russkogo jazyka. SPb: Izd-vo S. Peterb. un-ta (2004). (in Russian)
17. Boyarsky, K.K., Kanevsky, E.A.: Predsintaksicheskiy modul' v analizatore SemSin. Internet i sovremennoe obshchestvo: sbornik nauchnyh statey. Trudy XVI Vserossiyskoy ob"edinennoy konferentsii «Internet i sovremennoe obshchestvo». SPb, pp. 280–286 (2013). (in Russian)
18. Boyarsky, K.K., Kanevsky, E.A.: Sistema produktsionnyh pravil dlya postroeniya sintaksicheskogo dereva predlozheniya. Prikladna lingvistika ta lingvistichni tekhnologii: MegaLing-2011.Kiev, pp. 73–80 (2012). (in Russian)
19. Boyarsky, K.K., Kanevsky, E.A.: Yazyk pravil dlya postroeniya sintaksicheskogo dereva // Internet i sovremennoe obshchestvo: Materialy XIV Vserossiyskoy ob"edinennoy konferentsii «Internet i sovremennoe obshchestvo». SPb, pp. 233–237 (2011). (in Russian)
20. Nivre, J., Boguslavsky, I.M., Iomdin, L.L.: Parsing the SynTagRus treebank of Russian. In: Proceedings of the 22nd International Conference on Computational Linguistics, vol. 1, pp. 641–648. Association for Computational Linguistics (2008)
21. Avdeeva, N., Boyarsky, K., Kanevsky, E.: Extraction of low-frequent terms from domain-specific texts by cluster semantic analyses. In: Proceedings of the ISMW-FRUCT 2016, Saint-Petersburg, Russia, FRUCT Oy, Finland, pp. 86–89 (2016)

22. Artemova, G., Boyarsky, K., Gouzévitch, D., Gusarova, N., Dobrenko, N., Kanevsky, E., Petrova, D.: Text categorization for generation of a historical shipbuilding ontology. In: Klinov, P., Mouromtsev, D. (eds.) KESW 2014. CCIS, vol. 468, pp. 1–14. Springer, Cham (2014). https://doi.org/10.1007/978-3-319-11716-4_1

23. Pronoza, E., Yagunova, E., Pronoza, A.: Construction of a Russian paraphrase corpus: unsupervised paraphrase extraction. In: Braslavski, P., Markov, I., Pardalos, P., Volkovich, Y., Ignatov, Dmitry I., Koltsov, S., Koltsova, O. (eds.) RuSSIR 2015. CCIS, vol. 573, pp. 146–157. Springer, Cham (2016). https://doi.org/10.1007/978-3-319-41718-9_8

24. Avdeeva, N., Artemova, G., Boyarsky, K., Gusarova, N., Dobrenko, N., Kanevsky, E.: Subtopic segmentation of scientific texts: parametr optimisation. Commun. Comput. Inf. Sci. **518**, 3–15 (2015)

25. Russkiy semanticheskiy slovar'. Tolkovyy slovar', sistematizirovannyy po klassam slov i znacheniy/Rossiyskaya akademiya nauk. In-t rus. yaz. im. V.V. Vinogradova; Pod obshchey red. N. YU. SHvedovoy. – M.: "Azbukovnik" (1998). (in Russian)

26. Azarova, I.V., Mitrofanova, O.A., Sinopal'nikova, A.A.: Komp'yuternyy tezaurus russkogo yazyka tipa WordNet. Komp'iuternaia Lingvistika i Intellektual'nye Tekhnologii: Trudy Mezhdunarodnoi Konferentsii "Dialog 2003", pp. 168–177 (2003). (in Russian)

27. Lukashevich, N.V.: Tezaurusy v zadachah informatsionnogo poiska. – M. MGU (2011). (in Russian)

28. http://www.paraphraser.ru/download/get?file_id=5

RuThes Thesaurus in Detecting Russian Paraphrases

Natalia Loukachevitch[(✉)], Aleksandr Shevelev, Valerie Mozharova,
Boris Dobrov, and Andrey Pavlov

Lomonosov Moscow State University, Moscow, Russia
louk_nat@mail.ru, alex.shevelev@hotmail.com, valerie.mozharova@gmail.com,
dobrov_bv@mail.ru, pavlov.andrew.m@gmail.com

Abstract. In this paper we study the contribution of semantic features to the detection of Russian paraphrases. The features were calculated on the Russian Thesaurus RuThes. First, we applied RuThes synonyms in clustering news articles, many of which had been created with rewriting (that is paraphrasing) of source news, and found significant improvement. Second, we applied several semantic similarity measures proposed for English thesaurus WordNet to RuThes thesaurus and utilized them for detecting Russian paraphrased sentences.

1 Introduction

Paraphrase detection algorithms should determine if two text fragments convey the same meaning. Identifying paraphrases is an important task, useful in information retrieval, question answering [1], document clustering [2], text summarization [3,4], plagiarism detection [5], machine translation evaluation [6] and others.

The paraphrase detection task is well-studied for English. For evaluating approaches, specialized databases with automatically collected paraphrases have been created [7,8]. Systems detecting English paraphrases are being evaluated at the SemEval conference since 2012 [9,10]. But for other languages, the number of such studies is much smaller.

It is natural that central components in the paraphrase tasks are semantic features, which can be calculated on manual resources, such as thesaurus WordNet [11] or Wikipedia. Other types of semantic features utilize distributional characteristics of words [12]. In this paper we study approaches for finding Russian paraphrases and describe the results of applying the Russian thesaurus RuThes [13] in two tasks.

The first task is the clustering of news articles. A news cluster grouping similar news articles is a basic unit of any modern news aggregators. Current news flows contain a lot of articles similar in content. Many of such articles are created with so-called rewriting technologies, which heavily exploits rephrasing.

A. Filchenkov et al. (Eds.): AINL 2017, CCIS 789, pp. 242–256, 2018.
https://doi.org/10.1007/978-3-319-71746-3_20

Paraphrases in news articles hinder the recognition of the similarity of news publications. In this task we study the contribution of RuThes synonyms.

The second task is detecting paraphrases using Russian Paraphraser corpus, prepared for the shared task on Russian paraphrasing [14,15]. We have already described initial experiments with the use of RuThes in this task in [16]. But in the current paper we give the more detailed view on contribution of specific thesaurus paths and path restrictions on Russian paraphrase detection.

The paper is organized as follows. In Sect. 2 we consider related work. Section 3 describes the RuThes thesaurus, its structure and possibilities to make inference using properties of the thesaurus relations. Section 4 is devoted to testing impact of using RuThes synonyms in news article clustering. In Sect. 5 we consider the contribution of the RuThes synonyms and relations in detection of paraphrased sentences.

2   Related Work

Most research on paraphrase detection was carried out for English. The proposed approaches exploit multiple groups of features (string intersections, machine translation and information retrieval similarities, syntactic structures, etc.) and combine them with machine learning methods [17–19].

Among these features, semantic features based on manual thesauri such as WordNet [11] have a prominent role. These features are calculated on the basis of word similarity measures based on paths between corresponding synsets in WordNet. Other semantic similarity measures additionally use frequencies in a large text corpus to estimate path-based word semantic similarity.

Kozareva, Montoyo [17] used a WordNet-based semantic similarity measure as one of multiple features in supervised machine learning. Mihalcea et al. [20] calculated text semantic similarity using maximum similarity of a word with another word in another text and summed up the obtained similarities also accounting the idf information-retrieval feature. Fernando and Stevenson [21] compared several different WordNet similarity measures with each other. They formed a similarity matrix based on WordNet measures between compared sentences. The matrix allowed summing up semantic similarity values of a word with several words in another sentence.

At the SemEval evaluations, the best system of SemEval-2012 combined WordNet word similarity measures with Explicit Semantic Analysis (ESA) [22]. The best model of SemEval-2013 exploited Latent Semantic Analysis and modified its results using the WordNet semantic information [12]. The best approach of SemEval-2016 employs two important components: the unfolding recursive autoencoder and the penalty-award weight system based on WordNet [23]. First, recursive autoencoder is used to perform unsupervised learning on parse trees, then the WordNet module adjusts the distances of vectors using awards and penalties based on semantic similarities of words.

For German, Gurevych and Niederlich [24] studied the calculation of word semantic relatedness on GermaNet (German wordnet) [25] using semantic measures earlier proposed for WordNet. Muller et al. [26] integrated GermaNet-based word similarity measures into two tasks: query expansion in information retrieval and text similarity detection.

For Russian, Pronoza and Yagunova [18] studied various factors of paraphrase detection on the Russian paraphrase corpus including shallow measures based on word or characters overlap, dictionary-based measures and distributional semantic measures based on finding context similarity between words in a text corpus. They experimented on the Russian paraphrase corpus containing 6281 sentence pairs (1482 precise, 3247 loose and 2209 non-paraphrases). Altogether more than 80 features of sentences were calculated and combined with the Gradient Boosting classifier. The similarity between synonyms in a dictionary was based on calculating the probability to meet the words in the same set of synonyms.

In 2016 the shared task on evaluation of methods for detecting Russian paraphrases has been organized [15].

3   RuThes Thesaurus

The thesaurus of the Russian language RuThes [13] is a linguistic ontology for natural language processing, i.e. an ontology, where the majority of concepts are introduced on the basis of actual language expressions. RuThes is a hierarchical network of concepts. Each concept has a name, relations with other concepts, a set of language expressions (words, phrases, domain terms) whose senses correspond to the concept, so called ontological synonyms.

Ontological synonyms of a concept can comprise words belonging to different parts of speech; language expressions relating to different linguistic styles, genres; idioms and even free multiword expressions, for example, synonymous to single words): красный, краснота, красный цвет (*red, redness, red color*).

The experts have been encouraged to create rich rows of ontological synonyms. After a concept has been introduced, an expert searches for all possible synonyms or orthographic variants, single words, and phrases that can be associated with this concept. The publicly available version of the RuThes thesaurus, RuThes-lite 2.0 [27] comprises 31.5 thousand concepts, 115 thousand Russian words and expressions, which means that each thesaurus entry has almost four ontological synonyms on the average.

The relations in RuThes are only conceptual, not lexical (as antonyms or derivational links in wordnets). They are constructed as more formal, ontological relations originated from traditional information-retrieval thesauri. The set of conceptual relations includes:

- the class-subclass relation;
- the part-whole relation applied with the following restriction: the existence of the concept-part should be strictly attached to the concept-whole (so tree can grow in many places therefore concept *tree* cannot be directly linked to concept *forest* with the part-whole relation, the additional concept *forest tree* should be introduced). This restriction gives the possibility to utilize the transitivity property of the part-whole relation;

– the external ontological dependence when the existence of a concept depends on the existence of another concept. In such a way, the concept *forest* depends on the existence of the concept *tree* [28]. In RuThes this relation is denoted as association with indexes: asc_1 is directed to the main concept, asc_2 indicates the dependent concept;
– symmetric associations between much related concepts can be established in a restricted number of cases (denoted as asc).

Several properties are defined over RuThes relations. These properties give the possibility to make logical inference, in particular, to find semantic relatedness between text entries that are not directly connected with each other by the thesaurus relations. These properties include:

– transitivity of class-subclass relations:

$$class(c_i, c_j) \land class(c_j, c_k) \rightarrow class(c_i, c_k)$$

– transitivity of part-whole relations. It should be noted that the transitivity of the part-whole relation is an often discussed issue in computational applications (see more detailed discussion in [29]) but the rules of establishing those relations in RuThes allows exploiting this property:

$$whole(c_i, c_j) \land whole(c_j, c_k) \rightarrow whole(c_i, c_k)$$

Also the following inheritance rules are valid in RuThes:

– whole relations are inherited to subclasses:

$$class(c_i, c_j) \land whole(c_j, c_k) \rightarrow whole(c_i, c_k)$$

– asc_1 relations are inherited to subclasses and parts:

$$class(c_i, c_j) \land asc_1(c_j, c_k) \rightarrow asc_1(c_i, c_k)$$
$$whole(c_i, c_j) \land asc_1(c_j, c_k) \rightarrow asc_1(c_i, c_k)$$

– asc relations are inherited to subclasses and parts:

$$class(c_i, c_j) \land asc(c_j, c_k) \rightarrow asc(c_i, c_k)$$
$$whole(c_i, c_j) \land asc(c_j, c_k) \rightarrow asc(c_i, c_k)$$

Considering all possible relation paths existing between two thesaurus concepts c_1 and c_2, it was supposed that those paths that can be reduced to a single relation with the application of the above-mentioned rules of transitivity and inheritance indicate semantic relatedness between concepts c_1 and c_2, so called semantic paths. Word and phrases presented as thesaurus entries assigned to the concepts c_1 and c_2 are also considered semantically related even if the length of the path is quite large (five and more relations). Such defined semantic similarity between words and phrases included in RuThes is used for query expansion in information retrieval, thematic text representation [4], representation of categories in text categorization [29], and others. The thesaurus RuThes-lite 2.0[1] can be obtained in form of xml files for noncommercial use.

4   Using RuThes Synonyms in News Article Clustering

The approaches in paraphrase detection can be employed in the clustering of news articles. Such clustering is a necessary component of news aggregators such as Yandex.news, Google.news, etc. Thousand of news agencies generate news reports on similar topics or rewrite news articles from authoritative sources. To present the news in a more readable way, online news services group similar news in clusters, which are main units for visualization of current news flows [2].

News articles describing the same topic or event contain a lot of paraphrases, which can be used for extracting and storing them [8] or creating evaluation datasets [18]. But at the same time these synonyms and paraphrases can hamper correct clustering of news flows because they greatly increase the difference between similar texts. Thus, it is interesting to estimate the impact of synonym accounting on the performance of news clustering methods.

For experiments with news clustering in Russian, the news collection of Russian seminar on information retrieval ROMIP [30] can be used. This collection contains news stories from Yandex.news service for three weeks of 2003–2004[2]. The collection was gathered for evaluation of information extraction approaches but also can be used for evaluating document clustering because it contains multiple news articles created on the same dates (Table 1). For experiments, wednesdays of these weeks were chosen.

Table 1. Dataset for clustering evaluation

Weeks	Days	Number of documents
Shevardnadze week	2003-11-20	1752
Ordinary week	2003-12-03	1715
Election week	2004-04-02	1809

The clustering of documents is usually carried out on the basis of their vector representations when every document is described as a vector of $|V|$ dimensions, where $|V|$ is the size of the collection vocabulary. Each component of a vector corresponds to a specific word and calculated as tf.idf value where tf is the frequency of a word in a document and idf is the inverse document frequency of a word [31]. For Russian, document vectors are usually constructed using lemmas (or dictionary forms of words) obtained after morphological processing of texts.

---

[1] http://www.labinform.ru/ruthes/index.htm.
[2] http://romip.ru/ru/collections/news-collection.html.

Conventional news texts have the predefined structure, which determines the important contribution of the heading in the text content. Besides, the main contents of a news article is usually conveyed in the beginning of the text. It means that it is useful to have additional vector representation for text titles, which should be combined with the vector representation of the whole text.

To estimate possible contribution of thesaurus information into news clustering, concept vectors were generated. With this aim, news texts were processed with ALOT (Automatic Linguistic Text processing tool), which recognizes thesaurus concepts in the text [29]. In such processing, synonymous words, term variants, multiword expressions are gathered to the single concept; ambiguous words are disambiguated and assigned to different concepts. Thus, for every document, so called concept index is created.

Three representation for each news text (lemma (L), heading (H) and concept (C) representations) allowed us to create a combined representation, which can be used for calculating text similarity as a basis for document clustering as follows:

$$\mu_{lch}(d_i, d_j) = \alpha_L \mu_L(d_i, d_j) + \alpha_C \mu_C(d_i, d_j) + \alpha_H \mu_H(d_i, d_j)$$

where $\alpha_L + \alpha_C + \alpha_H = 1$

To estimate the quality of clustering, the gold standard clusters for news data were created. For evaluation, F1-measure was calculated on document pairs as follows:

- N_{11} is the number of document pairs for that the expert and the system agreed to assign them to the same cluster,
- N_{10} is the number of document pairs that the expert assigned to the same cluster, but the system assigned them to different clusters,
- N_{01} is the number of document pairs that the expert assigned to different clusters, but the system assigned then to the same cluster.

Then

$$R = Recall = \frac{N_{11}}{N_{11} + N_{10}}$$

$$P = Precsion = \frac{N_{11}}{N_{11} + N_{01}}$$

$$F1 = \frac{2PR}{P + R}$$

We experimented with four clustering methods: k-means clustering, agglomerative clustering and all its variants (single-link, complete-link, centroid, and average-link) [32], DBSCAN [33], and FOREL [34]. The Table 2 describes the results of applying clustering methods to news sets. For each method, the type of vectorization is given. For each method, only the best achieved results are shown.

It can be seen that for all clustering methods, the achievement of the best result is based on combination of all three indexes: Lemma, Heading, and Concepts. The best average result (0.7243 F-measure) on three days under consideration is obtained with the centroid variant of the agglomerative clustering, which combines the indexes in relatively equal proportions.

The Table 3 shows the contribution of each index in obtaining the best result. It can be seen that clustering using only titles (0, 0, 1) gives very low results (0.4972), using only the lemma index (1, 0, 0) is much better (0.5767), the combination of both indexes leads to significant improvement (0.6866). At last, the accounting of the concept index leads to 5.5% improvement of F-measure (0.7243).

Thus, we can see that the use of the RuThes synonyms allowed improving the news article clustering (Table 3).

Table 2. Comparison of clustering methods (LCH, means of vectorization, is given in parentheses)

Method	2003-11-21	2003-12-03	2004-04-02	Average
FOREL (0.6:0.2:0.2)	0.5282	0.8383	0.7364	0.6890
DBSCAN (0.6:0.2:0.2)	0.5173	0.8648	0.7504	0.6879
K-Means (0.6:0.15:0.25)	**0.5767**	0.8515	0.7616	0.7141
Agglomerative (0.4:0.3:0.3)	0.5716	**0.8685**	**0.7904**	**0.7243**

Table 3. Comparison of different vectorizations for agglomerative clustering

LCH = (0, 0, 1)	LCH = (1,0, 0)	LCH = (x,0,1-x)	bf LCH = (x,y,1-x-y)
Result = 0.4972	Result = 0.5767	Result = 0.6866	**Result = 0.7243**
		LCH = 0.7:0.0:0.3	**LCH = 0.4:0.3:0.3**
Method = center	Method = min	Method = center	bf Method = center

5 RuThes in Russian Paraphrasing Task

In this section we consider the use of RuThes thesaurus in the detection of paraphrased sentences. The approach is tested on the data prepared for the Russian shared task organized in 2016 [15].

5.1 Russian Paraphrasing Task

The evaluation on Russian paraphrasing included two tasks: three-way classification of sentence pairs (precise paraphrases, loose paraphrases and non-paraphrases) and binary classification: sentence pairs should be classified to paraphrases or non-paraphrases.

The participants could submit "standard" runs that utilize only the Para-Phraser corpus as training data and (or) manual dictionary resources, and "non-standard" runs that may use any other data. "Standard" and "non-standard" run have been evaluated separately.

The datasets were formed on the basis of news story headlines. The training collection contains about 7000 sentence pairs. Each candidate pair was manually annotated by three native speakers with the use of crowdsourcing. The test dataset (Gold standard set) contains 1924 sentence pairs.

The quality of submitted results has been assessed with Accuracy and macro F-measure for the three-class task. Accuracy and F-measure of detecting the paraphrase class were used for the two-class task.

5.2 Evaluating Thesaurus-Based Features in Paraphrase Detection

To apply RuThes to the Russian paraphrasing shared task, we calculated several lexical similarity measures proposed for Princeton WordNet. These measures exploit paths between concepts where words under comparison were assigned. The measures include Leacock-Chodorow measure (Lch), its variant without logarithm (Path), Lin measure (Lin), and Jiang-Conrath measure (Jcn) [35].

The Lch measure estimates the similarity of two nodes by finding the path length between them in the hierarchy ofthesaurus relations. It is computed as:

$$sim_{lch} = -log\frac{N_p}{2D}$$

where N_p is the distance between nodes and D is the maximum depth in the taxonomy. The distance is calculated in nodes, that is the distance between synonyms is equal 1, and the distance between a node and its hypernym is equal 2. The logarithm base is equal to 2D. In RuThes-lite, the maximum depth of the ontology accounting both types of relations is equal 14.

We used three variants of calculation of this measure:

- using only hyponym-hypernym relations;
- using hyponym-hypernym and part-whole relations exploting RuThes transitivity of both types of relations and its allowed combinations;
- using all relations according the defined relations' properties (Sect. 3).

We also used a modification of the lch-measure without logarithm:

$$sim_{path} = 1 - \frac{N_p}{2D}$$

Other two measures (Lin and Jcn) are calculated on the basis of word probabilities and so called information content (IC) [36]. For every word, the probability to meet this word in a corpus is calculated:

$$P(w) = \frac{Freq_w}{N}$$

where N is the size of a corpus in words. The probability of a concept is the sum of probabilities of all text entries assigned to this concept.

The information content of a concept is an estimate of how informative the concept is. It is supposed that frequently occurring concepts have low information content and rarely occurring concepts have high information content.

$$IC(C) = -log(P(C))$$

In calculating information content, probabilities of all lower concepts in the hierarchy should be summed up. For RuThes, lower concepts include subclasses, parts, and dependent concepts, which indicated directly or if such relations can be inferred using the relation properties (see Sect. 3).

The Lin measure is calculated as follows:

$$sim_{lin} = \frac{2IC(LCS(C_1, C_2))}{IC(C_1) + IC(C_2)}$$

where LCS is the least common subsumer of C_1 and C_2.

The Jcn measure combines the same values in another way:

$$sim_{jcn} = \frac{1}{IC(C_1) + IC(C_2) - 2IC(LCS(C_1, C_2))}$$

For Lin and Jcn measures, three variants were also calculated (only hypernyms, hypernyms and wholes, all relations) according to the RuThes relations' properties.

To estimate word frequencies for IC calculation, an additional news corpus was used. Therefore according to the evaluation rules, when we use the Lch or Path measures, the runs could be considered as standard. But when we utilize the Lin or Jcn similarity measures, these runs should be categorized as non-standard due to the use of the additional corpus.

The results of calculating above-mentioned similarity measures for words *Berlin* and *Germany* are shown in Table 4. It can be seen that the calculated measures give very different results of semantic similarity for these two words. The relation between the corresponding concepts is the part-whole relation therefore the only_hypernym path between these two words is long and gives relatively low values of the similarity measures. Lin and Jcn measures operate with the same values in different ways and the obtained similarity values differ significantly.

Table 4. Semantic similarity measures calculated for words *Berlin* and *Germany*

Type of path	Dist. in nodes	Lch	Path	Lin	Jcn
Only_hypernyms	12	0.254	0.57	0.382	0.064
Hypernyms and Wholes	2	0.792	0.928	0.845	0.358
All relations	2	0.792	0.928	0.845	0.358

Comparing sentences on the basis of the thesaurus-calculated word similarity, we used the approach proposed in [21], which allows summing the similarity of a word in one sentence with several words from another sentence. Sentences in this approach are represented as binary vectors $\overrightarrow{a}$ and $\overrightarrow{b}$. The similarity between the sentences is calculated as follows:

$$sim(\overrightarrow{a}, \overrightarrow{b}) = \frac{\overrightarrow{a} W \overrightarrow{b}}{|\overrightarrow{a}||\overrightarrow{b}|}$$

where W is a square matrix of the calculated similarities between words and expressions found in both sentences.

Each w_{ij} in W represents the similarity of words w_i and w_j according to some lexical similarity measure. In our case the measures are symmetric, i.e. $w_{ij} = w_{ji}$ and the matrix is also symmetric. Diagonal elements represent self similarity and have the greatest values equal to 1 [16].

As preprocessing, before thesaurus features calculating, sentences are lemmatized, function words are removed, numbers mentioned in sentences are substituted with corresponding words. Words not found in the thesaurus but met in both sentences have maximal similarity 1.

It is worth noting that Fernando and Stevenson [21] calculating such a matrix took into consideration only the values of word similarities more than 0.8 and nullified all other values of similarity. In our case we did not nullified any values but calculated variants of all measures without any restrictions on the path length and for paths with the restricted length.

As an example of the sentence matrix, we can consider two sentences:

(s1) Около 15 тысяч человек пришли к памятнику Воину-освободителю в ФРГ (About 15 thousand people came the monument to the Soviet soldier in FRG);

(s2) Около 15 тысяч человек пришли поклониться памятнику Воину-освободителю в Берлине (About 15 thousand people came to worship the monument to the Soviet soldier in Berlin).

The matrix according the Jcn measure (all relations, length of the path 3) for these sentences is presented in Table 5. Besides related words Берлин and ФРГ, also the relation between words человек and воин was found.

Table 5. Matrix of Jcn similarity for the example sentences

	человек	тысяча	прийти	поклониться	воин	Берлин	ФРГ
человек	1	0	0	0	0.245	0	0
тысяча	0	1	0	0	0	0	0
прийти	0	0	1	0	0	0	0
поклониться	0	0	0	1	0	0	0
воин	0.245	0	0	0	1	0	0
Берлин	0	0	0	0	0	1	0.358
ФРГ	0	0	0	0	0	0.358	1

5.3 Finding the Best Thesaurus Feature

We calculated similarities between sentences in pairs using all above-mentioned semantic measures separately. To determine the best single semantic measure, we trained linear SVM classifier on the training data using each single thesaurus feature and applied the classifier to the test data. In fact, in such a way we tried to find the best threshold dividing thesaurus feature values to two or three classes according to the task.

The Table 6 presents the results of our evaluation for all feaures and their variants based on the used relations. Using this evaluation we could choose the paths of the best length for each thesaurus feature. The parentheses contain the path length with the best achieved result.

It can be seen that the results obtained for specific features in the binary task are very similar and locate in the interval (0.78, 0.80). These values are very close to the results of [21] for the English paraphrase corpus where the F-measure for WordNet features belongs to (0.80, 0.825) F-measure.

It can also be seen that relatively short paths gave the best results. Longer paths can revealed erroneous similarities. It seems that such results agree with the paper by Fernando and Stevenson [21], who nullified the values of word similarities less than 0.8 in the sentence matrixes.

Table 6. The best results (F1-measure) achieved by single Thesaurus measures. In parentheses, the path length for the best result is indicated

Method	Relations	2-class best result	3-class best result
Lch	Only Hyper	0.784 (6)	0.541 (3)
	Hyp+Wholes	0.788 (5)	0.545 (5)
	All	0.789 (5)	0.549 (5)
Path	Only Hyper	0.784 (3)	0.542 (5)
	Hyp+Wholes	0.788 (4)	0.543 (4)
	All	0.788 (5)	0.542 (2)
Lin	Only Hyper	**0.795 (2)**	0.558 (2)
	Hyp+Wholes	**0.793 (2)**	0.550 (2)
	All	**0.794 (2)**	0.548 (2)
Jcn	Only Hyper	**0.795 (5)**	**0.562 (2)**
	Hyp+Wholes	**0.794 (7)**	**0.564 (3)**
	All	0.786 (3)	0.555 (3)

5.4 Combining Thesaurus Features with Other Features

To understand the contribution of the thesaurus features, we combined the best semantic features with other features and trained a classifier. We used the following types of features: string-based features, information-retrieval features (BM-25 and Idf for word difference), pos-features for words different in both sentences, as described in [16]. In preliminary experiments, we chose Random Forest as a basic machine learning method. We used the implementation from scikit-learn package with grid parameter tuning[3].

Table 7. Contribution of the best thesaurus features into Russian paraphrase detection

Feature set	2-class task	3-class task
	Acc/F1	Acc/MacroF
(1) String-based	73.80/79.00	60.03/57.99
(2) String-based+BM25	74.06/79.18	60.96/58.99
(3) String-based+BM25+5 POS features	74.42/79.32	61.07/59.03
(4) String-based+BM25+5 POS features+BestThes	**77.33/81.71**	**62.57/60.93**
Loukachevitch (2017) [16] results		
(3)+$Thes_{lch}$	–	61.48/59.33
(3)+$Thes_{lch} + Thes_{jcn}$	–	62.00/60.03
Best results of the shared task		
Standard runs	74.59/80.14	59.01/56.92
Non-standard runs	77.39/81.10	61.81/58.38

Table 7 contains the results of the classification. Here BM25 is an average value between BM25 from the first sentence to the second sentence and vice versa, calculated on the training collection. The BestThes feature is the best set of the thesaurus features with the most contribution to the overall result. In this case, the BestThes contains two variants of lch semantic features: with only hyponym-hypernym paths and paths containing hyponym-hypernym and part-whole relations.

For the best run, we did not use any additional corpora. Thus, we can qualify our results as results of a standard run and can see considerable improvement over the shared task results. The obtained results are better than the results presented in [16] because we additionally applied restrictions on the RuThes path length.

6 Conclusion

In this paper we studied semantic features calculated on the Russian language Thesaurus RuThes in processing of Russian paraphrases. First, we applied RuThes synonyms in clustering of news articles, many of which are created with rewriting (that is paraphrasing) of source news and found significant improvement.

---

[3] http://scikit-learn.org/stable/index.html.

Second, we tested several semantic similarity measures proposed for English thesaurus WordNet on the material of RuThes and utilized them for detecting Russian paraphrases. We experimented with several variants of paths allowed by RuThes relations' properties and their lengths. We found that the best results in the paraphrase detection based on RuThes semantic features are obtained on relatively short thesaurus paths. The combination of RuThes semantic features with other features gave high results in detecting Russian paraphrases.

Acknowledgments. This work was partially supported by Russian National Foundation, grant N16-18-02074.

References

1. Fader, A., Zettlemoyer, L.S., Etzioni, O.: Paraphrase-driven learning for open question answering. In: Proceedings of ACL-2013, pp. 1608–1618 (2013)
2. Vossen, P., Rigau, G., Serafini, L., Stouten, P., Irving, F., van Hage, W.R.: NewsReader: recording history from daily news streams. In: Proceedings of LREC-2014, pp. 2000–2007 (2014)
3. Nenkova, A., McKeown, K.: A survey of text summarization techniques. In: Aggarwal, C., Zhai, C. (eds.) Mining Text Data Book, pp. 43–76. Springer, Boston (2012). https://doi.org/10.1007/978-1-4614-3223-4_3
4. Loukachevitch, N., Alekseev, A.: Summarizing news clusters on the basis of thematic chains. In: Proceedings of LREC-2012, pp. 1600–1607 (2012)
5. Clough, P., Gaizauskas, R., Piao, S., Wilks, Y.: METER: MEasuring TExt reuse. In: Proceedings of the 40th Anniversary Meeting for the Association for Computational Linguistics (ACL 2002), pp. 152–159 (2002)
6. Marton, Y., Callison-Burch, C., Resnik, P.: Improved statistical machine translation using monolingually-derived paraphrases. In: Proceedings of the 2009 Conference on Empirical Methods in Natural Language Processing, EMNLP-2009, pp. 381–390 (2009)
7. Dolan, W.B., Quirk, C., Brockett, C.: Unsupervised construction of large paraphrase corpora: exploiting massively parallel news sources. In: Proceedings of the 20th International Conference on Computational Linguistics, Coling-2004, Geneva, Switzerland (2004)
8. Pavlick, E., Rastogi, P., Ganitkevitch, J., Durme, B., Callison-Burch, C.: PPDB 2.0: better paraphrase ranking, fine-grained entailment relations, word embeddings, and style classification. In: Proceedings of ACL-2015 and the 7th International Joint Conference on Natural Language Processing, vol. 2, pp. 425–430 (2015)
9. Agirre, E., Diab, M., Cer, D., Gonzalez-Agirre, A.: Semeval-2012 task 6: a pilot on semantic textual similarity. In: Proceedings of the Sixth International Workshop on Semantic Evaluation, pp. 385–393. Association for Computational Linguistics (2012)
10. Agirre, E., Banea, C., Cer, D., Diab, M., Gonzalez-Agirre, A., Mihalcea, R., Wiebe, J.: Semeval-2016 task 1: semantic textual similarity, monolingual and cross-lingual evaluation. In: Proceedings of SemEval, pp. 497–511 (2016)
11. Fellbaum, C. (ed.): WordNet: An Electronic Lexical Database. The MIT Press, Cambridge (1998)

12. Han, L., Kashyap, A., Finin, T., Mayfield, J., Weese, J.: UMBC EBIQUITY-CORE: semantic textual similarity systems. In: Second Joint Conference on Lexical and Computational Semantics (*SEM), Volume 1: Proceedings of the Main Conference and the Shared Task: Semantic Textual Similarity, Atlanta, Georgia, USA, June, pp. 44–52. Association for Computational Linguistics (2013)

13. Loukachevitch, N., Dobrov, B.: RuThes linguistic ontology vs. Russian wordnets. In: Proceedings of Global WordNet Conference GWC-2014, pp. 154–162 (2014)

14. Pronoza, E., Yagunova, E., Pronoza, A.: Construction of a Russian paraphrase corpus: unsupervised paraphrase extraction. In: Braslavski, P., Markov, I., Pardalos, P., Volkovich, Y., Ignatov, D.I., Koltsov, S., Koltsova, O. (eds.) RuSSIR 2015. CCIS, vol. 573, pp. 146–157. Springer, Cham (2016). https://doi.org/10.1007/978-3-319-41718-9_8

15. Pivovarova, L., Pronoza, E., Yagunova, E., Pronoza, A.: ParaPhraser: Russian paraphrase corpus and shared task. In: Filchenkov, A., et al. (eds.) AINL 2017. CCIS, vol. 789, pp. 211–225. Springer, Cham (2018)

16. Loukachevitch, N., Shevelev, A., Mozharova V.: Testing features and methods in Russian Paraphrasing Task. In: Proceedings of International Conference on Computational Linguistics and Intellectual Technologies Dialog 2017, vol. 1, pp. 135–145 (2017)

17. Kozareva, Z., Montoyo, A.: Paraphrase identification on the basis of supervised machine learning techniques. In: Salakoski, T., Ginter, F., Pyysalo, S., Pahikkala, T. (eds.) FinTAL 2006. LNCS (LNAI), vol. 4139, pp. 524–533. Springer, Heidelberg (2006). https://doi.org/10.1007/11816508_52

18. Pronoza, E., Yagunova, E.: Low-level features for paraphrase identification. In: Sidorov, G., Galicia-Haro, S.N. (eds.) MICAI 2015. LNCS (LNAI), vol. 9413, pp. 59–71. Springer, Cham (2015). https://doi.org/10.1007/978-3-319-27060-9_5

19. Brockett, C., Dolan, W.B.: Support vector machines for paraphrase identification and corpus construction. In: Proceedings of the 3rd International Workshop on Paraphrasing, pp. 1–8 (2005)

20. Mihalcea, R., Corley, C., Strapparava C.: Corpus-based and Knowledge-based measures of text semantic similarity. In: Proceedings of the American Association for Artificial Intelligence (2006)

21. Fernando, S., Stevenson, M.: A semantic similarity approach to paraphrase detection. In: Proceedings of the 11th Annual Research Colloquium of the UK Special Interest Group for Computational Linguistics, pp. 45–52 (2008)

22. Bar, D., Biemann, C., Gurevych, I., Zesch, T.: UKP: computing semantic textual similarity by combining multiple content similarity measures. In: Proceedings of the 6th International Workshop on Semantic Evaluation, Held in Conjunction with the 1st Joint Conference on Lexical and Computational Semantics, pp. 435–440 (2012)

23. Rychalska, B., Pakulska, K., Chodorowska, K., Walczak, W., Andruszkiewicz, P.: Samsung Poland NLP team at SemEval-2016 Task 1: necessity for diversity; combining recursive autoencoders, wordnet and ensemble methods to measure semantic similarity. In: Proceedings of the 10th International Workshop on Semantic Evaluation (SemEval 2016), San Diego, CA, USA (2016)

24. Gurevych, I., Niederlich, H.: Computing semantic relatedness in German with revised information content metrics. In: Proceedings of OntoLex 2005 - Ontologies and Lexical Resources, IJCNLP 2005 Workshop (2005)

25. Kunze, C., Lemnitzer, L.: GermaNet-representation, visualization, application. In: LREC-2002 (2002)

26. Muller, C., Gurevych, I., Muhlhauser, M.: Integrating semantic knowledge into text similarity and information retrieval. In: International Conference on Semantic Computing, ICSC 2007, pp. 257–264. IEEE (2007)
27. Loukachevitch, N.V., Dobrov, B.V., Chetviorkin, I.I.: Ruthes-lite, a publicly available version of thesaurus of Russian language ruthes. In: Computational Linguistics and Intellectual Technologies: Papers from the Annual International Conference Dialogue-2014, Bekasovo, Russia, pp. 340–349 (2014)
28. Guarino, N.: The ontological level: revisiting 30 years of knowledge representation. In: Borgida, A.T., Chaudhri, V.K., Giorgini, P., Yu, E.S. (eds.) Conceptual Modeling: Foundations and Applications. LNCS, vol. 5600, pp. 52–67. Springer, Heidelberg (2009). https://doi.org/10.1007/978-3-642-02463-4_4
29. Loukachevitch, N., Dobrov, B.: The Sociopolitical Thesaurus as a resource for automatic document processing in Russian. Terminology **21**(2), 238–263 (2015). Special issue Terminology across languages and domains
30. Dobrov, B.V., Kuralenok, I., Loukachevitch, N.V., Nekrestyanov, I., Segalovich, I.: Russian information retrieval evaluation seminar. In: LREC-2004 (2004)
31. Manning, C.D., Raghavan, P., Schütze, H.: Introduction to Information Retrieval. Cambridge University Press, Cambridge (2008)
32. Rokach, L., Maimon, O.: Clustering Methods. Data Mining and Knowledge Discovery Handbook, pp. 321–352. Springer, New York (2005)
33. Ester, M., Kriegel, H., Sander, J., Xu, X.: A density-based algorithm for discovering clusters in large spatial databases with noise. In: Simoudis, E., Han, J., Fayyad, U.M., (eds.) Proceedings of the Second International Conference on Knowledge Discovery and Data Mining (KDD 1996), pp. 226–231. AAAI Press (1996)
34. Zagoruiko, N.G.: Intellectual data analysis based on a rival similarity function. Optoelectron. Instrum. Data Process. **44**(3), 211–217 (2008)
35. Budanitsky, A., Hirst, G.: Evaluating wordnet-based measures of lexical semantic relatedness. Comput. Linguist. **32**(1), 13–47 (2006)
36. Resnik, P.: Using information content to evaluate semantic similarity in a taxonomy. arXiv preprint cmp-lg/9511007 (1995)

Knowledge-lean Paraphrase Identification Using Character-Based Features

Asli Eyecioglu[1(✉)] and Bill Keller[2]

[1] Bartin University, 74100 Bartin, Turkey
aozmutlu@bartin.edu.tr
[2] University of Sussex, Brighton BN19QJ, UK
billk@sussex.ac.uk

Abstract. The paraphrase identification task has practical importance in the NLP community because of the need to deal with the pervasive problem of linguistic variation. Accurate methods should help improve the performance of NLP applications, including machine translation, information retrieval, question answering, text summarization, document clustering and plagiarism detection, amongst others. We consider an approach to paraphrase identification that may be considered "knowledge-lean". Our approach minimizes the need for data transformation and avoids the use of knowledge-based tools and resources. Candidate paraphrase pairs are represented using combinations of word- and character-based features. We show that SVM classifiers may be trained to distinguish paraphrase and non-paraphrase pairs across a number of different paraphrase corpora with good results. Analysis shows that features derived from character bigrams are particularly informative. We also describe recent experiments in identifying paraphrase for Russian, a language with rich morphology and free word order that presents a particularly interesting challenge for our knowledge-lean approach. We are able to report good results on a three-way paraphrase classification task.

Keywords: Paraphrase identification · Paraphrase corpora · Character N-grams
Lexical overlap · Support vector machines

1 Introduction

The work described in this paper is concerned with natural language processing (NLP) approaches to the task of paraphrase identification (PI henceforth). According to Lintean and Rus [31], PI may be defined as "the task of deciding whether two given text fragments have the same meaning". PI has elsewhere also been referred to as paraphrase detection [49, 52] or paraphrase recognition [2]. PI can be viewed as a classification task: in the simplest case, given a pair of texts, a binary decision is made as to whether they mean the same thing or not. It should be noted, however, that the notion that two texts "have the same meaning" requires further elaboration and does not in itself offer an effective definition of the paraphrase relation. In practice, human intuition about the semantic equivalence of texts provides a proxy for such a definition. Current NLP approaches to PI typically make use of machine learning and aim to mimic human intuition.

© Springer International Publishing AG 2018
A. Filchenkov et al. (Eds.): AINL 2017, CCIS 789, pp. 257–276, 2018.
https://doi.org/10.1007/978-3-319-71746-3_21

Although PI aims to identify sentences that are semantically equivalent, a number of researchers have shown that classifiers trained on lexical overlap features may attain relatively good performance. This performance is achieved without the use of knowledge-based semantic features or other external knowledge sources such as parallel corpora [7, 31]. Such approaches may be characterized as 'knowledge-poor' or 'knowledge-lean'.

As far as the authors are aware, the term 'knowledge-poor' was first introduced by Hearst and Grefenstette [24]. Their motivation is the use of knowledge-poor corpus-based approaches for the automatic discovery of lexical relations. They argue that adopting a knowledge-poor corpus-based approach has advantages and can yield good results without the need for more complex, knowledge-based methods and resources. The term 'knowledge-lean' is used in the same way by Pedersen and Bruce [40] in order to draw attention to the significance of corpus-based measures compared to knowledge-based measures for the word-sense disambiguation task.

Methods are considered knowledge-lean if they make use only of the text data at hand, and avoid the use of external, knowledge-based processing tools and other resources. A knowledge-lean approach is a strategy that generally requires less alteration of experimental data. An important consequence of this is that knowledge-lean approaches will generally preserve information or features of the language data that might otherwise be thrown out or lost during the transformation process. This in turn may offer a number of advantages over knowledge-based approaches. Knowledge-lean methods may be more readily applied to less well resourced languages. They may offer a reduction in manual annotation effort, and provide the potential to learn from samples of the actual data to be processed rather than from possibly unrepresentative approximations. The potential benefits have led the NLP community to investigate techniques that require less knowledge, thereby avoiding the costly and time-consuming construction of knowledge-rich resources.

During the last two decades, the PI task has gained considerable significance in NLP. The PI task has practical importance in the NLP community because of the need to deal with the pervasive problem of linguistic variation. Accurate methods for PI should help improve the performance of NLP applications, including machine translation, information retrieval, question answering, text summarization, document clustering and plagiarism detection, amongst others. Question Answering (QA) systems, for example, must deal with cases where the answer to a question may not always be in the same form as the question. The alternative answer can be retrieved from a text by rephrasing the sentence. Producing several variants of a question also increases the possibility of finding the right answer to the question. Mckeown [37] addresses solutions to the problems in natural language systems by using a paraphraser mechanism. Ravichandran and Hovy [45] explore paraphrase patterns in order to improve QA systems. Barzilay, Mckeown, and Elhadad [5] show that paraphrases increase the chances of finding the right brief statement of a text for a summarization task of multiple sources such as news articles. Different phrases explaining the same event create alternatives for generating a summarized sentence from a text.

Acquired paraphrases have also been shown to improve the performance of Statistical Machine Translation (SMT) systems [33, 39]. SMT systems are trained on 'bitexts', bilingual parallel corpora [3]; augmenting such data with paraphrases can significantly improve translation quality and coverage [8, 23]. Recent attempts to use paraphrases to augment the coverage of SMT include filtering out the out-of-vocabulary words [36] and identifying particular words such as negators and antonyms [35].

Information Retrieval (IR) and Information Extraction (IE) methods may benefit from paraphrase patterns. An early paper on IR argued for the utility of paraphrase [10] and paraphrases of words in IE patterns can be identified so as to extract the required information from stored text [48]. Barron-Cedeno et al. [4] address the difficulty of detecting paraphrases for an automatic plagiarism detection system. They suggest that identifying paraphrases improves the performance of plagiarism detection systems. Ganitkevitch et al. [22] present state-of-the-art results for compression systems, and benefit from the extraction of sentential paraphrases. Also, natural language generation systems may benefit from paraphrasing for a sentence re-writing task [43].

A notable aspect of the knowledge-lean approach to PI described here is the use of character-based features to represent texts. We describe a number of experiments that explore the use of both lexical- and character-based features in order to represent candidate paraphrase pairs. It is shown that features based on character n-grams are particularly important in providing a basis for reliably identifying paraphrases. This observation is robust across corpora for English, Turkish and Russian, as well as for a highly non-standard form of English represented by Twitter. It is hypothesized that character n-grams are useful for capturing lexical similarity, even in the presence of morphological processes or other changes such as non-standard spellings. Thus, representations that use character n-grams as the basis for defining features may offer some advantages over representations simply utilising word-based features.

In Sect. 5 we first report on PI experiments previously conducted with several different paraphrase corpora: the Microsoft Research Paraphrase Corpus [14], the Plagiarism Detection Corpus [32], the Twitter Paraphrase Corpus [51] and a recently constructed Turkish Paraphrase Corpus [17]. Our approach uses combinations of features based on lexical- and character-based feature overlap, together with Support Vector Machine (SVM) classifiers. Good results have already been reported using such methods with the Twitter Paraphrase Corpus [16]. The focus here, however, is on identifying individual features or combinations that provide for robust PI across the different corpora.

Next, in Sect. 6 we report on new experimental results for ParaPhraser [42, 44], a Russian paraphrase corpus. Russian, with its rich morphology and free word order, presents a particularly interesting challenge for our knowledge-lean approach to PI. A further novel aspect of this work is the extension of our approach to a multi-class classification problem. We train classifiers to distinguish between precise paraphrase, near paraphrase and non-paraphrase and are able to show that character-based features also work well in this context.

2 Related Work

Much PI research makes use of existing NLP tools and other knowledge-base resources. For example, Duclaye et al. [15] exploits the NLP tools of a question-answering system, while Finch et al. [20], Mihalcea et al. [38], Fernando and Stevenson [19], Malakasiotis [34], and Das and Smith [11] all employ lexical semantic similarity information based on lexical knowledge bases such as WordNet [18].

In contrast to this, however, a number of researchers have investigated whether near state-of-the-art PI results might be obtained without the use of such tools and external resources. Blacoe and Lapata [7], for example, use distributional methods to find compositional meaning of phrases and sentences. They find that the performance of their approach is comparable to methods that rely on knowledge-based resources. Lintean and Rus [31] consider the use of overlap methods based on word unigrams and bigrams. Word bigrams, in particular, have the potential to capture word order information, which can in turn capture syntactic similarities between two text fragments. SMT metrics, alone and in combination, can be used to identify sentence-level paraphrases [20, 32, 39]. Ji and Eisenstein [27] attains state-of-the-art results based on latent semantic analysis and a new term-weighting metric, TF-KLD.

A variety of classifier models have been employed for the purpose of identifying paraphrases. Kozarova and Montoyo [29] measure lexical and semantic similarity with the combination of different classifiers: k-NN, SVM, and Maximum Entropy. SVM classifiers remain the most applicable in recent research, whether applied solely [20, 50] or as part of combined classifiers [29, 31, 32].

In the current work, SVM classifiers are trained to provide a binary classification between paraphrase and non-paraphrase pairs. In addition, for the Russian ParaPhraser corpus, SVM classifiers are combined to provide a three-way classification that also distinguishes between precise paraphrase and near paraphrase.

3 Paraphrase Corpora

A paraphrase corpus is a collection of monolingual, text pairs (typically, pairs of sentences), where the text pairs have been annotated according to whether they represent paraphrases or not. In the absence of an effective linguistic definition of the notion "have the same meaning", paraphrase corpora provide a basis for the training and evaluation of models of paraphrase. The construction of paraphrase corpora has therefore been central to the development of methods for identifying paraphrase. In the later sections we report on experiments conducted using a number of different paraphrase corpora. We outline each of these corpora below.

3.1 The Microsoft Paraphrase Corpus

The Microsoft Research Paraphrase Corpus (MSRPC) [14] has been used for a decade as the standard for comparison of results. The initial dataset was comprised of comparable corpora, constructed by collecting comparable newswire articles, using an approach similar to that of [6, 47]. Unlike these approaches, however, Dolan et al. [13] use broad-domain news agencies, clustering them to align pairs of sentences referring to the same event. A simple measure of string distance supported by heuristics is used to extract sentential paraphrases. Dolan and Brockett [14] further develop this initial dataset, training SVMs on the data and constructing a monolingual parallel corpus: MSRPC. There are 5,802 sentence pairs in the MSRPC. Paraphrase pairs (3,900) are scored as 1 and non-paraphrase pairs (1,901) are scored as 0. The MSRPC is split into two chunks; a training set and a test set, containing 1,725 and 4,076 sentence pairs, respectively. Both sets consist of randomly chosen paraphrase and non-paraphrase pairs. The MSRPC has been widely used for the development and comparison of PI methods.

3.2   The Plagiarism Detection Corpus

The Plagiarism Detection Corpus (PAN) [32] has been constructed by deriving aligned, corresponding sentences from 41,233 plagiarised documents. PAN has been made available for use in the development and testing of plagiarism detection and PI methods, and its authors published initial results experimental results. PAN is comprised of 13,000 sentence pairs in total: 10,000 reserved as a training set and 3,000 for a test set. The data contain equal numbers of paraphrase and non-paraphrase pairs in both the training and test sets. Candidate paraphrase pairs are labelled in the same way as for the MSRPC: paraphrase pairs scored 1, non-paraphrase pairs 0.

3.3   The Twitter Paraphrase Corpus

The Semeval-2015 Task1, "Paraphrase and Semantic Similarity in Twitter" involved predicting whether two tweets have the same meaning. Training and test data were provided in the form of a Twitter Paraphrase Corpus (TPC) [51]. The TPC is constructed semi-randomly and annotated via Amazon Mechanical Turk by 5 annotators. It consists of around 35% paraphrases and 65% non- paraphrases. Training and development data consist of 18,000 tweet pairs and test data 1,000 tweet pairs. The test data are drawn from a different time period to the training and development data and annotated by an expert. A novel aspect of the TPC compared to other paraphrase corpora is the inclusion of topic information, which is also used during the construction process.

3.4   A Turkish Paraphrase Corpus

Although a paraphrase corpus for Turkish has previously been reported [12] the corpus data are not widely available and currently do not provide any negative instances or scoring scheme. To address this, a small Turkish Paraphrase Corpus (TuPC) [17][1] was constructed from news items extracted from online Turkish newspapers. The method of construction was adapted from that used for the MSRPC. A relatively fine-grained, semantic similarity score is assigned to candidate paraphrase pairs following guidelines provided for the SemEval 2012 pilot semantic textual similarity task [1]. Four native

---

[1] In order to download TuPC: https://osf.io/wp83a/.

speakers of Turkish annotated each candidate pair. Annotation made use of a six-point scale, ranging from 0 (completely unrelated in meaning) to 5 (identical in meaning). A similar scheme was adopted for expert annotation of the TPC test data. For the purpose of experimenting with simple PI, the assigned scores were also converted to binary labels. Pairs scored as 5, 4 and 3 were labeled paraphrase, while those marked 2, 1 and 0 were labeled non-paraphrase. Inter-annotator agreement was measured using Fleiss's Kappa [21]. This showed 'moderate agreement' (0.42) for the coarser-grained, binary labels. After converting scores to binary labels, we obtained 563 paraphrase, 285 non-paraphrase and 154 debatable pairs[2]. Excluding the 154 debatable pairs, TuPC has 848 sentence pairs that can be used for the PI task.

3.5   A Russian Paraphrase Corpus

A Russian Paraphrase Corpus (ParaPhraser) [44] was constructed from data collected in real-time from daily news headlines sourced via a number of different online, Russian news agencies. An unsupervised, matrix similarity metric was used in order to obtain candidate paraphrase pairs. Candidate pairs were further evaluated and labelled by crowdsourcing using a user-friendly, online interface. Labelling of candidate pairs in the corpus is three-way, distinguishing between precise paraphrases, near paraphrases and non-paraphrases. The ParaPhraser corpus data used in the experiments reported in Sect. 6 were provided for a shared paraphrase detection task[3] [42] at the conference Artificial Intelligence and Natural Language (AINL-FRUCT) 2016. The training set for the shared task consisted of 7227 sentence pairs. It is reported in Pronoza and Yagunova [44] that most of the negative instances (approximately half of the data) are those pairs labelled non-paraphrase by the annotators. However, to keep the corpus balanced, some candidate pairs rejected by the unsupervised similarity metric are also included in the data. A test set, made available later, consisted of 1925 sentence pairs.

4   Knowledge-Lean Paraphrase Identification

Text pre-processing is essential to many NLP applications. It may involve tokenisation, removal of punctuation, part-of-speech tagging, morphological analysis, lemmatisation, and so on. For identifying paraphrases, this may not always be appropriate. Removal of punctuation and stop words, and word lemmatisation, for example, can all result in a loss of information that may be critical in terms of PI. In keeping with our knowledge-lean approach, in all of the reported experiments we keep text pre-processing to a minimum. In general, punctuation is retained and no lexical normalization, such as correction of spelling errors or stripping of morphological affixes, is carried out. For all corpus data, text is tokenized by splitting at white space and lowercasing is also performed.

---

[2] A debatable pair arises where the decisions of the four annotators are equally divided between "paraphrase" and "non-paraphrase".

[3] http://ainlconf.ru/paraphraser.

4.1   Representing Paraphrase Pairs

As the basis for deriving a number of overlap features, we begin by considering different representations of a text as a set of tokens. In general, a token may be either a word n-gram or a character n-gram. In practice, for all of the work described here we restrict attention to word and character unigrams, bigrams and trigrams. Use of a variety of machine translation techniques [32] that utilise word n-grams originally motivated their use in our work for representing candidate paraphrase pairs for the PI task. Word n-grams (for n > 1) have the potential to provide useful syntactic information about a text. Character bigrams and trigrams, on the other hand, can prove beneficial in capturing similarity between related word forms. This may be particularly useful where languages are morphologically rich, as for example, in the case of Turkish and Russian, or where non-standard spellings or abbreviations are encountered (e.g. for Twitter).

Possible features are constructed using the following basic set-theoretic operations. For a given type of token we may define:

- **Size of union (U):** the size of the union of the sets of tokens representing the two texts of a candidate paraphrase pair.
- **Size of intersection (N):** the number of tokens common to the sets tokens representing the two texts of a candidate paraphrase pair.

In addition we consider text size (S). For a given pair of texts, feature S_1 represents the size of the set of tokens representing the first text of a candidate paraphrase pair. Similarly S_2 is the size of the second text. The four features U, N, S_1 and S_2 are each computed for word and character unigrams, bigrams and trigrams, respectively. It may be noted that, in general, just knowing about the union, intersection or size features in isolation from one another is unlikely to be particularly informative. However, for a given token type, the four derived features in combination may provide very useful information about the similarity of texts.

In order to visualize how these individual features are separable with a hyper-plane, Fig. 1 shows plots of features derived from character bigrams from the Twitter test data set using Matplotlib[4] software [26]. Note that numbers used to label the axes for the plots represent scaled and normalized features values (see Sect. 4.4). The overall pattern shown in the figure is representative of the patterns seen in similar plots for the other corpus data used in our studies (See Sect. 5). The scatter plots in A and B of Fig. 1 show the results of classification on the basis of just the union and intersection features, using either a linear kernel (A) or an RBF kernel (B). It is clear that both of the classifiers are readily able to find decision boundaries that generally succeed in separating paraphrases from non-paraphrases based on the union and intersection features. We can conclude that these features are highly informative when used in combination. This is not altogether surprising, as the ratio of these features corresponds to the well-known Jaccard similarity coefficient. In contrast, it is clear from plots C and D of Fig. 1 that using just the size features S_1 and S_2 is not sufficient to separate the instances. Intuitively, this is because the features S_1 and S_2 only relate to the individual texts of a candidate pair, rather than providing information about the pair as a whole. Nevertheless, our

---

[4] http://matplotlib.org/.

experiments have shown that information about the size of the individual texts is useful in the context of the other two features and we have consistently found that better performance is obtained when all four types of feature are present. This observation can also be related to commonly used measures of similarity. In particular, the ratio between the size of intersection (dot product) of two vectors and some function of their sizes (norms) is the basis of measures including both the Dice similarity coefficient and the cosine measure. This helps to explain why it is important to include the additional features S_1 and S_2.

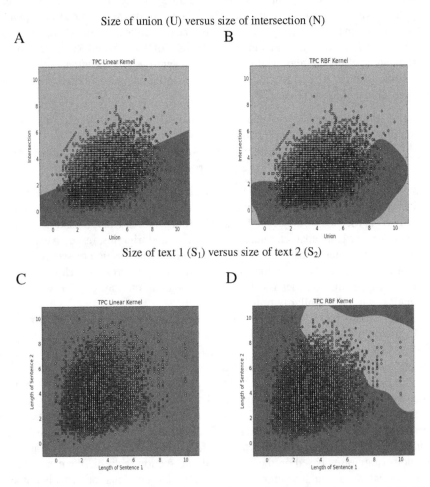

Fig. 1. Plotting features derived from character-bigrams for data from the test set of the TPC.

A given candidate paraphrase pair is represented as a vector of features derived from lexical and character n-grams, either alone or in combination. In the results reported later, we denote character-based features as C1, C2 and C3, representing the four features corresponding to character unigrams, bigrams and trigrams, respectively. Similarly, W1, W2 and W3 each denote four features generated by word unigrams, bigrams and trigrams, respectively. A combination such as C3W2 comprises the four features derived from character trigrams plus the four derived from word bigrams, totaling eight features in all. The notation C12 is used as convenient to abbreviate the combination of both C1 and C2, and so on.

4.2   Classifier Training

A Support Vector Machine (SVM) classifier maps the feature vectors corresponding to training instances into high dimensional vector space and constructs a maximum margin, linear decision boundary separating this space into two categories. New data points are then classified according to which side of the decision boundary they fall. The applicability of SVMs has been proven for a range of different NLP tasks and applications, including PI.

In all of the experiments reported later we make use of SVM implementations from scikit-learn (Pedregosa et al., 2001) and have experimented with a variety of classifiers. For all experiments, the results were obtained using Support Vector Classifier (SVC), which was adapted from *libsvm* [9] by embedding different kernels. In particular, we experimented with linear and Radial Basis Function (RBF) kernels. Linear kernels are known to work well with large datasets and RBF kernels are the first choice if a small number of features is applied [25]. Both cases apply to our experimental datasets.

We further note that in all experiments reported below, the linear and RBF classifiers are simply used with their default parameter settings: no attempt is made to tune parameters. Our rationale is that the main focus here is on the choice of features and representation of paraphrase pairs for classification and not on achieving absolute optimal performance, as such. Tuning of the SVM parameters to further enhance performance is therefore a secondary issue from our perspective.

4.3   Feature Scaling

Scaling is a smoothing technique for data distribution which transforms features into a new form without loss of their informative characteristics. It is stated [25] that feature scaling is an essential step for SVM classifiers. The SVM features that represent different properties of the data should be scaled (normalised or standardised) for better performance. In order to weight each feature equally, the numeric range difference of each feature should be scaled to fall within a standardized range (typically, [0,1] or [−1,1]). If this is not done, features that are greater in numeric range will tend to dominate the smaller ones.

Weighting of features with a scheme such as TF-IDF gives a value to each feature according to their significance to text. Ji and Eisentein's [27] attain new TF-KLD scheme is a modified form of TF-IDF which they use in conjunction with a linear SVM classifier, achieving good performance on the MSRPC. This also proves that although SVM classifiers are powerful for separating features linearly, a successful feature scaling and weighting process is vital to ensure good results on a given dataset.

In the present work we apply scaling but keep the scheme as simple as possible by applying a form of standardisation known as the z-score in statistics. Subtracting the mean, μ_x, from the feature vector, x, and dividing each of those features by its standard deviation, σ, scales features, and a new feature vector, $\hat{x}$, is obtained (Eq. 1). For the z-score, the transformed data variable has a mean of 0 and a variance of 1. Apart from this simple scaling method, all features are otherwise kept as they are.

$$\hat{x} = \frac{x - \mu_x}{\sigma} \tag{1}$$

4.4 Experiments

We have previously reported good results for the PI task with respect to the Twitter Paraphrase Corpus [51] and shown that our approach is also applicable to a highly inflected, agglutinative language such as Turkish [17]. In the following section, our focus is on identifying feature combinations that provide robust PI across different corpora. For this we present a comparison of experimental results previously reported separately for the MSRPC, PAN, TPC and TuPC. In these experiments the representations of candidate paraphrase pairs are constructed using features based on word and character n-grams. Although we have considered the use of character trigrams and even four-grams in earlier work, the results presented here are based only on unigrams and bigrams as these provide the most robust and comparable results across the different corpora.

New experimental results obtained using the Russian ParaPhraser corpus are then presented separately in Sect. 6.2. In this work, we also consider the utility of additional features based on word and character trigrams. One other notable aspect of our work here is the extension of our PI method to a three-way classification task (precise paraphrase, near paraphrase, non-paraphrase). We note that good results are obtained relative to other approaches on this task, suggesting that our approach is effective in distinguishing between the categories precise paraphrase and near paraphrase. In all of the experiments reported, 10-fold cross validation was applied.

5 Combination of Word- and Character-Based Features

Experiments were originally conducted with SVMs trained using both linear and RBF kernels and using features derived from character-based and word-based unigrams and bigrams. For simplicity in the following, results are only reported for experiments conducted with the RBF kernel. It is noted that the pattern of results observed for the linear kernel is very similar to that obtained using the RBF kernel. However, while the differences between the two kernels for the same set of features are marginal, the RBF kernel does generally outperform the linear kernel for these particular data sets. One exception to this is for the experiments conducted with the TuPC, where the linear kernel actually yields slightly better results than the RBF kernel. As already noted, our objective here is to gain insight into which feature combinations appear to be effective, and not to report the best results overall. For this purpose, it is the general pattern of results that is of interest, rather than individual best performances.

Table 1 presents Accuracy and F1 scores for experiments performed using a number of different feature combinations and across the different corpora. With the exception of the TuPC, F1 scores for informed baseline results are also provided. For the MSRPC, the baseline result is taken from Mihalcea et al. (2006) and represents a vector-based, cosine similarity measure, as traditionally used in information retrieval, with TF-IDF weighting; for PAN the baseline is a combination of the BLEU, NIST and TER machine translation metrics, as reported in Madnani et al. [32] and for TPC, the baseline is Logistic Regression as reported by Xu [51]. For the TuPC, on the other hand, the scores represent a naive baseline simply obtained by labeling every sentence pair as positive (i.e. paraphrase). Scores for current state-of-the-art results are also shown for MSRPC [27], PAN [32] and TPC [51]. For the latter, F1 scores only are shown, as accuracy scores were not reported.

Table 1. Comparing PI results obtained using various features and across the different corpora

Features	MSRPC		PAN		TPC		TuPC	
	Acc.	F1	Acc.	F1	Acc.	F1	Acc.	F1
C1	69.8	81.2	76.1	75.1	76.4	33.6	68.8	80.5
C2	73.0	**82.3**	90.9	90.8	*86.2*	*66.7*	*76.4*	*83.3*
W1	**73.1**	81.9	**92.0**	**91.9**	85.2	62.4	72.2	80.5
W2	69.6	80.9	90.9	89.9	85.0	61.8	71.3	80.5
C12	72.9	82.2	90.8	90.7	**86.0**	**66.5**	75.5	82.8
W12	73.1	81.9	92.1	92.0	85.9	63.6	73.7	81.4
C1W1	73.1	82.2	91.8	91.7	85.2	64.2	72.4	81.0
C2W2	74.0	*82.7*	91.9	91.8	85.3	64.4	*76.4*	*83.3*
C1W2	71.2	81.5	90.2	90.1	84.6	59.1	71.6	81.0
C2W1	*74.2*	*82.7*	*92.4*	*92.3*	85.4	65.1	76.3	83.2
Baseline	65.4	75.3	88.6	87.8	--	58.9	66.4	79.8
State-of-art	80.4	85.9	92.3	92.1	--	72.4	--	--

It is noted that the best results shown here for the different corpora comfortably outperform all of the given baselines. For PAN, the combination of C2 and W1 features already yields an F-Score of 92.3, which outperforms the state-of-the-art result reported by Madnani et al. [32]. For TPC and MSRPC, while the results shown fall some way below the state-of-the-art, they are nevertheless competitive with knowledge-based methods that make use of language-specific resources and processing tools. In the case of the results for Twitter, an optimized selection of features attained an F1 score of 67.4 on this test set and was ranked first for the binary PI task in the SemEval-2015 Task 1: Paraphrase and Semantic Similarity in Twitter (PIT) [16].

Considering features based just on a single token type (upper part of the table), it can be seen that character bigrams (C2) tend to perform well. For both TPC and TuPC, the C2 features clearly outperform W1. On the other hand, for both MSRPC and PAN, the performance is less marked. For MSRPC, features based on character bigrams attain the best F1 score (if not the highest accuracy) but for PAN it is clear that W1 actually outperforms C2.

Turning to combinations of features, for MSRPC and PAN, the combination of character bigram features with lexical features (C2W1) yields some gain over just the C2 or W1 features alone. For TPC and TuPC, on the other hand, results based solely on character-bigram features appear to perform best overall.

The performance differences that are observed may be explained in terms of differences in the language data as well as in the construction of the corpora. For both MSRPC and PAN, the construction methods and the relatively impoverished morphology of English mean that candidate pairs tend to exhibit a comparatively high degree of simple lexical overlap. The PAN corpus, for example, was derived from aligned pairs of sentences within plagiarized text. The alignment algorithm made use of information about bag-of-words overlap, resulting in positive examples with high lexical similarity. As a result, there is a tendency for paraphrase pairs to have in common the same lexical forms and for simple word overlap measures (W1) to perform well, leading to the high scores observed in experiments using this data set. This is borne out by the scatter plots presented in Fig. 2. The plots show union and intersection features based on word unigrams for the two corpora. Classifiers using an RBF kernel are readily able to separate paraphrases and non-paraphrases on the basis of these features and this is particularly noticeable in the case of the PAN corpus.

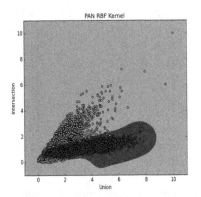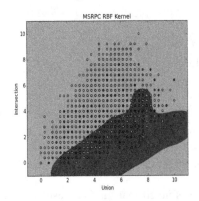

Fig. 2. Plotting features (union and intersection) of word unigrams for the test set of PAN Corpus and MSRPC

In contrast to MSRPC and PAN, both TPC and TuPC show a greater degree of variation in lexical form. For Twitter, the character limit on tweets notoriously means that words may be creatively shortened or abbreviated. Standard spelling rules are often ignored and additional characters may also be added to words for emphasis. Turkish, on the other hand, is a highly inflected and agglutinative language and the productive use of affixes is very typical, either to change the meaning or the stress of a word. In both cases, this means that candidate paraphrase pairs are less likely to share exactly the same word forms. Instead, they may share related forms and for this, character bigram features may be more applicable.

We conclude by noting some examples where pairs are incorrectly labeled on the basis of our overlap measures. This provides some indication of the limitations of our knowledge-lean approach. In particular, there are cases where high lexical overlap scores are misleading, so that sentence pairs cannot be correctly identified. For example, the sentence pair below is marked as a non-paraphrase in MSRPC. However, taken as strings of words the sentences are near identical, the only differences being the words marked in bold:

*S1: NBC will probably end the season as the second most popular network behind CBS, **although it's** first among the key 18-to-49-year-old demographic.*
*S2: NBC will probably end the season as the second most-popular network behind CBS, **which is** first among the key 18-to-49-year-old demographic.*

Table 2 presents some examples from the TPC that are incorrectly labeled by the system we used in SemEval-2015 Task 1. The pairs on the left are erroneously predicted as non-paraphrase. These examples are clear cases where simple overlap measures are unlikely to work well. Both require relatively deep understanding of the semantics of the sentences and indeed some inference (e.g. that "those last 3 battles in 8 mile" implies "the ending to 8 mile"). The pairs on the right of the table on the other hand, are examples of non-paraphrases that are erroneously labeled as paraphrases. In these cases, as for the MSRPC example shown above, high lexical overlap is misleading. The examples also indicate the limitations of a knowledge-lean approach that takes no account of word order or grammatical structure. Nevertheless, the results for the PI task obtained on the Twitter data outperformed all other systems submitted for SemEval-2015 Task 1.

Table 2. Twitter paraphrase pairs incorrectly labeled by our system using character bigram features

Paraphrase pairs that are predicted false	Non-paraphrase pairs that are predicted true
S1: the ending to 8 mile is my fav part of the whole movie *S2: those last 3 battles in 8 mile are the shit*	*S1: hahaha that sounds like me* *S2: sounds like a successful day to me*
S1:chris davis is putting the team on his back *S2: chris davis is so fucking good*	*S1: world of jenks is on at 11* *S2: world of jenks is my favorite show on tv*

6   The Russian Paraphrase Task

A shared task on sentence paraphrase detection for the Russian language was held as part of the Artificial Intelligence and Natural Language conference (AINL-FRUCT) in November 2016. A core data set for the task was the Russian Paraphraser corpus. Participants were invited to submit runs for two subtasks. For Task 1, given a candidate paraphrase pair, it was necessary to label the pair according to a three-way classification scheme. Task 2 on the other hand required a simple binary classification of a given pair of sentences as either paraphrase or non-paraphrase. Both tasks permitted runs in two categories (standard and non-standard) that differed in terms of the data allowed for the purposes of training and development. For the standard runs, the only data permitted for use in training were those provided by the workshop organisers. For the non-standard runs, on the other hand, the use of any external data sources was allowed. Note that, in keeping with our knowledge-lean philosophy, we have only submitted runs for the standard category and no additional data sources or processing tools were used in our experiments.

6.1 Three Class Versus Binary Classification

SVM classifiers were again trained using implementations from scikit-learn [41]. As for the earlier experiments, the results reported below were obtained using SVC adapted from libsvm [9] and experiments were conducted with both linear and RBF kernels.

Paraphrase corpora have generally been constructed to provide a simple binary distinction between paraphrases and non-paraphrases. However, Rus et al. [46] suggests that a simple binary distinction is not sufficient and that paraphrase corpora should allow for a more nuanced set of categories, representing gradations of paraphrase or similarity of meaning. Following this philosophy, for ParaPhraser, each candidate pair in the data is labeled as precise paraphrase (1), near paraphrase (0) or non-paraphrase (−1). Combining the precise and near paraphrase items also allows for a simple binary classification PI task between paraphrase and non-paraphrase categories.

We experimented with both the binary and three-class classification tasks using SVMs. SVM are binary classifiers, so in order to achieve three-way classification it is necessary to train and combine multiple binary classifiers in some way. The three-class classification method adopted here uses a standard "one-against-one" approach for multi-class classification. Each individual classifier distinguishes between two of the classes. In general, if there are n classes in the dataset, there will be $(n * (n − 1)/2)$ classifiers constructed for training. In our case, three different classifiers are constructed because there are three different classes in the dataset.

At prediction time, each classifier is used to "vote" for one of the classes and the class that receives the most votes is selected. If two or more classifiers receive the same number of votes, then classification confidence is taken into account and the class with the highest aggregate confidence is selected.

6.2 Results

We present here selected results submitted for both Task 1 (Table 3) and Task 2 (Table 4) using combinations of features based on character and word unigrams, bigrams and trigrams. Results in the tables are reported in terms of accuracy and F1 score as provided by the task organisers. Note that for the three-way classification problem, the reported F1 scores shown in the table are macro-averages of the F1 scores obtained against each of the distinct classes in the data.

Although we trained classifiers against all possible combinations of features, the submitted runs were just those that appeared to perform well on the training set. As noted previously, we only participated in the standard runs for Task 1 and Task 2. No external resources were used for these runs. Tables 3 and 4 also show baseline results provided by the organisers.

Table 3. Task 1-Standard results of combined features using two different SVM kernels

Task 1	Linear		RBF	
Features	Acc.	F1	Acc.	F1
C2	56.42	55.34	55.94	53.45
C13	56.95	55.08	57.11	54.40
C123	**57.32**	**55.57**	56.89	54.19
Baseline	34.39	33.41	34.39	33.41

Table 4. Task 2-Standard results of combined features using two different SVM kernels

Task 2	Linear		RBF	
Features	Acc.	F1	Acc.	F1
C2	**72.11**	78.73	71.37	**78.88**
C2W1	71.74	78.51	71.26	**78.88**
C13	69.78	77.23	68.62	77.38
C123	71.95	78.60	71.95	79.31
Baseline	49.66	54.03	49.66	54.03

Although we also experimented using word level features, both singly and also in combination with character level features, the best results for Task 1 were obtained using character level features alone. Table 3 therefore shows results for the best representations, based on the feature combinations C2, C13 and C123. We show results for both the linear and RBF kernels. As can be observed, while classifiers utilizing either kernel comfortably outperform the baselines, linear kernels generally perform better than RBF kernels for the ParaPhraser data on this task. A minor exception is that for the feature combination C13, the RBF kernel performs slightly better than the linear kernel in terms of accuracy (though not F1). The highest F1 score overall on Task 1 is obtained using a linear kernel and a combination of all character level features (C123). Ranking according to accuracy, this result was placed as the second highest result reported on this task for the shared paraphrase task, as shown in Table 3, above.

Table 4 presents the results for our approach on Task 2. The results show that the performance of the linear kernel is better than the RBF kernel in terms of accuracy for the same set of features. However, it is noted that F1 scores tend to be higher for the RBF kernel. In terms of the features employed, the highest results are again obtained from using character level features, although word unigram features in combination with character bigram features (C2W1) perform well. The highest accuracy is obtained from features constructed from character bigrams (C2) and using a linear kernel and this result is ranked as the 8th in Task 2 (ranking of results according to accuracy) as shown in Table 5. Interestingly, the highest F-score is 79.31, obtained from combined character level features (C123) using an RBF kernel. However, its accuracy is lower as compared to the other results shown in Table 5.

Table 5. Our highest results in comparison with other team results for the Task 1 and Task 2

Rank	Task 1-Standard	Acc.	F1 (macro)
1	Team3348	59.01	56.92
2	**C123 (Linear)**	**57.32**	**55.57**
3	Penguins	57.21	44.43
4	**C23 (Linear)**	**57.16**	**54.60**
5	Team3348	57.11	54.50
Rank	Task 2-Standard	Acc.	F1 (macro)
1	dups	74.59	80.44
2	Team3348	74.48	80.78
5	NLX	72.74	78.80
8	**C2 (Linear)**	**72.11**	**78.73**
9	**C123 (Linear)**	**71.95**	**78.60**

Intuitively, the three-way classification problem represents a more difficult task than the simple binary task of deciding between paraphrase and non-paraphrase. In the case of Russian PI, the experimental results show that three-class classification performs relatively well as compared to binary classification. Table 6 shows a confusion matrix for the C123 feature combination using a linear kernel on Task 1. As might be expected, the classifier is not doing a particularly good job of distinguishing precise paraphrases (1) from near paraphrases (0). Recall of precise paraphrase is particularly low at around 42%; precision for predictions of near paraphrase is around 50%. The classifier does somewhat better in terms of distinguishing non-paraphrases (−1) from paraphrases, however, with precision and recall for this class at around 68% and 60%, respectively.

Table 6. Confusion matrix for C123 features with Linear Kernel on Task 1.

True	Predicted C123 (Linear)		
	1	0	−1
1	**158**	177	39
0	115	**484**	179
−1	15	294	**463**

7   Discussion and Conclusions

Knowledge-lean PI techniques can perform comparably to methods that utilise external processing tools and other resources. An advantage of adopting such methods is that they make use of just the text at hand and as a consequence may be readily applied to languages that may lack processing tools and other knowledge-based resources. This may in turn lead to faster development of NLP applications.

We have reported the results of experiments that show that robust PI results can be attained across a number of different corpora. The corpora represent paraphrase data drawn from a several different language and text-types. MSRPC and PAN represent informal, though relatively standard forms of written (American) English. The English language Twitter data drawn from the TPC, on the other hand, is highly non-standard. Its short texts (tweets) widespread use of non-standard grammar, spelling and punctuation, as well as slang, abbreviations and neologisms make it particularly interesting in the context of evaluating a knowledge-lean approach to PI. Turkish and Russian, as languages with rich morphology, also present particular challenges for our approach.

Our results suggest that relatively simple overlap features based on character n-grams are informative for the PI task. This is especially noticeable where there is rich and productive morphology (Turkish or Russian) or non-standard orthographic conventions (Twitter). In such cases, candidate paraphrase pairs are less likely to share the same set of word forms, though they may share word forms that are related by morphological processes or non-standard spelling, for example. Features derived from character n-grams provide a way of detecting similarity of related word-forms, since different, but related forms may still share a relatively high proportion of character n-grams. Character bigrams appear particularly to be associated with good performance on the PI task, whether alone or in combination with word-based features.

We have extended our previous work on PI to a three-way classification scheme, as required by Task 1 of the shared paraphrase detection task using the Russian ParaPhraser corpus. Our approach involved training several different binary (SVM) classifiers, using a "one against one" strategy to achieve multi-class classification, and was ranked 2nd overall using a combination of character-based features only. Intuitively, making a three-way classification of candidate paraphrase pairs presents a more demanding task than a simple binary decision. However, training a single SVM classifier for the binary case did not perform as well as expected. Using the same combination of character-based features our approach was ranked 9th on Task 2. In fact, marginally better performance on this task was attained using features based just on character bigrams. Why this is so is not clear and we intend to analyse the performance of the classifiers more closely in order to investigate this further. For example, it is possible that using the SVM classifiers trained for Task 1 to simply distinguish between paraphrase and non-paraphrase might have attained better performance on Task 2 than training a single binary classifier.

Character level features are worth exploring further and there is growing interest in exploiting them to perform language processing tasks. For instance, the use of character level vectors has been proposed in a few of the latest studies [28, 30]. In this recent work, word representations are composed of characters of vectors. In the future, we also intend to explore the use of character-level vectors with convolutional neural networks for the paraphrase identification task.

Acknowledgements. The authors gratefully acknowledge the comments of our reviewers on earlier drafts of this paper.

References

1. Agirre, E., et al.: Semeval-2012 task 6: A pilot on semantic textual similarity. In: Proceedings of the 6th International Workshop on Semantic Evaluation, in Conjunction with the First Joint Conference on Lexical and Computational Semantics, pp. 385–393 (2012)
2. Androutsopoulos, I., Malakasiotis, P.: A survey of paraphrasing and textual entailment methods. Artif. Intell. Res. **38**(1), 135–187 (2010)
3. Bannard, C., Callison-Burch, C.: Paraphrasing with bilingual parallel corpora. In: Proceedings of the 43th Annual Meeting on Association for Computational Linguistics, pp. 597–604 (2005)
4. Barron-Cedeno, A., et al.: Plagiarism meets paraphrasing: insights for the next generation in automatic plagiarism detection. Comput. Linguist. **39**(4), 917–947 (2013)
5. Barzilay, R., et al.: Information fusion in the context of multi-document summarization. In: Proceedings of ACL, pp. 550–557 (1999)
6. Barzilay, R., Lee, L.: Learning to paraphrase: an unsupervised approach using multiple-sequence alignment. In: Naacl-2003, pp. 16–23 (2003)
7. Blacoe, W., Lapata, M.: A comparison of vector-based representations for semantic composition. In: Proceedings of the 2012 Joint Conference on Empirical Methods in Natural Language Processing and Computational Natural Language Learning (EMNLP-CoNLL 2012), pp. 546–556 (2012)
8. Callison-Burch, C., et al.: Improved statistical machine translation using paraphrases. In: Proceedings of the Main Conference on Human Language Technology Conference of the North American Chapter of the Association of Computational Linguistics (HLT-NAACL 2006), pp. 17–24 (2006)
9. Chang, C., Lin, C.: LIBSVM: a library for support vector machines. ACM Trans. Intell. Syst. Technol. **2**(3), 1–27 (2011)
10. Culicover, P.W.: Paraphrase generation and information retrieval from stored text. Mech. Transl. Comput. Linguist. **11**(1–2), 78–88 (1968)
11. Das, D., Smith, N.A.: Paraphrase identification as probabilistic quasi-synchronous recognition. In: Proceedings of the Joint Conference of the 47th Annual Meeting of the ACL and the 4th International Joint Conference on Natural Language Processing of the AFNLP: ACL-IJCNLP 2009, pp. 468–476 (2009)
12. Demir, S., et al.: Turkish paraphrase corpus. In: Proceedings of the Eight International Conference on Language Resources and Evaluation (LREC 2012), pp. 4087–4091 (2012)
13. Dolan, W.B., et al.: Unsupervised construction of large paraphrase corpora: exploiting massively parallel news sources. In: Proceedings of the 20th International Conference on Computational Linguistics, COLING 2004. Association for Computational Linguistics, Geneva (2004)
14. Dolan, W.B., Brockett, C.: Automatically constructing a corpus of sentential paraphrases. In: Proceedings of IWP, pp. 9–16. Asia Federation of Natural Language Processing (2005)
15. Duclaye, F., et al.: Using the web as a linguistic resource for learning reformulations automatically. In: Proceedings of the Third International Conference on Language Resources and Evaluation (LREC 2002), Las Palmas, Canary Islands, Spain, pp. 390–396 (2002)
16. Eyecioglu, A., Keller, B.: ASOBEK: Twitter paraphrase identification with simple overlap features and SVMs. In: Proceedings of the 9th International Workshop on Semantic Evaluation (SemEval 2015), Denver, Colorado, pp. 64–69 (2015)

17. Eyecioglu, A., Keller, B.: Constructing a Turkish corpus for paraphrase identification and semantic similarity. In: Proceedings of the 17th International Conference on Intelligent Text Processing and Computational Linguistics. LNCS, pp. 562–574 (2016)
18. Fellbaum, C.: WordNet. An Electronic Lexical Database. MIT Press, Cambridge (1998)
19. Fernando, S., Stevenson, M.: A semantic similarity approach to paraphrase detection. In: Proceedings of the 11th Annual Research Colloquium of the UK Special Interest Group for Computational Linguistics, pp. 45–52 (2008)
20. Finch, A., et al.: Using Machine translation evaluation techniques to determine sentence-level semantic equivalence. In: Proceedings of the Third International Workshop on Paraphrasing (IWP 2005), pp. 17–24 (2005)
21. Fleiss, J.L.: Measuring nominal scale agreement among many raters. Psychol. Bull. **76**, 378–382 (1971)
22. Ganitkevitch, J., et al.: Learning sentential paraphrases from bilingual parallel corpora for text-to-text generation, pp. 1168–1179. Computational Linguistics (2011)
23. He, W., et al.: Enriching SMT training data via paraphrasing. In: Proceedings of the 5th International Joint Conference on Natural Language Processing, IJCNLP 2011, pp. 803–810. Asian Federation of Natural Language Processing (2011)
24. Hearst, M.A., Grefenstette, G.: Refining automatically-discovered lexical relations: combining weak techniques for stronger results. In: Statistically-Based Natural Language Programming Techniques, Papers from the 1992 AAAI Workshop, Menlo Park, CA, pp. 64–72 (1992)
25. Hsu, C.-W., et al.: A practical guide to support vector classification. BJU Int. **101**(1), 1396–1400 (2008)
26. Hunter, J.D.: Matplotlib: a 2D graphics environment. Comput. Sci. Eng. **9**(3), 90–95 (2007)
27. Ji, Y., Eisenstein, J.: Discriminative improvements to distributional sentence similarity. In: Proceedings of the 2013 Conference on Empirical Methods in Natural Language Processing, pp. 891–896. Association for Computational Linguistics, Seattle (2013)
28. Kim, Y., et al.: Character-aware neural language models. CoRR 1508.06615 (2015)
29. Kozareva, Z., Montoyo, A.: Paraphrase identification on the basis of supervised machine learning techniques. In: Salakoski, T., Ginter, F., Pyysalo, S., Pahikkala, T. (eds.) FinTAL 2006. LNCS (LNAI), vol. 4139, pp. 524–533. Springer, Heidelberg (2006). https://doi.org/10.1007/11816508_52
30. Ling, W., et al.: Finding function in form: compositional character models for open vocabulary word representation. CoRR 1508.02096 (2015)
31. Lintean, M., Rus, V.: Dissimilarity kernels for paraphrase identification. In: Proceedings of the 24th International Florida Artificial Intelligence Research Society Conference, Palm Beach, FL, pp. 263–268 (2011)
32. Madnani, N., et al.: Re-examining machine translation metrics for paraphrase identification. In: Proceedings of the 2012 Conference of the North American Chapter of the Association for Computational Linguistics: Human Language Technologies (NAACL-HLT 2012), PA, USA, pp. 182–190 (2012)
33. Madnani, N., et al.: Using paraphrases for parameter tuning in statistical machine translation. In: Proceedings of the Second Workshop on Statistical Machine Translation (WMT 2007), Prague, Czech Republic (2007)
34. Malakasiotis, P.: Paraphrase recognition using machine learning to combine similarity measures. In: Proceedings of the ACL-IJCNLP 2009 Student Research Workshop, Suntec, Singapore, pp. 27–35 (2009)

35. Marton, Y., et al.: Filtering antonymous, trend-contrasting, and polarity-dissimilar distributional paraphrases for improving statistical machine translation. In: Proceedings of the Sixth Workshop on Statistical Machine Translation, pp. 237–249. Association for Computational Linguistics, Edingburgh, Scotland (2011)
36. Marton, Y., et al.: Improved statistical machine translation using monolingually-derived paraphrases. In: Conference on Empirical Methods in Natural Language Processing (EMNLP), Singapore (2009)
37. Mckeown, K.R.: Paraphrasing questions using given and new information. Comput. Linguist. **9**(1), 1–10 (1983)
38. Mihalcea, R., et al.: Corpus-based and knowledge-based measures of text semantic similarity. In: Proceedings of the 21st National Conference on Artificial Intelligence, vol. 1, pp. 775–780. AAAI Press (2006)
39. Owczarzak, K., et al.: Contextual bitext-derived paraphrases in automatic MT evaluation. In: StatMT 2006, Stroudsburg, PA, USA, pp. 86–93 (2006)
40. Pedersen, T., Bruce, R.: Knowledge lean word-sense disambiguation. In: Proceedings of the Fifteenth National Conference on Artificial Intelligence, pp. 800–805. AAAI Press (1998)
41. Pedregosa, F., Varoquaux, G., Gramfort, A., Michel, V., et al.: Scikit-learn: Machine Learning in Python. http://scikit-learn.org/stable/
42. Pivovarova, L., et al.: ParaPhraser: Russian paraphrase corpus and shared task. In: Filchenkov, A., et al. (eds.) AINL 2017, CCIS, vol. 789, pp. 211–225. Springer, Cham (2018)
43. Power, R., Scott, D.: Automatic generation of large-scale paraphrases. In: Proceedings of the 3rd International Workshop on Paraphrasing (IWP2005), Jeju, Republic of Korea, pp. 33–40 (2005)
44. Pronoza, E., Yagunova, E.: Comparison of sentence similarity measures for Russian paraphrase identification. In: Proceedings of the AINL-ISMW FRUCT 2015, pp. 74–82. IEEE (2015)
45. Ravichandran, D., Hovy, E.: Learning surface text patterns for a question answering system. In: Proceedings of the 40th Annual Meeting of the Association for Computational Linguistics. (2002)
46. Rus, V., et al.: On paraphrase identification corpora. In: Proceedings of the Ninth International Conference on Language Resources and Evaluation (LREC 2014). European Language Resources Association (ELRA), Reykjavik, Iceland (2014)
47. Shinyama, Y., et al.: Automatic paraphrase acquisition from news articles. In: Proceedings of the Second International Conference on Human Language Technology Research, pp. 313–318 (2002)
48. Shinyama, Y., Sekine, S.: Paraphrase acquisition for information extraction. In: Proceedings of the second International Workshop on Paraphrasing - Volume 16 (PARAPHRASE 2003), vol. 16, pp. 65–71. Association for Computational Linguistics, Stroudsburg (2003)
49. Socher, R., et al.: Dynamic pooling and unfolding recursive autoencoders for paraphrase detection. In: Advances in Neural Information Processing Systems, pp. 801–809 (2011)
50. Wan, S., et al.: Using dependency-based features to take the 'Para-farce' out of paraphrase. In: Proceedings of the Australasian Language Technology Workshop, Sydney, Australia, pp. 131–138 (2006)
51. Xu, W.: Data-driven approaches for paraphrasing across language variations. New York University (2014)
52. Zhang, Y., Patrick, J.: Paraphrase identification by text canonicalization. In: proceedings of the Australasian Language Technology Workshop, Sydney, Australia, pp. 160–166 (2005)

Paraphrase Detection Using Machine Translation and Textual Similarity Algorithms

Dmitry Kravchenko[✉]

Department of Computer Science, Ben-Gurion University of the Negev,
Beer Sheva, Israel
`to.dmitry.kravchenko@gmail.com`

Abstract. I present experiments on the task of paraphrase detection for Russian text using Machine Translation (MT) into English and applying existing sentence similarity algorithms in English on the translated sentences. But since I use translation engines - my method to detect paraphrases can be applied to any other languages, which translation into English is available on translation engines. Specifically, I consider two tasks: given pair of sentences in Russian – classify them into two (non-paraphrases, paraphrases) or three (non-paraphrases, near-paraphrases, precise-paraphrases) classes. I compare five different well-established sentence similarity methods developed in English and three different Machine Translation engines (Google, Microsoft and Yandex). I perform detailed ablation tests to identify the contribution of each component of the five methods, and identify the best combination of Machine Translation and sentence similarity method, including ensembles, on the Russian Paraphrase data set. My best results on the Russian data set are an Accuracy of 81.4% and F1 score of 78.5% for an ensemble method with the translation using three MT engines (Google, Microsoft and Yandex). This compares favorably with state of the art methods in English on data sets of a similar size which are in the range of Accuracy 80.41% and F1-score of 85.96%. This demonstrates that, with the current level of performance of public MT engines, the simple approach of translating/classifying in English has become a feasible strategy to address the task. I perform detailed error analysis to indicate potential for further improvements.

Keywords: Paraphrase detection · Semantic similarity algorithms
Machine translation · Supervised classification

1 Introduction

1.1 Motivation

Paraphrase identification is useful in many natural language applications such as search engines (to calculate relevance of one sentence to the other), in plagiarism detection systems, authorship identification, patents and copyright detection systems, question-answering bots (to compute the semantic similarity between a

© Springer International Publishing AG 2018
A. Filchenkov et al. (Eds.): AINL 2017, CCIS 789, pp. 277–292, 2018.
https://doi.org/10.1007/978-3-319-71746-3_22

sentence given by a human and sentences stored in a corpus database), and text summarization. This task consists of determining whether two sentences convey similar content to the extent that one can held as a re-statement of the other.

The task has been well studied in English, and methods have been developed that reach pretty good results. These methods are not easily applicable directly to other languages, because they rely on rich lexical resources such as thesauri and large-scale word embedding which are not available in many other languages.

In this paper, I report on experiments to assess the feasibility of a simple strategy to adapt existing techniques in English to a Russian data set of paraphrase sentences: I first translate the Russian sentences into English sentences using publicly available Machine Translation (MT) engines (I test Google Translate, Microsoft Bing and Yandex) and then apply a variety of English techniques on the translated sentences to establish their paraphrase relation.

I find that this simple strategy provides "good enough" results on the data set for little effort, especially when compared with the complexity of acquiring large coverage thesauri in Russian and/or training statistical models on large amounts of Russian text.

1.2 Objective

My objective is to establish the feasibility of applying the strategy of MT as a preprocessing step to address the paraphrase detection task. Clearly, using a translation engine introduces noise because existing MT engines have limited accuracy. I compare three different MT engines to control for this aspect.

1.3 Task Description

Given the Russian paraphrases data set – the goal is to compute sentence similarity. The task is cast as two distinct classification tasks: (1) separate the list of sentence pairs into three classes: non-paraphrases, near-paraphrases and precise-paraphrases; (2) classify the pairs into two classes: non-paraphrases and paraphrases. Results are measured by two scores: F1 score and Accuracy. I use the shared task data set distributed at the Workshop of the International conference in Artificial Intelligence and Natural Language - AINL 2016.

2 Related Work

For English - there are two main paraphrase data sets: the Microsoft Research Paraphrase Corpus (MSRP) [12] and PPDB: the Paraphrase Database [11]. MSRP contains 5801 pairs of sentences (4076 are in the training set, and 1725 are in the test set), which are classified by humans into two classes: paraphrases and non-paraphrases. The highest achievement on MSRP is recorded by Ji and Eisenstein (2013), who used a method of matrix factorization with supervised reweighting, and which achieved an Accuracy of 80.41% and an F1 score of 85.96%. Another notable result was achieved by Socher et al. (2011), which achieved an Accuracy of 76.8% and an F1 score of 83.6%. Recent application of deep learning to the task is reported in Yin and Schutze (2015), where they applied convolutional neural networks and achieved an Accuracy of 78.4% and an F1 score of 84.6%.

For Russian corpus in one of the recent publications [10] Pronoza et al. (2015) achieved F1 score of 82.46% on binary classification task in paraphrase detection.

Madnani et al. (2012) [19] in their paper *Re-examining Machine Translation Metrics for Paraphrase Identification*, for solving the task of this paper used only Machine Translation metrics like BLEU(1-4), MAXSIM, BADGER, SEPIA, TER, NIST(1-5), METEOR, TERp, and has achieved on MSRP corpus 77.4% in Accuracy and 84.1% in F1 score, proving by this the effectiveness of these metrics. In my paper I will use only BLEU scores, and all rest would be semantic similarity algorithms scores. And machine translation I would use to prepare input data for these similarity algorithms.

Paraphrase detection is closely related to the task of textual entailment (TE) identification [17]. TE is a directed relationship between text and hypothesis. Bidirectional TE have not reached the same level of performance as direct paraphrase detection in English.

3   Data Set

Input data is a list of pairs of sentences in Russian which are collected from news headlines. The training set includes 7,227 pairs of sentences, which are classified by humans into three classes: 2,582 non-paraphrases, 2,957 near-paraphrases, and 1,688 precise-paraphrases.

Experimental settings: 14,454 sentences with approximately 117,000 words, in which approximately 23,000 words have unique forms. Sentence length is 8 words on average.

Output data is the list of predicted classes for each one of two tasks (described above), which is assigned to each pair of the sentences of the test part of cross-validation test.

4   Baseline

4.1   Algorithm

As a baseline algorithm I use the standard BLEU sentence similarity metric which can get input in Russian, and doesn't require translation into English.

BLEU scores with smoothing methods are from [2] with word n-grams. It is mentioned in [2] that original BLUE scores required no smoothing, as they were developed for document-level classification. But for the sentence-level classification they used these smoothing techniques.

These two smoothing techniques work as follows: assume that I match word n-grams for $n = 1...N$ (usually, $N = 4$). Let m_n be the count of the matching words in both sentences, and let $\widehat{m_n}$ be the modified n-gram match count.

Smoothing 1 is defined as follows: if there are no matched words in n-grams, then I use a small positive value ϵ to replace the 0 for n $\in$ [1..N]. if $m_n = 0$ then $\widehat{m_n} = \epsilon$.

Smoothing 2 (proposed by Lin and Och, 2004) is defined as follows: I add 1 to the matched n-gram count and the total n-gram count for n ranging from 2 to N.

Formally: for $n \in [2..N]$ I calculate: $\widehat{m_n} = m_n + 1$, and $\widehat{l_n} = l_n + 1$

4.2   Results

Table 1 contains execution results of BLEU algorithms on First Task (3-way classification) and Second task (2-way classification) with word n-grams.

Table 1. BLEU of two smoothing types

BLEU smoothing type	First task		Second task	
	Accuracy	F1 score	Accuracy	F1 score
Type 1 (1-g)	**57.43**	**55.07**	**76.64**	**71.33**
Type 1 (2-g)	55.76	52.83	73.72	70.76
Type 1 (3-g)	55.27	51.54	73.66	70.10
Type 1 (4-g)	49.50	45.50	64.28	41.70
Type 2 (1-g)	57.43	55.07	76.64	71.33
Type 2 (2-g)	56.81	53.44	76.54	70.57
Type 2 (3-g)	56.45	52.71	76.26	70.57
Type 2 (4-g)	56.33	52.20	76.17	70.67

5   Algorithm

5.1   Brief Explanation

In Appendix A, you can see the figure of algorithm data-flow.

Many sentences in the data set (which are news feeds) contain acronyms. I substitute acronyms to their full names using acronyms list derived from www. wiktionary.org. Acronym expansion is performed on the Russian sentences before they are translated.

I then translate the sentences with expanded acronyms from Russian to English. I used translation engines APIs to do so.

Finally, I construct a feature vector for the classifier - using a variety of sentence similarity algorithms which compute similarity scores on pairs of sentences.

These feature vectors I use as an input to GradientBoosting classifier and after computation it gives the class prediction for each pair of sentences.

5.2   Detailed Description

Step 1: Preprocessing I substitute all of the acronyms in the sentences to their full names. This step is very important because all of the toolkits which I use do not recognize Russian acronyms, particularly after they are translated to English. Expanding acronyms helps the MT engine and the sentence similarity methods process all words with better access to the meaning as opposed to the acronym. I used as an acronyms dictionary - online thesaurus https://www.wiktionary. org/. On the training set it had coverage of 47% of the acronyms.

Step 2: After substituting acronyms, I translate Russian to English. I used 3 online translation engines: Google, Microsoft and Yandex. Each of these MT engines has it's own APIs to receive an original sentences, and and send back a translated ones. Each of the MT engines gave its own translation, most of the time slightly different from one another. Each of the translations gave different score results when passed to the sentence similarity toolkits. On the given corpus of sentences, Yandex and Google showed the most accurate translation as measured by the classification performance downstream (higher F1 score and Accuracy scores).

Step 3: Running sentence similarity toolkits on the pairs of sentences, translated to English, and getting scores on each pair, saving them into a json data set file. I use six distinct semantic similarity toolkits for first task, and five - for second task, which are described below.

Step 4: Train a classifier: I train a Gradient Booster classifier algorithm, fed with the vectors of sentence similarity measures. This classifier is comparable to Support Vector Machines in its method, and it gives better results both in F1 measure and Accuracy on our corpus. This method consists of learning an ensemble classifier which combines the similarity scores of five English sentence similarity methods.

I use the Scikit-learn implementation of Gradient Boosting. The following Python code shows the specific parameters I used:

```
import sklearn.ensemble
clf=sklearn.ensemble.GradientBoostingClassifier(n_estimators=100,
max_depth=3)
```

I chose Gradient Boosting classifier, since it gave more accurate classification than either SVM with Gaussian Kernel, or Random Forest. In both cases it gave more then 1 percent to F1 score and Accuracy, then two mentioned classifiers.

My feature vector includes 77 features (which would be described in detail further) for the First Task (3-way classification: 77 features = 23 features * 3 translate engines + 8 BLEU features) and 69 features (described in detail further) for the Second Task (2-way classification: 69 features = 23 features * 3 translations).

The difference between the number of features used in the two tasks is because I did not include BLEU scores to solve the Second Task since these scores worsened class recognition and led to lower F1 score and Accuracy. BLEU scores did help on the First Task, and improved classification.

5.3 Feature Vector Structure for Each One of the Three Translations

Toolkits:

1. SEMILAR [4–7]
   - Number of used features: 6
   - Feature names: bleuComparer, cmComparer, dependencyComparerWn-LeskTanim, greedyComparerWNLin, lsaComparer, optimumComparerL-SATasa
2. DKPro Similarity [9]
   - Number of used features: 13
   - Feature names: CosineSimilarity, ExactStringMatchComparator, GreedyStringTiling 2-g, GreedyStringTiling 4-g, JaroSecondStringComparator, JaroWinklerSecondStringComparator, normalized Levenshtein-Comparator, LongestCommonSubsequenceNormComparator, Substring-MatchComparator, WordNGramContainmentMeasure, WordNGram JaccardMeasure 2-g, WordNGramJaccardMeasure 3-g, WordNGramJaccardMeasure 4-g
3. Python difflib
   - Number of used features: 1
   - Feature name: difflib SequenceMatcher comparator

   Example of code using it:
   ```
   import difflib
   sm = difflib.SequenceMatcher(None)
   sm.set_seq1('sentence one')
   sm.set_seq2('sentence two')
   print sm.ratio()
   ```
4. Algorithm of [1]
   - Number of used features: 2
   - Feature names: Sentence similarity scores
5. Swoogle [3]
   - Number of used features: 1
   - Swoogle comparator
6. BLEU scores (on the Russian version of the sentences, no need for English translation) [2]
   - Number of used features: 8
   - Feature names: BLEU with smoothing method number 1 (described in [2]) (1/2/3/4-g), BLEU with smoothing method number 2 (described in [2]) (1/2/3/4-g)

6   Comparison of Toolkits on First Task (3-Way Classification)

6.1   Results

To understand which of the scores separately gave more recognition rate - I did 5-fold cross-validation for First Task (3 class classification) on the training set, on all three translations (Google, Microsoft, Yandex) and the results are in Table 2.

Table 2. Toolkits results

Toolkit	Accuracy	F1 score
SEMILAR	**62.26**	**60.15**
DKPro Similarity	61.14	59.30
Python difflib	57.07	53.51
Algorithm of [1]	60.02	57.66
Swoogle	59.15	55.52
BLEU	57.34	54.96
All six toolkits together	**64.14**	**62.46**

6.2   Confusion Matrix

In Table 3 is a confusion matrix, which I got by running First Task using all six toolkits together:

Table 3. Confusion matrix

	Non-paraphrases	Near-paraphrases	Precise-paraphrases
Non-paraphrases	1751	798	33
Near-paraphrases	444	2164	349
Precise-paraphrases	74	893	721

As it can be seen from the matrix that two major mistakes in classification are:

– of non-paraphrases, which were incorrectly classified as near-paraphrases (798 pairs), which is 30.9% of total amount. Cases for such an errors are described in 'error analysis' section;

– of precise-paraphrases, which were incorrectly classified as near-paraphrases (893 pairs). Also can be noticed that only 721 precise-paraphrases were classified correctly, which is only 42.71% of total amount. This shows us that for semantic similarity algorithm it is hard to distinguish between near-paraphrases and precise-paraphrases, which could be explained that in fact precise-paraphrases have just a light semantic difference from near-paraphrases.

7   Ablation Test and Its Analysis on Second Task (2-Way Classification)

7.1   Results

To understand which of the scores gave more effect to the result - I did 5-fold cross-validation for Second Task (2 class classification) on the training set, on all three translations (Google, Microsoft, Yandex).

For each on the following experiments I combined feature vectors for the classifier as a concatenation of feature vectors of relevant toolkits.

Since it is the Second task I did not include BLEU scores (because they are not improving the results).

The following result are random combinations of toolkits, chosen in such a way - to cover most of the cases Tables 10, 11, 12 and 13 in Appendix B.

It can be seen from the Table 10 (appendix) that all of the toolkits give Accuracy between 75.92% and 80.13% and F1 score between 71.36% and 77.02%. By combining scores (Tables 11, 12 and 13) of these toolkits together I achieve maximum of **81.41%** in Accuracy and **78.51%** in F1 score. If I take as the basis scores from SEMILAR toolkit - by adding other scores to feature vector I achieve improvement of 1.28% in Accuracy and 1.49% in F1 score. Note that each time by adding more scores from one more toolkit to the feature vector - I improve the result, hence I chose optimal scores.

7.2   Confusion Matrix

In Table 4 is a confusion matrix, which I got by running Second Task using all five toolkits together:

Table 4. Confusion matrix

	Non-paraphrases	Paraphrases
Non-paraphrases	1612	970
Paraphrases	373	4272

The matrix indicates that the major source of error is the classification of non-paraphrases, which were mistakenly classified as paraphrases (970 pairs), which is 37.57%. The sources for such an errors are described in 'error analysis' section.

7.3   Identifying Best SEMILAR Toolkit Score

To detect the best SEMILAR toolkit score, I ran each of them separately, with the following results (Table 5):

Table 5. Score from SEMILAR toolkit

Score name	Accuracy	F1 score
bleuComparer	66.30	65.50
cmComparer	78.42	74.53
dependencyComparerWnLeskTanim	75.77	70.57
greedyComparerWNLin	**79.18**	**75.74**
lsaComparer	70.11	64.82
optimumComparerLSATasa	78.87	75.06

It can be seen that the highest results both in Accuracy and F1 score gave greedyComparerWNLin score. Let's recall that all 13 scores from DKPro Similarity toolkit gave us approximately the same result: Accuracy of 79.52 and F1 score of 75.78. So on the given corpus this one SEMILAR score alone gives approximately the same recognition rate as 13 scores from DKPro, which is impressive.

GreedyComparerWNLin score refers to a sentence to sentence similarity method which greedily aligns words between two sentences. The word alignment method used is WordNet based method proposed by Lin (1998) [16]. The method is described in [4].

8   Comparison of Translation Engines for Second Task (2-Way Classification)

In this section, I compare all three translation engines (Google, Microsoft and Yandex), and conclude which of them gave better F1 score and Accuracy (Table 6):

Table 6. Translation engines comparison

Toolkits	Google		Microsoft		Yandex	
	Accuracy	F1 score	Accuracy	F1 score	Accuracy	F1 score
SEMILAR	78.95	75.19	78.41	74.71	78.40	74.67
DKPro Similarity	78.38	74.30	77.84	73.83	78.74	74.79
Python difflib	74.88	70.04	74.05	68.70	75.05	71.40
Algorithm of [1]	76.72	72.15	76.32	72.39	77.05	72.82
Swoogle	77.94	73.12	77.95	72.80	77.69	73.02
All five toolkits together	**79.90**	**76.52**	**79.25**	**75.76**	**79.93**	**76.53**

Google and Yandex give similar results in translating from Russian to English on our corpus, both better than Microsoft's MT.

9   Error Analysis

Let's examine common mistakes our scores (algorithms) make in giving higher/lower values, causing the classifier to mistakenly predict the wrong class. Provided below sentences in Russian are from AINL 2016 paraphrase shared task corpus (available on http://www.paraphraser.ru/download/).

9.1   False Positive: Mistakenly Predicted as 'Paraphrase'

Such pairs of sentences typically contain for the most part *the same words* (or similar in the meaning), except for a few words which make all the difference, changing the meaning of the sentence completely.

The following tables are different cases of such a words (Tables 7 and 8):

Table 7. Different words are antonyms or different in the meaning words

Sentence pairs (with corresponding English translations) and explanations
Gogol center will show a video of the controversial performance of Tannhauser./ "Гоголь-центр"покажет видеозапись скандального спектакля "Тангейзер"
Kekhman banned Gogol-center to show the video recording of Tannhauser./ Кехман запретил «Гоголь-центру» показывать видеозапись «Тангейзера»
in first sentence: will show video in second sentence: banned to show video
The leader of the British labour party retained the seat in Parliament./ Лидер британских лейбористов сохранил место в парламенте.
The leader of the British labour party resigned because of the defeat in the elections./ Лидер британских лейбористов подал в отставку из-за поражения на выборах.
in first sentence: retained the seat in second sentence: resigned

Idea How to Solve Such Cases

(1) *Difference feature:* Different words are antonyms or different in the meaning words.

   Idea for solution: Create a score (algorithm) which checks to which objects (or concepts) are related those antonyms (or different in meaning words), and if they are related to the same objects (or concepts) - this is a sign that different meaning in the pair of sentences exists, so the algorithm should take them into account in our score, so that afterwards classifier can pick such a cases.

(2) *Difference feature:* Different words make the subject of each sentence different.

   Idea for solution: create a score (algorithm) which gets the main subject of the sentence, and checks if both of the sentences are of the same subject, so that afterwards classifier can pick such a cases.

Table 8. Different words make the subject of each sentence different

Sentence pairs and explanations
Agent: again the RFU delays the salary of Fabio Capello./
Агент: РФС вновь задерживает зарплату Фабио Капелло.
Mass media: Agent of Fabio Capello is threatening to sue the RFU./
СМИ: Агент Фабио Капелло грозится подать в суд на РФС.
in first sentence: RFU delays salary
in second sentence: Fabio Capello is threatening to sue.
The results of the elections in the UK has strengthened the pound./
Результаты выборов в Великобритании укрепили фунт.
Became known the final results of the elections in the UK./
Стали известны окончательные результаты выборов в Великобритании.
in first sentence: results strengthened the pound
in second sentence: results became known

9.2 False Negatives: Mistakenly Predicted as 'Non-paraphrase'

Rule for Such Pairs, with Corresponding Examples. Such pairs of sentences typically contain for the most part *different words* but the meaning of the whole sentence is the the same or closely related (since I have only one paraphrase class for second task, I can combine closely related sentences to paraphrase class): Table 9.

Table 9. Sentence pairs

Apakan called on the parties of the Ukrainian conflict to observe a truce on 9 may./
Апакан призвал стороны украинского конфликта соблюдать перемирие 9 мая.
The OSCE has called for peace in Ukraine on May 9./
В ОБСЕ призвали к миру на Украине 9 Мая.
Media: the United States are planning to deploy Osprey
convertiplane in the suburbs of Tokyo./
СМИ: США планируют разместить конвертопланы Osprey в пригороде Токио.
The Pentagon has decided to place in the suburbs of Tokyo ten convertiplanes./
Пентагон решил разместить в пригороде Токио десять конвертопланов
Media: police of United States of America is looking for
unknowns who have stolen the clothes of Demi Moore./
СМИ: полиция США ищет неизвестных, укравших одежду Деми Мур.
Demi Moores' clothes for 200 thousand dollars were stolen./
У Деми Мур украли одежду на 200 тысяч долларов.

Idea How to Solve Such Cases. Pay attention that the word Apakan in the *first pair* is the name of the current Chief Monitor of the OSCE Special Monitoring Mission to Ukraine. So if I would have knowledge base, which associated Apakan with OSCE - and gave that knowledge base to those scores (algorithms), that would have helped to identify the similarity of these two sentences.

The same goes with the *second pair* since Pentagon is United States *state level* organization.

In the *third pair* I need sophisticated understanding since 'clothes of Demi Moore' mentioned in first sentence can be expensive enough, since she is a celebrity, and can afford to buy something expensive. So because in the second sentence is mentioned the price of the clothes, and due to this amount of money ('200 thousand dollars') - we understand that is being talked about expensive clothes - we can conclude that the subject of both sentences in this pair is the same (e.g. it is about 'expensive clothes that were stolen').

Such a sophisticated AI can be created, but it remains an open challenge.

10   Conclusions and Future Work

In this work I address the paraphrase classification problem by combining Machine Translation and using an ensemble of semantic similarity algorithms. The resulting F1 score and Accuracy metrics have shown the effectiveness of such an algorithm. I achieved recognition of **64.14%** in Accuracy and **62.46%** in F1 score for the First Task (3-way classification), and **81.41%** in Accuracy and **78.51%** in F1 score for the Second Task (two-way classification). Results on both tasks are significantly better than the baseline BLEU algorithm (ran on Russian source, without translation).

I completed ablation test to detect which of those algorithms gave more effect in gaining the correct answer. I observe that the best toolkit is SEMILAR. Interesting enough, python difflib showed pretty good result for just a regular python library, which results in only about 5% less than our winner - SEMILAR. The best score of all used, for given corpus, is greedyComparerWNLin, which is a part of the SEMILAR toolkit. The best translation engines, for our corpus in Russian, are Yandex and Google.

I used most popular semantic similarity algorithms (toolkits) however more algorithms are available. In future work I intend to use additional algorithms to improve paraphrase recognition. Additionally I plan to develop scores which will allow us to cover the cases mentioned in the error analysis section, to let the classifier to include these cases into the correct classes.

Although the corpus was in Russian and I translated it to English using automated translation engines (Google, Microsoft and Yandex), and semantic similarity algorithms I launched on translations which were not 100% correct in the suitability of translated words and syntactic structure, since these translations engines can give only approximation to the most fitted translation. So F1 score and Accuracy could be higher if the corpus was originally in English.

11   Repository

Python code sources are available on:
https://github.com/dmikrav/paraphraser.ainlconf.2016
Web-site is: https://dmikrav.github.io/paraphraser.ainlconf.2016/.

A   Appendix: Algorithm Data-Flow

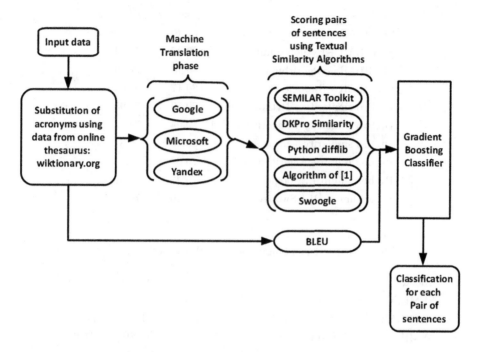

B   Appendix: Ablation Test Tables

Table 10. Each toolkit launched separately

Toolkit	Accuracy	F1 score
SEMILAR	**80.13**	**77.02**
DKPro Similarity	79.52	75.78
Python difflib	75.92	71.36
Algorithm of [1]	78.76	75.02
Swoogle	78.94	75.03

Table 11. Combinations by two toolkits

Toolkits	Accuracy	F1 score
SEMILAR + DKPro similarity	**80.86**	**77.89**
SEMILAR + Python difflib	80.37	77.34
SEMILAR + Algorithm of [1]	80.18	77.10
SEMILAR + Swoogle	80.17	77.05
DKPro similarity + Python difflib	79.59	75.95
DKPro similarity + Algorithm of [1]	80.04	76.62
DKPro similarity + Swoogle	80.26	76.87
Python difflib + Algorithm of [1]	79.09	75.44
Python difflib + Swoogle	79.42	75.69
Algorithm of [1] + Swoogle	80.10	76.66

Table 12. Combinations by three toolkits

Toolkits	Accuracy	F1 score
SEMILAR + DKPro similarity + Python difflib	80.66	77.66
SEMILAR + DKPro similarity + Algorithm of [1]	80.55	77.59
SEMILAR + DKPro similarity + Swoogle	**80.89**	**77.89**
SEMILAR + Python difflib + Algorithm of [1]	80.43	77.45
SEMILAR + Python difflib + Swoogle	80.61	77.66
SEMILAR + Algorithm of [1] + Swoogle	80.73	77.77
DKPro similarity + Python difflib + Algorithm of [1]	80.37	77.14
DKPro similarity + Python difflib + Swoogle	79.93	76.55
Python difflib + Algorithm of [1] + Swoogle	80.03	76.60

Table 13. Combinations by four and five toolkits

Toolkits	Accuracy	F1 score
SEMILAR + DKPro similarity + Python difflib + Algorithm of [1]	80.94	77.99
SEMILAR + DKPro similarity + Python difflib + Swoogle	**81.36**	**78.39**
SEMILAR + Python difflib + Algorithm of [1] + Swoogle	80.75	77.79
DKPro similarity + Python difflib + Algorithm of [1] + Swoogle	80.72	77.56
All five toolkits together	**81.41**	**78.51**

References

1. Li, Y., McLean, D., Bandar, Z., O'Shea, J., Crockett, K.: Sentence similarity based on semantic nets and corpus statistics. IEEE Trans. Knowl. Data Eng. **18**, 1138–1150 (2006)
2. Chen, B., Cherry, C.: A systematic comparison of smoothing techniques for sentence-level BLEU. In: WMT@ACL, pp. 362–367 (2014)
3. Ding, L., Finin, T., Joshi A., Peng, Y., Cost, S., Sachs, J., Pan, R., Reddivari, P., Doshi, V.: Swoogle: a semantic web search and metadata engine. In: CIKM, pp. 652–659 (2004)
4. Rus, V., Lintean, M.: A comparison of greedy and optimal assessment of natural language student input using word-to-word similarity metrics. In: BEA@NAACL-HLT, pp. 157–162 (2012)
5. Ştefănescu, D., Banjade, R., Rus, V.: Latent semantic analysis models on Wikipedia and TASA. In: LREC, pp. 1417–1422 (2014)
6. Lintean, M., Rus, V.: Paraphrase identification using weighted dependencies and word semantics. Informatica (Slovenia) **34**, 19–28 (2010)
7. Rus, V., Lintean, M., Banjade, R., Niraula, N., Stefanescu, D.: SEMILAR: the semantic similarity toolkit. In: ACL (Conference System Demonstrations), pp. 163–168 (2013)
8. Banjade, R., Niraula, N., Maharjan, N., Rus, V., Stefanescu, D., Lintean, M., Gautam, D.: NeRoSim: a system for measuring and interpreting semantic textual similarity. In: Proceedings of the 9th International Workshop on Semantic Evaluation, SemEval@NAACL-HLT, pp. 164–171 (2015)
9. Bar, D., Zesch, T., Gurevych, I.: DKPro similarity: an open source framework for text similarity. In: ACL (Conference System Demonstrations), pp. 121–126 (2013)
10. Pronoza, E., Yagunova, E.: Low-level features for paraphrase identification. In: Sidorov, G., Galicia-Haro, S.N. (eds.) MICAI 2015. LNCS (LNAI), vol. 9413, pp. 59–71. Springer, Cham (2015). https://doi.org/10.1007/978-3-319-27060-9_5
11. Ganitkevitch, J., Van Durme, B., Callison-Burch, C.: PPDB: the paraphrase database. In: HLT-NAACL, The Association for Computational Linguistics, pp. 758–764 (2013)
12. Dolan, B., Brockett, C., Quirk, C.: Microsoft Research Paraphrase Corpus (2005)
13. Ji, Y., Eisenstein, J.: Discriminative improvements to distributional sentence similarity. In: Proceedings of the 2013 Conference on Empirical Methods in Natural Language Processing, EMNLP, pp. 891–896 (2013)
14. Socher, R., Huang, E., Pennington, J., Ng, A., Manning, C.: Dynamic pooling and unfolding recursive autoencoders for paraphrase detection. In: Advances in Neural Information Processing Systems 24: 25th Annual Conference on Neural Information Processing Systems, pp. 801–809 (2011)
15. Yin, W., Schutze, H.: Convolutional neural network for paraphrase identification. In: Proceedings of the 2015 Conference of the North American Chapter of the Association for Computational Linguistics: Human Language Technologies, HLT-NAACL, pp. 901–911 (2015)
16. Lin, D.: An information-theoretic definition of similarity. In: Proceedings of the Fifteenth International Conference on Machine Learning (ICML), pp. 296–304 (1998)
17. Dagan, I., Dolan, B., Magnini, B., Roth, D.: Recognizing textual entailment: rational, evaluation and approaches - Erratum, In: Natural Language Engineering, vol. 16 (2010)

18. Madnani, N., Tetreault, J., Chodorow, M.: Re-examining machine translation metrics for paraphrase identification. In: Human Language Technologies: Conference of the North American Chapter of the Association of Computational Linguistics, Proceedings, pp. 182–190 (2012)
19. Chitra, A., Rajkumar, A.: Plagiarism detection using machine learning-based paraphrase recognizer. J. Intell. Syst. **25**, 351–359 (2016)
20. Dey, K., Shrivastava, R., Kaushik, S.: A paraphrase and semantic similarity detection system for user generated short-text content on Microblogs. In: COLING, 26th International Conference on Computational Linguistics, Proceedings of the Conference, pp. 2880–2890 (2016)
21. Pivovarova, L., Pronoza, P., Yagunova, E., Pronoza, A.: ParaPhraser: Russian paraphrase corpus and shared task. In: Filchenkov, A., et al. (eds.) AINL 2017. CCIS, vol. 789, pp. 211–225. Springer, Cham (2018)

Character-Level Convolutional Neural Network for Paraphrase Detection and Other Experiments

Vladislav Maraev[(✉)], Chakaveh Saedi, João Rodrigues, António Branco, and João Silva

Department of Informatics, Faculty of Sciences,
University of Lisbon, Lisbon, Portugal
{vlad.maraev,chakaveh.saedi,joao.rodrigues,antonio.branco,
jsilva}@di.fc.ul.pt

Abstract. The central goal of this paper is to report on the results of an experimental study on the application of character-level embeddings and basic convolutional neural network to the shared task of sentence paraphrase detection in Russian. This approach was tested in the standard run of Task 2 of that shared task and revealed competitive results, namely 73.9% accuracy against the test set. It is compared against a word-level convolutional neural network for the same task, and varied other approaches, such as rule-based and classical machine learning.

Keywords: Paraphrase detection · Word embeddings
Character embeddings · Convolutional neural networks
Distributional semantics

1 Introduction

The Russian language is a morphologically rich language with free word order and can be an interesting workbench for testing different models of paraphrase detection, which have been studied mostly against English datasets.

In this paper, we report on addressing this task by using a system that we developed and showed competitive results in the standard run Task 2 of Russian paraphrase detection shared task,[1] where participating systems cannot resort to data other than the ones provided for the shared task. This system is based on a character-based convolutional neural network.

We report also on the results obtained with the application of other approaches that we developed and tested initially for the task of duplicate question detection [12,14].

Paraphrase detection belongs to a family of semantic text similarity tasks, which have been addressed in SemEval challenges since 2012, and which in the

---

[1] http://www.paraphraser.ru/contests/result/?contest_id=1.

© Springer International Publishing AG 2018
A. Filchenkov et al. (Eds.): AINL 2017, CCIS 789, pp. 293–304, 2018.
https://doi.org/10.1007/978-3-319-71746-3_23

last SemEval-2016, for instance, included also tasks like the degree of similarity between machine translation output and its post-edited version, among others.

Semantic textual similarity assesses the degree to which two textual segments are semantically equivalent to each other, which is typically scored on an ordinal scale ranging from semantic equivalence to complete semantic dissimilarity.

Paraphrase detection is a special case of semantic textual similarity, where the scoring scale is reduced to its two extremes and the outcome for an input pair of textual segments is yes/no.

The present paper is organized as follows. In the next Sect. 2, the conditions of and the results for the shared task are discussed. The character-level convolutional neural network and respective results are discussed in Sect. 3. Sections 4, 5 and 6 present the experimental results of a range of other approaches, respectively, rule-based, supervised classifiers and other deep neural networks. In Sect. 7, the results obtained are discussed. Sections 8 and 9 discuss the related work and present the conclusions.

2 Dataset and Results of Participation

For the experimental results reported in the present paper, we resorted to the shared task's ParaPhraser dataset [11], a freely available corpus of Russian sentence pairs manually annotated as precise paraphrases, near-paraphrases and non-paraphrases. Each pair was collected from news headlines and then manually annotated by three native speakers.

The size of the training set is 7,000 pairs and the test set contains 1,924 pairs. The number of tokens, the number of types and average sentence length in the training set are presented in the Table 1.

Table 1. Quantitative attributes of the training set.

Pairs	7,000
Total tokens	126,303
Lowercased types	20,252
Average sentence length (words)	8.7

The shared task consists of two subtasks: one for three-class classification, and another for binary classification. We have tackled the second one (Task 2) which is defined as follows:

Given a pair of sentences, to predict whether they are paraphrases (whether precise or near paraphrases) or non-paraphrases.

There were two types of shared settings: the *standard run* where only the ParaPhraser corpus could be used for training, and the *non-standard run* where any other corpora could be also used. We participated in both types of submissions.

According to the results obtained by submitting the output to the shared task organisation: (i) our system *CNN-char*, which participated in standard run obtained a competitive accuracy score of 72.7%, which stands just 1.9% points below the best system's score; (ii) our system *CNN-word*, which participated in the non-standard run obtained an accuracy score of 69.9%, which is quite lower than the best system's accuracy of 77.4%.

Below we will discuss also the results obtained a posteriori in our lab once the test sets were released, which are slightly different from the ones above reported by the shared task organization, due to the random initialization of the weights of the neural network.

3 Convolutional Neural Network

The architecture of convolutional neural network (CNN) used to address the paraphrase detection task was introduced by Bogdanova et al. [3] for the task of detecting semantically equivalent questions in online question answering forums. It also takes advantage of the approach introduced by Kim [7] for sentence classification task using a set of convolutional filters of an arbitrary length.

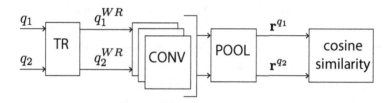

Fig. 1. CNN architecture.

Figure 1 shows the layers of the CNN: token (word or character) representation layer (TR), convolution layer(s) (CONV), pooling layer (POOL) and cosine similarity measurement step.

To obtain the representation of a sentence, it is pipelined along these major steps:

1. Obtaining token representations;
2. Applying a set of convolutional filters;
3. Concatenating the results of convolution;
4. Pooling the product of convolution filters.

We resort to two variants[2] for paraphrase detection using a convolutional neural network.

---

[2] Source code is available as a part of Vladislav Maraev's MA dissertation at: https://github.com/vladmaraev/msrdsdl.

The first one uses randomly initialized character representations on a token representation layer that are further passed as input to a set of convolutional filters.

The second one follows Bogdanova et al. [3] and relies on pre-trained word embeddings for the initial token representation.

3.1   Character Embeddings

In the first variant, referred to as *CNN-char*, we split sentences into characters instead of tokenizing them into words. The main reason to have followed this route is that character-level embeddings are reported to be good in capturing morphological information [8,15], which is important for a morphologically rich language like Russian.

In terms of preprocessing, a few basic procedures were applied, namely, lowercasing the input and removing non-word characters.

Table 2 summarizes the hyper-parameters that were used for this run.

Table 2. Hyper-parameters of CNN-char.

Parameter	Value	Description
k	$\{2, 3, 5, 7, 9, 11\}$	Sizes of k-grams
l_u	100	Size of each convolutional filter
d	100	Size of character representation
epochs	20	Number of training epochs
pooling	MAX	Pooling layer function
optimizer	SGD	Stochastic Gradient Descent
loss	MSE	Mean Squared Error

Results. This approach leads to the highest accuracy of 73.9%, reported in this work regarding Russian paraphrase detection task.

3.2   Pre-trained Word Embeddings

In this other variant, referred to as *CNN-word*, the approach adopted by Bogdanova et al. [3] for the task of duplicate question detection was followed here for paraphrase detection, where word embeddings were pre-trained.

We employed word2vec word embeddings from Kutuzov and Andreev [9].[3]

In order to preprocess the input sentences, these were lowercased, lemmatised and PoS-tagged using MyStem [16], which is the same tool that was reported by the authors of RusVectores model [9].

The Table 3 summarizes the hyper-parameters that were used for this run.

---

[3] These word embeddings for Russian are available from: http://rusvectores.org/ru/ models/, *ruscorpora_2015* model.

Table 3. Hyper-parameters of CNN-word.

Parameter	Value	Description
k	3	Size of k-gram
l_u	300	Size of convolutional filter
d	300	Size of word representation
epochs	5	Number of training epochs
pooling	MAX	Pooling layer function
optimizer	SGD	Stochastic Gradient Descent
loss	MSE	Mean Squared Error

Results. This variant leads to an accuracy score of 70.6%, which is 3.3 pp. lower than the score obtained by the character-based model in spite of the usage of external resources.

4   Rule-Based

A rule-based approach, referred to as *Jaccard*, was used to establish a baseline. We used the Jaccard Coefficient over n-grams (n ranging from 1 to 4), inspired by the usage of this coefficient in [17].

Before applying this technique, the textual segments were preprocessed by submitting them to lowercasing, tokenization and lemmatisation using the MyStem tool [16].

To find the best threshold, the training set was used in a series of trials. This led to the thresholds of 0.13 for the English dataset, and 0.1 for the Russian dataset.

Results. This system achieves the accuracy score of 67.0%. This result is lower than ones obtained by *CNN-char* and described above. It is in line tough with the scores obtained in other experiments that were carried out for another task, namely duplicate question detection [12,14].

5   Classic Machine Learning Approaches

5.1   SVM with Basic Features

To set up a paraphrase detection system based on a supervised machine learning classifier, we resorted to support vector machines (SVM), following its acknowledged good performance in many NLP tasks. We employed SVC (Support Vector Classification) implementation from the sklearn support vector machine toolkit [10].

For the first version of the classifier, a basic feature set (FS) was created. N-grams, with n ranging from 1 to 4, were extracted from the training set.

Afterwards, among those extracted n-grams, the ones with at least 10 occurrences were selected to support the FS. We tried thresholds ranging from 5 to 15 and the best result was achieved when the threshold was set to 10.

For each textual segment in a pair, a vector of size k was generated, where k is the number of n-grams included in the FS. Each vector encodes the occurrences of the n-grams in the corresponding segment, where vector position i will be 1 if the i-th n-gram occurs in the segment, and 0 otherwise. Then a feature vector of size $2k$ is created by concatenating the vectors of the two segments. This vector is further extended with the scores of the Jaccard coefficient determined over 1, 2, 3 and 4-grams. Hence, the final feature vector representing the pair to the classifier has the length $2k + 4$.

Results. This system achieves 70.4% accuracy when trained over the Russian dataset, which suggests that the result is comparable with *CNN-word* that also uses external language resources.

5.2 SVM Classifier with Advanced Features

In order to get an insight on how strong an SVM-based system for paraphrase detection resorting to a basic FS like the one described above may be, we proceeded with further experiments, by adding more advanced features.

Lexical Features. The vector of each segment was extended with an extra feature, namely the number of negative words, e.g.: ("nothing"), ("never"), etc. occurring in it. And, to the concatenation of segment vectors, one further feature was added, the number of nouns that are common to both segments, provided they are not already included in the FS. Any pair was then represented by a vector of size $2(k + 1) + 4 + 1$.

Semantic Features. Eventually, any pair was represented by a vector of size $2(k+1)+4+2$, with its length being extended with yet an extra feature, namely the value of the cosine similarity between the embeddings of the segments in the pair.

For a given segment, its embedding, or distributional semantic vector, was obtained by summing up the embeddings of the nouns and verbs occurring in it, as these showed to support the best performance after experiments that have been undertaken with all parts-of-speech and their subsets. We employed word2vec word embeddings from Kutuzov and Andreev [9] the same ones that we used in the experiment discussed in Sect. 3.2.

Results. The resulting system permitted an improvement of over 1% points with respect to its previous version trained with basic features, scoring 71.7% accuracy, thus being slightly superior to our *CNN-word* system above, with pretrained word embeddings.

6   Deep Neural Network Architectures

In this section we discuss the experiments that were carried out in order to assess the performance, in the paraphrase detection task, of the deep neural network architectures that were able to achieve very high performance in the duplicate question detection task [12,14].

We begin by applying the architecture of MayoNLP, the system that was the top scoring system in SemEval-2016 Task 1 [2]. We will then proceed with discussing a hybrid approach that combines convolutional and fully-connected layers in a neural network.

The same preprocessing used on the convolutional neural networks (lower-casing, lemmatization, and PoS-tagged) was used in these models.

6.1   Deep Neural Network (MayoNLP)

We implemented a deep neural network (DNN) based on MayoNLP [1]. This system follows the architecture of Deep Structured Semantic Models, introduced by Huang et al. [5], which consists of a multi-layer neural architecture of feed-forward and fully connected layers. The neural network has as input a 30k neurons dense layer followed by two hidden multi-layers with 300 neurons each and finally a 128 neuron output layer.

MayoNLP also implemented a preprocessing dimension reduction with a word hashing method which creates trigrams for every word in the input sentence.

Given that we did not face the same dimension problem, we implemented a one-hot encoding process, which eventually ended up reducing even further, from an original 30k dimension in Mayo to 10k for the ParaPhrase dataset.

The MayoNLP system also differs from the Deep Structured Semantic Models by adopting a 1k neuron layer instead of two hidden layers in its architecture.

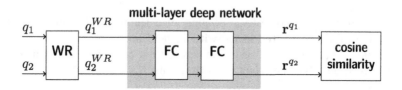

Fig. 2. DNN architecture: word representation layer (WR), fully connected layers (FC) and cosine similarity measurement layer.

A diagram of the implemented neural network is presented in Fig. 2. The Table 4 summarizes the hyper-parameters that were used.

Results. The model obtained a 59.9% accuracy, scoring the worst result in comparison with the results of the models experimented and reported in this paper. This is mainly due to the lack of sufficient data and the overwhelming complexity of the neural network for the given dataset.

Table 4. DNN-word approach hyper-parameters.

Parameter	Value	Description
lr	0.01	Learning rate
hidden neurons	728	Hidden layer neurons
epochs	20	Training epochs
pooling	MAX	Pooling layer function
optimizer	SGD	Stochastic Gradient Descent
loss	MSE	Mean Squared Error

6.2   Deep Convolutional Neural Network

Finally, we also experimented with a deep convolutional neural network (DCNN) model with which we obtained the best accuracies in a related semantic similarity task [12]. This model is a combination of the convoluted and dense models previously described. A lite version of the original model was deployed given the decrease in the available dataset when compared with the originally designed dataset. We resorted to Keras and Tensorflow for its implementation.

Both input sentences are fed to the neural network, both pass the same neural network layers in parallel and are compared before the output result, in a so-called Siamese architecture.

A vectorial representation for words is used at the beginning of the model with a layer that acts as a distributional semantic space and learns a vector for each word in the training dataset.

That vectorial representation is fed to a convolutional layer with 50 neurons and a window with size 15.

This convolutional layer is then combined with a pooling layer that resorts to a max filter.

With the resulting vector of the pooling layer the network connects to three dense layers of fully connected layers with 15 neurons each.

In a final step, the output of the layers is then computed by means of the cosine distance between the result of both inputs.

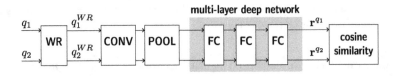

Fig. 3. DCNN architecture.

A diagram of this hybrid neural network is presented in Fig. 3. The Table 5 summarizes the hyper-parameters that were used.

Table 5. DCNN-word approach hyper-parameters.

Parameter	Value	Description
lr	0.01	Learning rate
epochs	20	Training epochs
d	50	Size of word representation
l_u	50	Size of convolutional filter
k	5	Size of convolutional kernel
hidden neurons	45	Hidden layer neurons
pooling	MAX	Pooling layer function
optimizer	SGD	Stochastic Gradient Descent
loss	MSE	Mean Squared Error

Results. The DCNN model obtained 70.0% accuracy, which is in line with the results of other models such as SVM and Jaccard. This is mainly due to it being a lite version of the original neural network. As it is common with neural networks, the more data the better, which makes us believe higher accuracies can be obtained with a larger dataset.

7   Discussion

The experimental results reported in the previous sections are summarized in Table 6.

Table 6. Accuracy of the 7 systems plus the majority class baseline over the Russian paraphrases dataset.

System	Accuracy (%)
Majority class	49.7
Jaccard*	67.0
SVM-bas*	70.4
CNN-word*	70.6
SVM-adv*	71.7
DNN	59.9
DCNN	70.0
CNN-char	73.9
Best system in shared task*	77.4
Best system in shared task	74.6

In this table, the star (*) superscript indicates systems that use resources other than just the ParaPhraser dataset distributed by the shared task organizers.

At the bottom of the table, the best results obtained by systems that participated in the shared task are displayed.

8   Related Work

The best three systems in the SemEval-2016 Task 1 are the following: Rychalska et al. [13], which employs autoencoders, WordNet and SVM; Brychcín and Svoboda [4], which combines various meaning representation algorithms and different classifiers; and the MayoNLP system [1], whose architecture is adopted in one of our experiments and was presented in Sect. 6.1.

The competitor non-NN-based system [6] uses discriminative term-weighting (TF-KLD) and matrix factorisation.

The work on CNNs reported in this paper was inspired by the work of Bogdanova et al. [3] that employ Siamese CNN with shared weights for detecting semantically equivalent question. It also takes advantage of the approaches introduced by Kim [7] for concatenating convolutional filters of various lengths and [8] for employing character embedding for morphologically rich languages.

9   Conclusions

This paper has presented the results of a range of experiments to address the task of paraphrase detection for Russian under the conditions and with the datasets of the respective shared task organized in 2016.

The application of the convolutional neural network model to this task showed the best results. In particular, the character-based convolutional neural network model achieves competitive performance for the task of detecting if two sentences are paraphrases without using any external resources.

Acknowledgements. The present research was also partly supported by the CLARIN and ANI/3279/2016 grants.

References

1. Afzal, N., Wang, Y., Liu, H.: MayoNLP at SemEval-2016 task 1: semantic textual similarity based on lexical semantic net and deep learning semantic model. In: Proceedings of the 10th International Workshop on Semantic Evaluation (SemEval-2016), pp. 1258–1263 (2016)
2. Agirre, E., Banea, C., Cer, D.M., Diab, M.T., Gonzalez-Agirre, A., Mihalcea, R., Rigau, G., Wiebe, J.: SemEval-2016 task 1: Semantic textual similarity, monolingual and cross-lingual evaluation. In: Bethard, S., Cer, D.M., Carpuat, M., Jurgens, D., Nakov, P., Zesch, T. (eds.) Proceedings of the 10th International Workshop on Semantic Evaluation, SemEval@NAACL-HLT 2016, San Diego, CA, USA, 16–17 June 2016, pp. 497–511. The Association for Computer Linguistics (2016). http://aclweb.org/anthology/S/S16/S16-1081.pdf

3. Bogdanova, D., dos Santos, C.N., Barbosa, L., Zadrozny, B.: Detecting semantically equivalent questions in online user forums. In: Alishahi, A., Moschitti, A. (eds.) Proceedings of the 19th Conference on Computational Natural Language Learning, CoNLL 2015, Beijing, China, 30–31 July 2015, pp. 123–131. ACL (2015). http://aclweb.org/anthology/K/K15/K15-1013.pdf

4. Brychcín, T., Svoboda, L.: UWB at SemEval-2016 task 1: semantic textual similarity using lexical, syntactic, and semantic information. In: Bethard, S., Cer, D.M., Carpuat, M., Jurgens, D., Nakov, P., Zesch, T. (eds.) Proceedings of the 10th International Workshop on Semantic Evaluation, SemEval@NAACL-HLT 2016, San Diego, CA, USA, 16–17 June 2016, pp. 588–594. The Association for Computer Linguistics (2016). http://aclweb.org/anthology/S/S16/S16-1089.pdf

5. Huang, P.S., He, X., Gao, J., Deng, L., Acero, A., Heck, L.: Learning deep structured semantic models for web search using clickthrough data. In: Proceedings of the 22nd ACM International Conference on Information and Knowledge Management, pp. 2333–2338. ACM (2013)

6. Ji, Y., Eisenstein, J.: Discriminative improvements to distributional sentence similarity. In: Proceedings of the 2013 Conference on Empirical Methods in Natural Language Processing, EMNLP 2013, 18–21 October 2013, Grand Hyatt Seattle, Seattle, Washington, USA, A meeting of SIGDAT, a Special Interest Group of the ACL, pp. 891–896. ACL (2013). http://aclweb.org/anthology/D/D13/D13-1090.pdf

7. Kim, Y.: Convolutional neural networks for sentence classification. In: Moschitti, A., Pang, B., Daelemans, W. (eds.) Proceedings of the 2014 Conference on Empirical Methods in Natural Language Processing, EMNLP 2014, 25–29 October 2014, Doha, Qatar, A meeting of SIGDAT, a Special Interest Group of the ACL, pp. 1746–1751. ACL (2014). http://aclweb.org/anthology/D/D14/D14-1181.pdf

8. Kim, Y., Jernite, Y., Sontag, D., Rush, A.M.: Character-aware neural language models. In: Schuurmans, D., Wellman, M.P. (eds.) Proceedings of the Thirtieth AAAI Conference on Artificial Intelligence, February 12–17, 2016, Phoenix, Arizona, USA, pp. 2741–2749. AAAI Press (2016). http://www.aaai.org/ocs/index.php/AAAI/AAAI16/paper/view/12489

9. Kutuzov, A., Andreev, I.: Texts in, meaning out: Neural language models in semantic similarity tasks for Russian, vol. 2, pp. 133–144 (2015)

10. Pedregosa, F., Varoquaux, G., Gramfort, A., Michel, V., Thirion, B., Grisel, O., Blondel, M., Prettenhofer, P., Weiss, R., Dubourg, V., Vanderplas, J., Passos, A., Cournapeau, D., Brucher, M., Perrot, M., Duchesnay, E.: Scikit-learn: machine learning in Python. J. Mach. Learn. Res. **12**, 2825–2830 (2011)

11. Pronoza, E., Yagunova, E., Pronoza, A.: Construction of a Russian paraphrase corpus: unsupervised paraphrase extraction. In: Braslavski, P., Markov, I., Pardalos, P., Volkovich, Y., Ignatov, D.I., Koltsov, S., Koltsova, O. (eds.) RuSSIR 2015. CCIS, vol. 573, pp. 146–157. Springer, Cham (2016). https://doi.org/10.1007/978-3-319-41718-9_8

12. Rodrigues, J.A., Saedi, C., Maraev, V., Silva, J., Branco, A.: Ways of asking and replying in duplicate question detection. In: Ide, N., Herbelot, A., Màrquez, L. (eds.) Proceedings of the 6th Joint Conference on Lexical and Computational Semantics, *SEM @ACM 2017, Vancouver, Canada, 3–4 August 2017, pp. 262–270. Association for Computational Linguistics (2017). https://doi.org/10.18653/v1/S17-1030

13. Rychalska, B., Pakulska, K., Chodorowska, K., Walczak, W., Andruszkiewicz, P.: Samsung Poland NLP team at semeval-2016 task 1: necessity for diversity; combining recursive autoencoders, wordnet and ensemble methods to measure semantic similarity. In: Bethard, S., Cer, D.M., Carpuat, M., Jurgens, D., Nakov, P., Zesch, T. (eds.) Proceedings of the 10th International Workshop on Semantic Evaluation, SemEval@NAACL-HLT 2016, San Diego, CA, USA, 16–17 June 2016, pp. 602–608. The Association for Computer Linguistics (2016). http://aclweb.org/anthology/S/S16/S16-1091.pdf

14. Saedi, C., Rodrigues, J., Silva, J., Branco, A., Maraev, V.: Learning profiles in duplicate question detection. In: Proceedings of IEEE IRI 2017 conference (2017, in press)

15. dos Santos, C.N., Gatti, M.: Deep convolutional neural networks for sentiment analysis of short texts. In: Hajic, J., Tsujii, J. (eds.) COLING 2014, 25th International Conference on Computational Linguistics, Proceedings of the Conference: Technical Papers, 23–29 August 2014, Dublin, Ireland, pp. 69–78. ACL (2014). http://aclweb.org/anthology/C/C14/C14-1008.pdf

16. Segalovich, I.: A fast morphological algorithm with unknown word guessing induced by a dictionary for a web search engine. In: Arabnia, H.R., Kozerenko, E.B. (eds.) Proceedings of the International Conference on Machine Learning; Models, Technologies and Applications. MLMTA 2003, 23–26 June 2003, Las Vegas, Nevada, USA, pp. 273–280. CSREA Press (2003)

17. Wu, Y., Zhang, Q., Huang, X.: Efficient near-duplicate detection for Q&A forum. In: Fifth International Joint Conference on Natural Language Processing, IJCNLP 2011, Chiang Mai, Thailand, 8–13 November 2011, pp. 1001–1009. The Association for Computer Linguistics (2011). http://aclweb.org/anthology/I/I11/I11-1112.pdf

Author Index

Printed in the United States
By Bookmasters